GROUND WATER

GROUND WATER

Editor

Guntram Pawlitzki

scitus
academics

Ground Water

Edited by **Guntram Pawlitzki**

This Book is Printed on Acid-Free Paper. ♾

ISBN: 9781681179681

Library of Congress Control Number: 2017949095

SCITUS Academics LLC,

616, Corporate Way, Suite 2, 4766,

Valley Cottage, NY 10989

www.scitusacademics.com

Printed in United States of America.

Preface

The most considerable water problem facing the world is the shortage of water. Scarcity is expected to strengthen during the 21st century due to global population growth, economic growth and the need to protect environmental assets. Groundwater is becoming gradually more popular resource because of the relative ease and flexibility with which it can be tapped. While developing groundwater resource promises to help alleviate poverty in many areas, the most formidable challenge are its sustainable use and management in regions where it is under threat. In many regions, climate change may cause scarcity to become more acute. Groundwater has the capacity to buffer extreme hydrologic events because recharge is not directly tied to precipitation and run-off events. Groundwater storage offers many opportunities to increase storage and enhance the general water supply. The value of groundwater will increase as scarcity intensifies and effective groundwater management will be required if groundwater values are to be reaped. Effective groundwater management will require attention to certain well-established economic principles. Groundwater monitoring helps us understand and make decisions about the quantity and quality of water beneath the earth's surface (such as in aquifers). Groundwater management schemes appear to be most effective when they are locally developed and managed

This book covers groundwater resources management, the socio-economic impact of its intensive use, and the physical, institutional and policy options for its management. The book provides an analysis of the socio-ecology of groundwater, based on a synthesis of macro and micro level data on the hydrological, social, economic and institutional parameters. It aims at enhancing progress towards access to safe drinking water through consolidating knowledge in groundwater fluoride occurrence, effects of fluoride on human health, and technologies available for water defluoridation. This book is anticipated to be useful to researchers, policy-makers, and non-governmental organizations working on water related projects in countries worldwide.

Table of Contents

CHAPTER 1

Groundwater and Contaminant Hydrology

Zulfiqar Ahmad[1], Arshad Ashraf[2], Gulraiz Akhter[1] and Iftikhar Ahmad[3]

[1] Department of Earth Sciences, Quaid-i-Azam University, Islamabad, Pakistan

[2] National Agricultural Research Center,Islamabad, Pakistan

[3] College of Earth and Environmental Sciences, Punjab University, Lahore, Pakistan

1. INTRODUCTION

Groundwater and Contaminant Hydrology has a range of research relating to the transport and fate of contaminants in soils and groundwater. The scope of the center includes: 1) the development of new sampling and site characterization techniques; and 2) other improved groundwater remediation techniques.

Contaminant hydrology is the study of processes that affect both ground and surface water pollution. It draws on the principles of hydrology and chemistry. Contaminant hydrology and water quality research seeks to understand the role of soil properties and hydrologic processes on ground and surface water pollution and develop strategies to mitigate their impacts. Research is done at all scales varying from soil pore to basin scale and covers both traditional and emerging contaminants. Groundwater and contaminant hydrology studies include fate and transport of jet fuel leakages from oil depots, producing water injection in shallow wells from the oil and gas exploration field concession areas, veterinary pharmaceuticals from land-applied manure, pathogen losses from manure application, fate and transport of disposal wastes in unlined evaporation ponds from pharmaceutical industries, impacts of tile drainage on sediment and nutrient pollution on Rivers, sediment-turbidity relationships, water quality modeling, and TMDL and paired watershed studies.

Some research institutes address national and international needs for subsurface contaminant characterization and remediation across a spectrum of approaches - laboratory experiments, field tests, and theoretical and numerical groundwater flow and transport investigations. Some of the developing countries most critical subsurface contamination issues, including the chemical evolution of highly alkaline radioactive waste in storage tanks; reduction, re-oxidation, and diffusion of uranium forms in sediments; hydraulic properties of unsaturated gravels; and the natural production of transport-enhancing mobile nanoparticles in the subsurface. Inverse modeling of reactive transport and joint hydrologic and geophysical inversion are investigated to develop new tools and approaches for estimating field-scale reactive transport parameters and characterizing contamination sites.

Contaminants can migrate directly into groundwater from below-ground sources (e.g. storage tanks, pipelines) that lie within the saturated zone. Additionally contaminants can enter the groundwater system from the surface by vertical leakage through the seals around well casings, through wells abandoned without proper procedures, or as a result of contaminant disposal of improperly constructed wells [1].

1.1. Governing processes of contaminant transport

Generally three processes can be distinguished which govern the transport of contaminants in groundwater: advection, dispersion and retardation. Dispersion and density/viscosity differences may accelerate contaminant movement, while retardation processes can slow the rate of movement. Some contamination problems involve two or more fluids. Examples include air, water and organic liquids in the unsaturated zone, or organic liquids and water in an aquifer. Tracers are useful for characterizing water flow in the saturated and unsaturated zone.

- Advection

The term advection refers to the movement caused by the flow of groundwater. Groundwater flow or advection is calculated based on Darcy's law. Particle tracking can be used to calculate advective transport paths [2]. Particle tracking is a numerical method by placing a particle into the flow field and numerically integrating the flow path.

- Dispersion

Dispersive spreading within and transverse to the main flow direction causes a gradual dilution of the contaminant plume. The dispersive spreading of a contaminant plume is due to aquifer heterogeneities. Dispersion on the macroscopic scale is caused by variations in hydraulic conductivity and porosity. Solute transport can be influenced by preferential flow-paths, arising from variations of hydraulic conductivity, at a decimetre scale.

- Retardation

Two major mechanisms that retard contaminant movement are sorption and biodegradation. If the sorptive process is rapid compared with the flow velocity, the solute will reach an equilibrium condition with the sorbed phase and the process can be described by an equilibrium sorption isotherm. The linear sorption isotherm can be described by the equation:

$$C^* = K_d C \quad (1)$$

Where C^*= mass of solute sorbed per dry unit weight of solid (mg/kg), C= concentration of solute in solution in equilibrium with the mass of solute sorbed onto the solid (mg/l) and Kd = distribution coefficient (L/kg)

- Non aqueous phase liquids (NAPL)

Organic liquids that have densities greater than water are referred to as DNAPL (dense nonaqueous phase liquids). Nonaqueous phase liquids that have densities less than water are called LNAPLs (light nonaqueous phase liquids). Contamination by LNAPL typically involves spills of fuels like gasoline or jet fuel.

1.2. Groundwater flow and contaminant transport modeling

The preliminary steps in modeling groundwater flow and contaminant transport include development of a conceptual model, selection of a computer code, and developing model design [3]. Defining a numerical groundwater flow model is based on parameters like (a) sources and sinks of water in the field system; (b) the available data on geohydrologic system; (c) the system geometry i.e. types and extent of model layers; (d) the spatial and temporal structure of the hydraulic properties; and (e) boundary condition. The widely used MODFLOW [4] and MT3D solute transport [5] numerical codes use finite differences schemes and are considered very reliable. MODFLOW is a three-dimensional modular finite-difference model of U.S. Geological Survey widely used for the description and prediction of the behavior of groundwater system. The program uses variable grid spacing in x and y directions. Parameter estimation can be approached to find the set of parameter values that provides the best fit of model results to field observations,. At first stage, the model computes drawdown, direction of flow and hydraulic heads on each nodal point using a finite difference grid system. Using the steady-state hydraulic heads calculated by the model as the initial condition, the MT3D model is run to simulate contaminant transport in a groundwater system. Once a model is calibrated, it can be used to make predictions for management or other purposes [6].

Two case studies including i) simulated transport of jet fuel leaking into groundwater, Sindh Pakistan; and ii) deep-seated disposal of hydrocarbon exploration produced water using three-dimensional contaminant transport model, Sindh Pakistan have been discussed to highlight the related issues, implications and concerns.

2. CASE STUDY — I

2.1. Introduction

Groundwater is a major source for domestic and industrial uses in many urban settlements of the world. Effluents from industrial areas as well as accidental spills and leaks from surface and underground storage tanks are the main sources of natural groundwater contamination. When such contamination is detected, it becomes essential to estimate the spatial extent of contamination. Conventionally, determination of the extent of contamination is undertaken by taking many samples within time and budgetary constraints from several points, which in general requires the installation of several observation wells. As a result, the cost of such operations can be very high, especially when measurements with higher resolution are required.

A three-dimensional model of the contaminant transport was developed to predict the fate of jet fuel, which leaked from above surface storage tanks in urban site of Karachi, Pakistan. Since the tanks were situated in a sandy layer, the dissolved product entered the groundwater system and started spreading beyond the site. The modelling process comprised of steady-state simulation of the groundwater system, transient simulation of the groundwater system in the period from January 1986 through December 2015, and calibration of jet fuel that was performed in context of different parameters in groundwater system. The fuel was simulated using a modular three-dimensional finite-difference groundwater model (PMWIN) ModFlow and solute transport model (MT3D) in the 1986-2001 periods under a hypothetical scenario. After a realistic distribution of piezometric heads within the aquifer system, calibration was achieved and matched to known conditions; the solute transport component was therefore coupled to the flow. Jet fuel concentration contour maps show the expanding plume over a given time, which become almost prominent in the preceding years.

Two-dimensional (2-D) solute transport models can be used to predict the effects of transverse dispersion of the contaminant plume (spreading). Additionally, 2-D models are appropriate where the contaminant source may lie within or near the radius of influence of a continuously pumping well. While three-dimensional (3-D) numerical models should only be used if extensive data are available regarding vertical and horizontal heterogeneity, and spatial variability in contaminant concentrations. A localized contaminant transport model for groundwater is developed to gain insight into the dynamics of the leakage of jet fuel from above-ground storage tanks in the metropolitan area of Karachi, Pakistan. Jet fuel consists of refined, kerosene-type hydrocarbons, which are mixtures of benzene, toluene, ethyl benzene, and isomers of xylene [7]. Hypothetical monitoring wells were established to estimate the concentrations of jet fuel over a stipulated time period as a result of continuous seepage from the storage depot. Although, no specific data on the history of seepage were available, in view of the results inferred from an electrical resistivity sounding survey (ERSS) regarding the nature of the subsurface lithologies coupled with the findings of previous investigators from Mott MacDonald Pakistan (MMP) [8], it is envisaged that seepage from the storage tanks occurred for more than a decade. ERSS is used to obtain the subsurface resistivity values that are assigned to different geological material. MMP [8] conducted the study on soil and groundwater assessment of environmental damage due to oil pollution and remedial measures were suggested for depots / installations / airfields of Shell Pakistan, scattered throughout the country. Appendix A provides the composition of jet-fuel (Table 1).

2.1.1. Site description

The project site is located between longitudes 67º 07′ 20" and 67º 10′ 30" and latitudes 24º 52′ 20" and 24º 54′ 20". The land surface elevation ranges from approximately 15 to 33 meters above mean sea level (masl). In the far south, the Malir River drains into the Arabian Sea (Figure 1). The typical lithology of the site is silty to sandy clay from 0 to 15 ft (4.6 m) bls, gravelly sand from 15 to 43 ft (4.6 to 13 m) bls, and clayey to silty sand from 43 to 80 ft (13 to 24 m) bls. The region is arid with an average annual rainfall of about 200 mm (7.9 in). Out of this, only 10% [9] is considered to recharge the aquifer system (6.34 x 10-9 m/sec).

The vadose zone is contaminated with up to 1300 ppm of total petroleum hydrocarbons (TPH) within the storage site. In the previous study by MMP [8], soil samples were collected from different locations with 3 feet (1 m) below land surface (bls) and analyzed to estimate the concentration of hydrocarbon compound. Soil samples associated with the storage area have indicated higher TPH concentrations. The contaminant plume follows the hydraulic gradient to the southwest.

2.1.2. Literature review

Kim and Corapcioglu [7] developed two-dimensional model to describe areal spreading and migration of light nonaqueous-phase liquids (LNAPLs) introduced into the subsurface by spills or leaks from underground storage tanks. The nonaqueous-phase liquids (NAPL) transport model was coupled with two-dimensional contaminant transport models to predict contamination of soil gas and groundwater resulting from a LNAPL migrating on the water table. Simulations were performed using the finite-difference method to study LNAPL migration and groundwater contamination. The model was applied to subsurface contamination by jet-fuel. Results indicated that LNAPL migration was affected mostly by volatilization. Further, the spreading and movement of the dissolved plume was affected by the geology of the area and the free-product plume. Most of the spilled mass remained as a free LNAPL phase 20 years after the spill. The migration of LNAPL for such a long period resulted in the contamination of both groundwater and a large volume of soil.

El-Kadi [10] investigated the US Navy's bulk fuel storage facility at Red Hill located in the island of Oahu. The facility consisted of 20 buried steel tanks with a capacity of about 12.5 million gallons each. The tanks contain jet-fuel and diesel fuel marine. The bottoms of the tanks are situated about 80 feet above the basal water table. The geology of the area is primarily basaltic lava flows. Investigations found evidence of releases from several tanks. Two borings were drilled to identify and monitor potential migration of contamination to the potable water source. A numerical model of the regional hydrogeology at the Red Hill Fuel Storage Facility (RHFSF) was developed to simulate the fate and transport of potential contamination from the jet-fuel tanks and the effect on the saltwater/freshwater transition zone of various pumping scenarios.

Periago et al. [11] investigated infiltration into soil of contaminants present in cattle slurry. Column experiments were performed in order to characterize the release of contaminants at the slurry-soil interface after surface application of slurry with subsequent rainfall or irrigation. The shape of the release curves suggests that the release of substances from slurry can be modeled by a single-parameter release function. They compared prediction of solute transport (a) with input defined by the release function and (b) assuming rectangular-pulse input.

Eric et al. [12] developed a parameter identification (PI) procedure and implemented with the United States Geological Survey's Method of Characteristics (USGS-MOC) model. The test results showed that the proposed algorithm could identify transmissivity and dispersivity accurately under ideal situations. Because of the improved efficiency in model calibration, extended application to field conditions was effective.

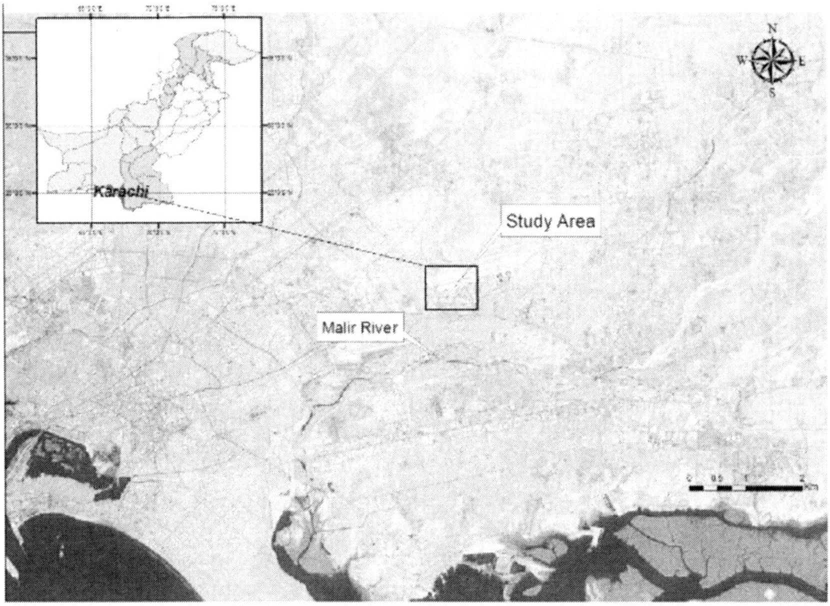

Figure 1. Location of the study area in Southern Pakistan

Jin et al. [13] investigated hydrocarbon plumes in groundwater through the installation of extensive monitoring wells. Electromagnetic induction survey was carried out as an alternative technique for mapping petroleum contaminants in the subsurface. The surveys were conducted at a coal mining site near Gillette, Wyoming, using the EM34-XL ground conductivity meter. Data from this survey used to validate with the known concentrations of diesel compounds detected in groundwater. Groundwater data correlated perfectly with the electromagnetic survey data, which was used to generate a site model to identify subsurface diesel plumes. Results from this study indicated that this geophysical technique was an effective tool for assessing subsurface petroleum hydrocarbon sources and plumes at contaminated sites.

2.2. Model conceptualization and simulation

The groundwater flow system is treated as a two layer. The upper layer (predominantly silty sand) is bounded above by the water table and is 15 ft

(4.6 m) thick, while the lower layer (predominantly clayey sand) is 65 ft (20 m) thick. These are unconfined and recharged from the surface by infiltrating rain, but only over permeable surfaces. A small stream runs along east of model area acts as a drain to groundwater, which flows from the northeast to the southwest. With the exception of the stream, all other boundaries are artificial, that is neither constant head, nor constant flow boundaries. The processes that control the groundwater flow are: (i) recharges from infiltrated rainfall; (ii) flow entering the model across the eastern boundary (also across the northern boundary); (iii) flow reaching the stream; (iv) flow leaving the model across the southern boundary; and (v) pumping from one well near tank no. 9. Values of hydraulic conductivities (K) for the layers are taken from the literature [14]. The K value for depth range 16-31 ft (4.9-9.4m) is taken as 1.7x10-6 ft/sec (5.2 x10^{-7} m/sec) and for depth range 31-97 ft (9.4-30 m) as 1.5x10^{-5} ft/sec (4.6 x10^{-6} m/sec).

The transport and fate of hydrocarbons depend on multi physical and chemical processes, including advection, dispersion, volatilization, dissolution, biodegradation, and sorption. When a solute undergoes chemical reactions, its rate of movement may be substantially less than the average rate of groundwater flow. In this study, retardation of the movement of dissolved hydrocarbons is simulated as a sorption process, which includes both adsorption and partitioning into soil organic matter or organic solvents. The MT3D software was used for simulation [5]. It uses a linear isotherm to simulate partitioning of a contaminant species between the porous media and the fluid phase due to sorption. This sorption process is approximated by the following equilibrium relationship between the dissolved and adsorbed phases:

$$S = K_d C \qquad (2)$$

Where *S* is the concentration of the adsorbed phase (M/M), *C* is the concentration of the dissolved phase (M/L3) and *Kd* is the sorption or distribution coefficient (L^3/M). *Kd* values for organic materials are commonly calculated as the product of the fraction of organic carbon in the soil, *foc*, and the organic carbon partitioning coefficient, *Koc*, or *Kd = foc Koc*. *Koc* values are contaminant specific and reported in various sources [15-17]. The *foc* in the uncontaminated soil was estimated to range from 0.001 to 0.02 based on guidelines by [18]. Assuming the linear isotherm, the retardation factor (*R*) is expressed as follows:

$$R = 1 + (r_b/n_{eff})\, K_d$$

Where r_b is the bulk density of the porous material (M/L3) and n_{eff} is the effective porosity.

For jet fuel, the distribution coefficient K_d is taken as 0.004415 ft^3/kg. With these values [7],

$$R = 1+ [(48\ kg/ft3)/0.25] \times 0.004415 = 1.848R = 1+ [(48\ kg/ft3)/0.25] \times 0.004415 = 1.848$$

2.2.1. Numerical ground water flow modeling

Processing ModFlow for Windows (PMWIN5), a modular 3-D finite-difference groundwater model, is used to configure the flow field [4]. The model consists of 41 columns and 39 rows in each layer (Figure 2). The size of cells is 410 ft x410 ft (125 m x 125 m) outside the fuel storage domain and 205 ft x 205 ft (62.5 m x 62.5 m) within the storage domain. Automatic calibration of the water table was made with algorithm - UCODE and a perfect match obtained with the known condition prior to developing the transport model [6]. Using the steady-state hydraulic heads calculated by PMWIN5 as the initial condition, the solute transport model MT3D was run to simulate the dispersion of the dissolved jet-fuel plume [5]. The parameters adjusted were the retardation factor R for each cell within the finite-difference grid, and the dispersion coefficient. Concentration-time curves have been calculated for ten monitoring wells. PMPATH [19] is used to retrieve the groundwater flow model and simulation result from PMWIN5. A semi-analytical particle-tracking scheme is used to calculate the groundwater flow paths, travel times, and time-related capture zones resulting from pumping a neighboring well at the storage facility [20]. As a preprocessor to modeling and creating input data files, the PMWIN5 utility package was used. Prior to initiating the modeling work, a groundwater information system was established with all data in binary and / or ASCII files that could be exported to other softwares.

2.2.2. Locations of hypothetical wells

The dissolved phase jet-fuel plume was traced using a combination of ten hypothetical monitoring wells (Figure 2) known as MW-1 through MW-10. The wells served to identify lithology, observe water levels, and monitor concentrations of organic compounds. The wells extend to a depth of 80 ft (24 m). In addition, actual well was completed to a depth of 100 ft (30.5 m) near storage tank no.9 in case of emergency need. In the modeling study, this well was used to track the time-related capture zone. The general layout of the storage tanks over the finite-difference grid is shown in Figure 3. The location of the pumping well is marked as a small red square in Figure 2 and Figure 3.

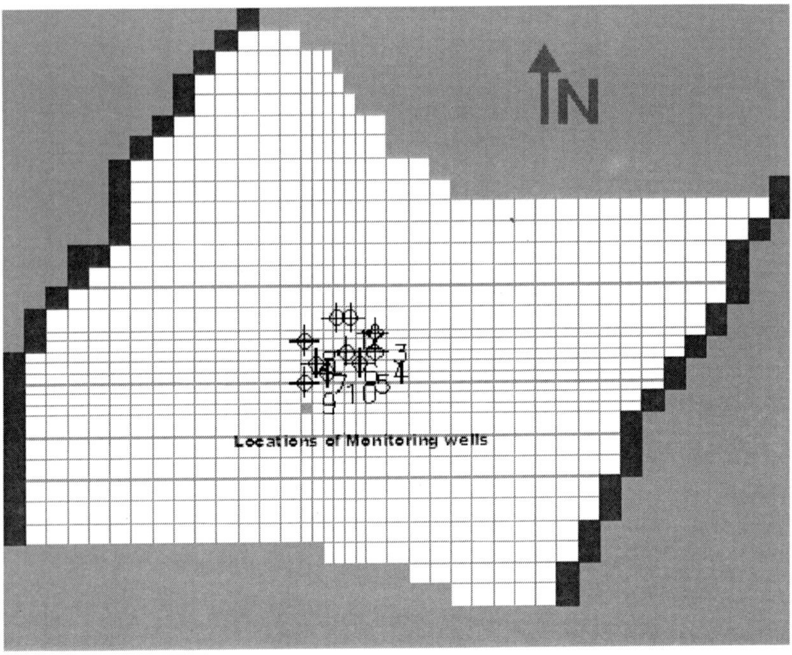

Figure 2. Model design indicating finite difference grid and locations of hypothetical monitoring wells

2.2.3. Model calibration

The model was calibrated for steady-state conditions. Since it was speculated that the seepage of the fuel might have started as early as 1986, simulation of the groundwater flow was begun in January 1971, the opening of storage facility. In the steady-state phase the only input comes from constant head boundaries (along the east and west of the model) and from infiltrated rainfall. All output goes into constant head boundaries (the stream and the southwest boundary of the model). The differences in the known and simulated heads were calibrated to less than 0.30 ft (0.091 m) by making slight adjustments in the *K* values of both the layers. Transient calibration of groundwater flow was accomplished using the time-variant hydraulic head values. Parameters such as recharge rates during each stress period, hydraulic heads in the stream and along the model boundaries, aquifer storage properties, pumping rates, and time-dependent capture zone were adjusted during the calibration. To be objective and consistent, the recharge from infiltration was made equal to 10% of rainfall in each month. Effective porosity of the aquifer was varied between 15% and 25% until the value of 25% was determined to be the best predictor for the model. Constant pumping rates of 1500 US gallons/hr (1.58×10^{-3} m^3/sec) and 1000 US gallons/hr (1.05×10^{-3} m^3/sec) were used in layer 1 and layer 2 respectively. The period from 1986 through 2000 was divided into seven stress periods, each of 2 years in duration. From 2000 to 2001, one stress period was assigned. The length of a stress period was made equal to the number of days in that month.

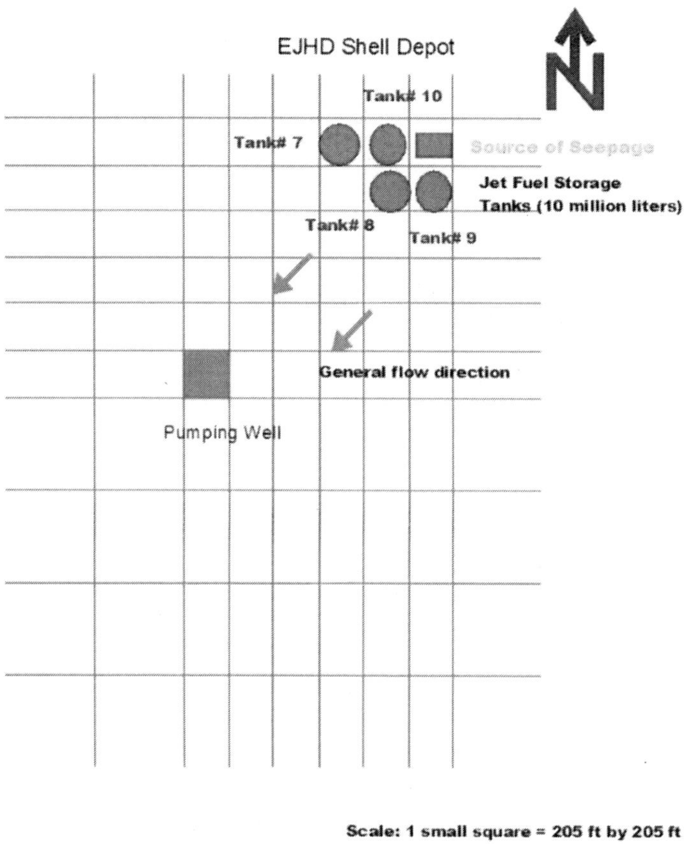

Figure 3. Layout of the storage tanks marked on a finite-difference grid

2.3. Results and discussion

The effect of the pumping well is clearly visible as a cone of depression (Figure 4). The drawdown was determined to be 14.0 ft (4.3 m) near the storage facility.

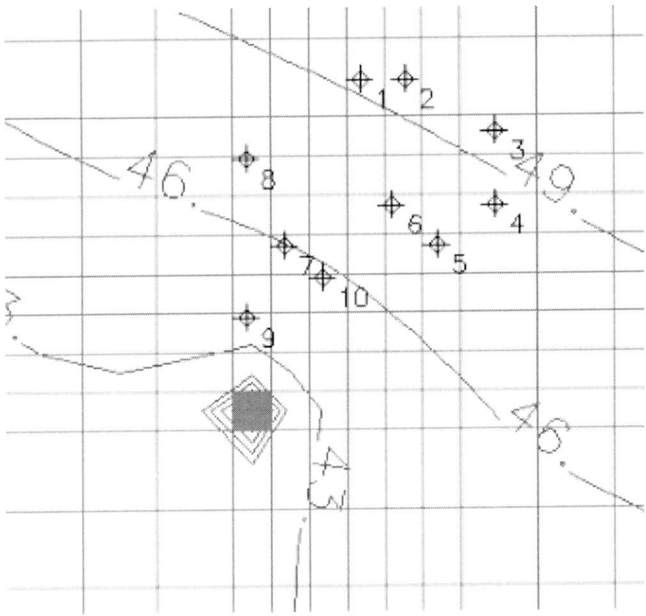

Figure 4. Cone of Depression visible around pumping well developed in layer 1

The model assumes uniform recharge from infiltrated rainfall to every "recharging" cell. Although effective porosity, hydraulic conductivity, and recharge may vary in space and time, the model is expected to have produced a reasonable configuration of the groundwater flow pattern throughout the whole period of simulation. The time-related capture zones produced due to constant pumping are shown in Figure 5 and Figure 6. Water balances, which was calculated for each year of the simulated period, showed a perfect match between the input and output components.

2.3.1. Calibration of plume dispersion

For simulation of the movement of dissolved jet fuel, lateral hydraulic conductivity values equal to 0.864 ft/day (0.263 m/day) for layer 1 and 1.29 ft/day (0.393 m/day) for layer 2 were accepted, while the vertical hydraulic conductivity were taken as 0.0864 ft/day (0.0263 m/day) and 0.129 ft/day (0.0393 m/day) for each layer, respectively [14]. The hydraulic gradient and flow-net were obtained by running the flow component of the model derived from water level information in the previous study.

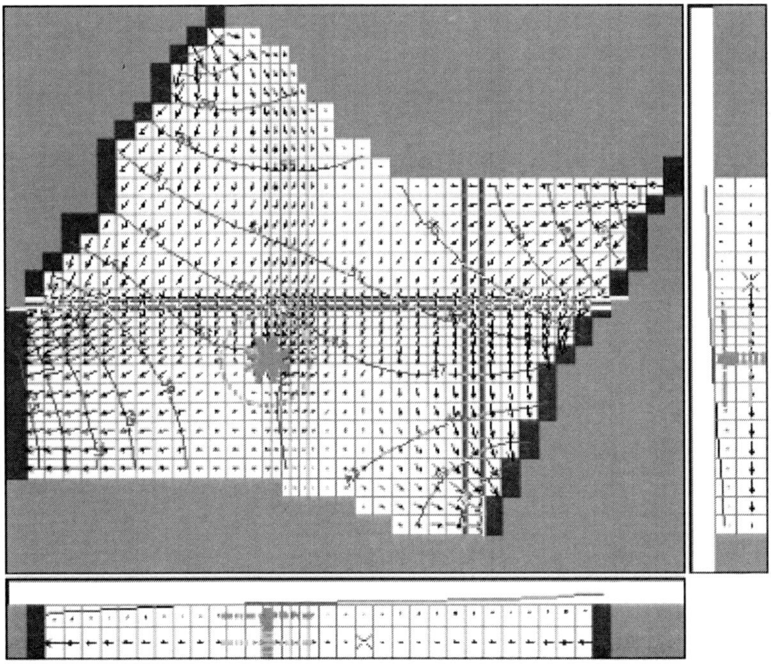

Figure 5. Capture zone of the pumping well with arrows indicating flow directions

The United States Environmental Protection Agency (USEPA) and the Georgia Environmental Protection Division (GAEPD) recommend that the value for longitudinal dispersion should be one-tenth of the distance from the place where a contaminant enters the groundwater system to the down-gradient receptor (a well, stream, or other point of compliance). The distance from the storage facility (tank no. 7) to the pumping well is approximately 100 ft (30.48 m). In all calibration runs, as recommended, the value for longitudinal dispersion was set at 10 ft (3 m). USEPA and GAEPD also recommend for a solute transport model that the value for transverse dispersion equal one-third for the longitudinal dispersion. For this model, transverse dispersion would equal 3.3 ft (1.0 m). In the simulation of the fate of jet fuel, the transverse dispersion coefficient was varied within a range of 2.0 ft to 3.3 ft (0.61 m to 1.0 m). In the model a value of 0.001 ft^2/day (1×10^{-5} cm^2/sec) was used for molecular diffusion. With a retardation factor of 1.80, dissolved jet fuel takes 1.33 years to travel a lateral distance of about 70 to 80 ft (21 to 24 m) in groundwater beneath tank no. 8. The best value of the microbial decay coefficient for jet fuel is estimated to be 1-10 / day with a microbial yield coefficient for oxygen of 0.52 [14].

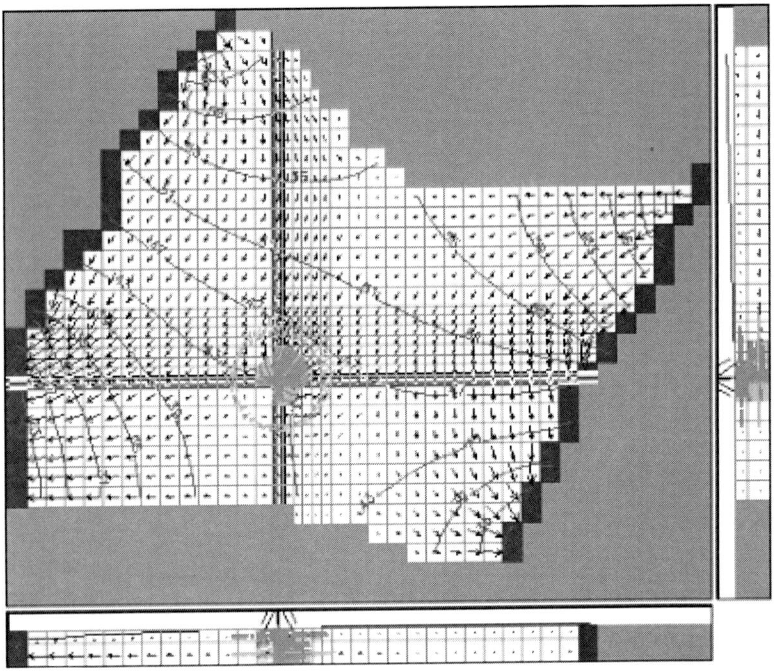

Figure 6. Days-capture zone calculated by PMPATH

2.3.2. Strategy development for release of jet fuel

The previous integrity test run on the storage tanks containing 10 million liters of jet fuel indicated no loss. The date when the leak initially began is unknown, although inventory records indicated that the leak was not present before tank integrity testing. The product has been detected in several hypothetical-monitoring wells (notably in MW-2 and MW-4) and in many soil samples taken within several tens of feet of the tank. The initial concentration of jet fuel entering the system is not of prime concern for the modeling. The product of the influx (in L^3/T) and the concentration (in M/L^3) gives the total mass of jet fuel entering the system in a certain time interval. For the purpose of calibrating the jet fuel input, the initial concentration used, based upon field data [8], varied from 0.095 to 0.19 g/ft^3 (0.0027 to 0.0054 g/m^3). The initial mass of jet fuel, as simulated by the model, was equal to each of four cells "injecting" at a mass rate of 95 to 190 g/ft^3 (2.7 to 5.4 g/m^3) following the initial period of 15 years during which no groundwater contamination was assumed (Table 1).

Table 1. Strategy developed for the plume modeling scenarios

Phase	**Stress period**	**Condition**
Safe Period	15 years (1971 to 1986)	No leakage
Hazardous Period	10 years (1986 to 1996)	Low to moderate leakage
Risk Assessment	5 years (1996-2001)	Moderate leakage
Future Prediction	14 years (2001 – 2015)	Accretion in leakage

Using steady-state hydraulic heads as initial conditions, the evolution of the plume was modeled over nine stress periods as a result of continuous seepage from cells (18,16,1; tank7), (19,16,1; tank 10), (19,17,1; tank 8), and (20,17,1; tank9) as shown in Table 2.

Table 2. Stress period used in time-dependent solute transport modeling of jet fuel

Stress Period	**Time interval (years)**	**Elapsed Time (sec)**	**Period**
1	2	6.30×10^7	1986 – 88
2	2	12.60×10^7	1988 – 90
3	2	18.92×10^7	1990 – 92
4	2	25.23×10^7	1990 – 94
5	2	3.15×10^8	1994 – 96
6	2	3.78×10^8	1996 - 98
7	2	4.41×10^8	1998 –00
8	1	4.73×10^8	2000 – 01
9	14	9.14×10^8	2001 – 15

2.3.3. Calibration scenario

Parameters describing various processes are used after calibration with different combination of parameters (Table 3).

Table 3. Preliminary and final values of parameters used in modeling

Parameters	Value
Longitudinal Dispersion	10 ft
Transverse Dispersion	3.3 ft
Molecular Diffusion	0.001 ft^2/day
Distribution Coefficient	0.004415 ft^3/kg
Retardation Factor (*R*)	1.80 to 1.848
Decay Coefficient	1 $x10^{-9}$ day^{-1}
Hydraulic Conductivity *K* (Layer-1)	$1x10^{-5}$ ft/sec (0.864 ft/day)
Hydraulic Conductivity *K* (Layer-2)	$1.49x10^{-5}$ ft/sec (1.29 ft/day)
Effective Porosity (Layer-1)	0.25
Effective Porosity (Layer-2)	0.30

The release of the jet fuel is simulated in four cells, all along columns 18 to 20 from row 16 to row 17. The area of injection is equal to 42025 ft^2 (3,904 m^2). The concentration of jet fuel at the source (95 to 190 mg/ft^3 [2.7 to 5.4 g/m^3]) maintained constant throughout the designated "hazardous period" simulation period (1986-1996). The concentration was increased slightly (about 0.01 %) from 1996 through 2011 and further up to longer time duration of 4 years i.e., up to 2015. The plume simulations are shown in Figure 7.

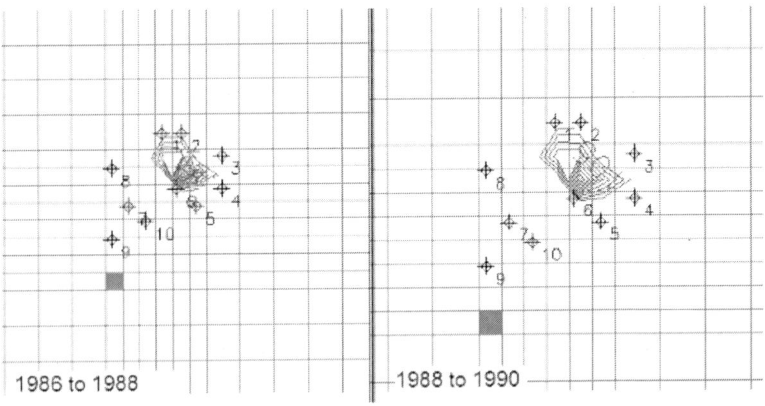

Figure 7. Simulated Jet-fuel plumes(1986 to1988 and 1988 to 1990)

The jet-fuel break-through curves for the hypothetical monitoring wells are shown in Figure 8. Conventionally, determination of the extent and level of contamination is undertaken by taking multiple measurements in wells [21-23]. However, higher spatial resolution generally requires installation of monitoring wells, which is costly [24].

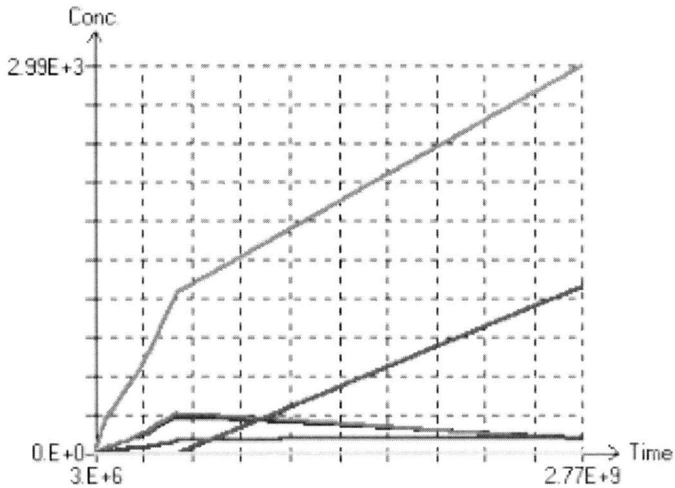

Figure 8. Concentration versus time based on data from 10 monitoring wells

The modeled concentration at MW-1, MW-2, and MW-6 was much higher than the concentrations in the remaining wells. The maximum level of concentrations recorded in MW-6 was 2990 μg/ft^3. Figure 9 reflects the plume spreading of year 2001.

The shape of the plume is elliptical, with the major axis in the direction of groundwater flow. This shape results from advection and longitudinal dispersion. The lateral spread of the plume results from transverse dispersion and molecular diffusion. Upgradient spread of the plume results from molecular diffusion [25-26]. The plume travels toward the stream, which is still far away in the west. By the end of 2015 the effect of the plume becomes evident and monitoring wells MW-4, MW-5, and MW-6 indicated increased concentration of jet fuel (Figure 10). The resultant plume appears to be spreading more in the elliptical path but in the direction of groundwater flow.

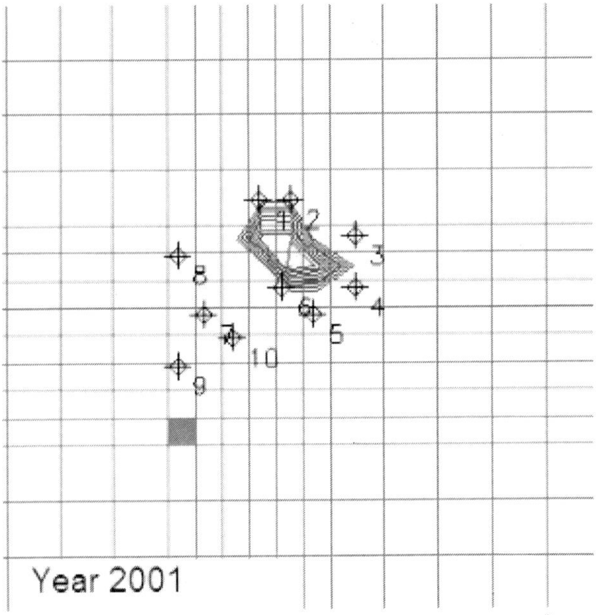

Figure 9. Extent of simulated plume in 2001

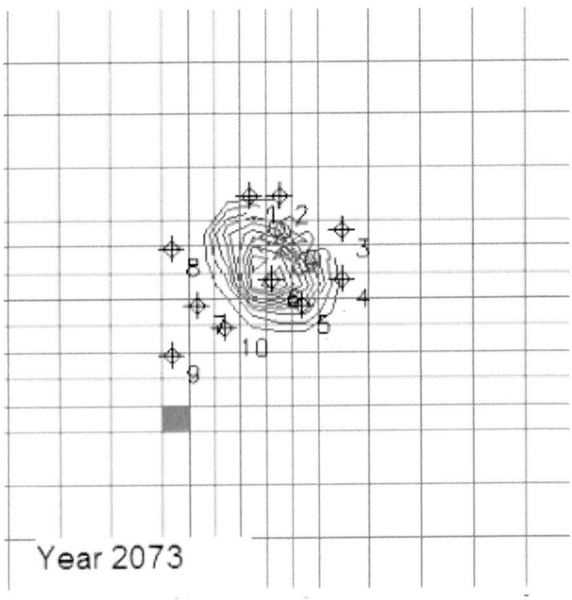

Figure 10. Extent of plume spreading in year 2015

2.4. Conclusions and recommendation

Based on the modeling, it is concluded that the jet-fuel plume has neither expanded nor moved considerably. It is less than 250 ft (76.2 m) beyond the storage tanks and is oriented northeast to southwest. The level of concentration found in the simulated monitoring wells is significant, but because groundwater is brackish and thus unlikely to be used, no harmful effects are expected. However, with the continuous process of leaking jet-fuel from the storage depot, the level of concentration is expected to increase over the period of time. Regionally, the jet-fuel expansion will have on prominent effect over a longer areal coverage and will be confined to a localized area.

An interdisciplinary investigation of the processes controlling the fate and transport hydrocarbons in the subsurface is needed. Concentrations should be performed in wells within and down-gradient of the plume, as field data would help develop a stronger argument for the fate of jet fuel in groundwater. Periodic observations need to be carried out in wells to have good control on the changes of groundwater chemistry. Defective storage depots need to be mended to stop the release of jet-fuel in future.

2.5. Appendix a

2.5.1 Composition of contaminant (Jet fuel)

Knowledge of the geochemistry of a contaminated aquifer is important to understand the chemical and biological processes controlling the migration of hydrocarbon contaminants in the subsurface. Originally, the jet fuel (kerosene oil) is the name assigned to a material with a biological origin, but now it is used to describe materials most of which contain carbon and hydrogen and which may contain oxygen, nitrogen, the halogens, and lesser amounts of other elements. The simplest of these are the hydrocarbons, molecules of hydrogen and carbon, many of which are the components of natural gas, petroleum, and coal. Petroleum, however, has a very large number of components ranging from methane to the high molecular weight materials asphalt and paraffin. Typical fractions into which crude oil is separated in an oil refinery and some principal molecular species are shown in Table 4.

Table 4. The Fractions and Representative Components obtained from Crude Oil

Fraction from distillation	Boiling range	Product of secondary treatment	Typical molecular components
Gas	Below 20°C	Gas Liquefied Pet. Gas (LPG)	CH_4methane, C_2H_6 ethane C_3H_8propane, C_4H_{10}butane
Naphtha	20° – 175°C	Naphtha gasoline	$C_{11}H_{24}$ $C_{18}H_{38}$
Kerosene	175° – 400° C	Kerosene diesel fuel	$C_{11}H_{24}$ $C_{18}H_{38}$
		Lubricating oil	$C_{15}H_{32}$ $C_{40}H_{82}$
Residue	above 400°C	Asphalt	Heavier hydrocarbons

3. CASE STUDY — II

3.1. Introduction

It is important to maintain the existing quality of groundwater because once contamination occurs; it is sometime difficult or rather impossible to clean the aquifer. There is high probability associated with certain landuses like agriculture, industrial/urban land and drainage wells for contaminating the groundwater. The early detection of such contamination can be executed through proper monitoring of the groundwater quality. Many oil & gas companies are disposing off their producing water into the deep Ranikot formation in Bhit oil-field area of southern Pakistan. The producing water contains Total dissolved solid (TDS) within range of 18,000 - 22,000 mg/l besides oil condensate. There were concerns that the producing water is affecting the fresh water aquifers belonging to the overlying Nari and Kirthar formations. This phenomenon has been studied by the utilization of groundwater contaminant transport model. Injection has been monitored at 2100 meters depth in the Pab sandstone formation. A three-dimensional contaminant transport model was developed to simulate and monitor the migration of disposal of hydrocarbon exploration produced water in Injection well at 2000 meters depth in the Upper Cretaceous Pab sandstone in the study area. Framework of regional stratigraphic and structural geology, landform characteristics, climate and hydrogeological setup were used to model the subsurface aquifer. The shallow and deep-seated characteristics of geological formations were obtained from electrical resistivity sounding surveys, geophysical well-logging information and available drilling data. The modeling process comprised of steady-state and transient simulations of the prolific groundwater system and, predictive simulation of contaminants transport after 1-, 10- and 30-year of injection. The contaminant transport was evaluated from the bottom of the injection well and its short and, long-term effects were determined on aquifer system lying in varying hydrogeological and geological conditions.

3.1.1. Description of study area

The study area of Bhit oil field is located about 43 km south of the Manchhar Lake within longitudes 67° 25' - 67° 48' E and latitudes 26° 01' - 26° 30' N in Dadu district in Sindh province of Pakistan (Figure 11). Manchhar Lake, one of the largest lake in Pakistan and in Asia, is formed in a depression in the western side of the Indus River in Sindh province. The total catchment area of the lake is about 97,125 km^2 [27]. The surface area of the lake fluctuates with the seasons from as little as 350 km^2 to as much as 520 km^2 [28]. The lake is fed by two canals, the Aral and the Danister emerging from the river Indus in their eastern side. Due to less rainfall and contamination of surface water, the Manchhar Lake contains brackish water. The elevation ranges between 45 m at the Manchhar Lake and 163 m towards Bhit study area. On regional scale the area is a part of Gigantic Indus river basin composed of alluvium transported by the river and its tributaries. The main surface water sources are Naig Nai stream originating from Bhit Mountain range in the west, Dhanar Dhoro stream passing close to the oil-field plan, besides other small intermittent streams which remain dry in most parts of the year. The discharge data of these streams were not available. The water supply for local communities is maintained by the springs originating from the nearby Limestone Mountains of the Kirthar Range. According to reconnaissance studies conducted in the area, fresh groundwater source is available at different locations in the alluvium. Presently, there is no significant groundwater development in the area. Only few water wells were constructed to fulfill the requirements of local communities. The major sources of potable water to humans and livestock and for irrigation are unconsolidated aquifers. A water supply scheme consisting of four tube wells at Jhangara village is providing water to the villages between Manchhar Lake and Jhangara. The rainfall is scanty. Average annual rainfall is about 200 mm. It is higher in summer months like July and August due to prevalence of monsoon conditions. The aquifer area is located in the alluvial deposits along the Naig Nai stream.

The water table is generally in phreatic to semi-unconfined conditions. The observation wells drilled in the area indicated water table depth of about 12 m and hydraulic head value of 148.8 meter above sea level (masl). The groundwater flow is generally from southwest towards northeast direction. The flow direction of groundwater is true replica of the flow direction of Naig Nai stream draining the area. The groundwater level is mainly influenced by seasonal floods, stream flows and tubewell (water well) discharge. The fluctuations are small in the deep water table in the piedmont plain.

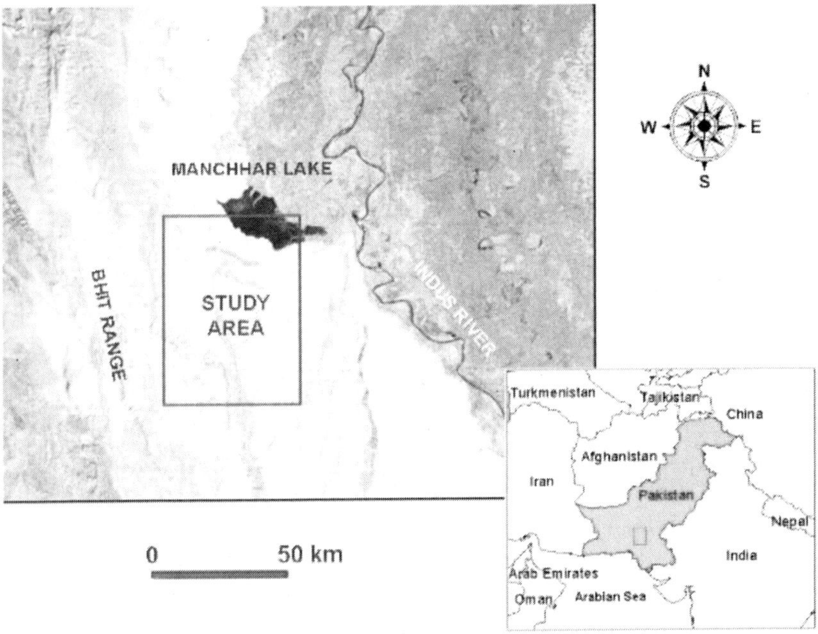

Figure 11. Location of study area in southern Pakistan

3.2. Material and methods

The data of subsoil properties, aquifer characteristics and existing groundwater conditions were collected through reconnaissance level field investigations including geophysical survey. The electrical resistivity soundings survey (ERSS) was conducted primarily to collect required input data for the modeling. The surface drainage and topographic information were extracted from the topo-sheets prepared by Survey of Pakistan on 1:50,000 scale. The climate data i.e. precipitation and temperature, of 1961-1976 period were acquired of nearby meteorological stations like Karachi, Nawabshah, Moenjodaro and Khuzdar from Meteorological Department and Water and Power Development Authority (WAPDA) Pakistan. The published literatures of the region i.e. see [29-32] had been used to firm up the study results.

3.2.1. Geophysical data analysis

The geophysical/drill logs of the injection well field suggest that the subsurface material is composed of layers of sandstone, limestone, dolomite and clay-stone of different formations. One of the interpretative seismic sections of the Bhit concession area (shown in Figure 12) indicates the deeper Chiltan limestone formation beyond the depth of injection well. The hard rock aquifers are mainly composed of partially fractured limestone and sandstone belonging to Nari and Kirthar formations. Limestone, which is the dominant formation, has solution channels due to water action having secondary permeability characteristics. Further, chances of transport contamination could take place much significantly through fracture zone of limestone and dolomitic formations. The hydraulic properties of the underlying overburden and rocks evaluated through geophysical well log data are shown in Table 5.

Table 5. Summary of the Aquifer characteristics of hard rock formations

S.No.	*Formation*	*Depth (m)*	*K (m/d)*	*Transmissivity T (m^2/d)*
1	Nari (sandstone)	113	5	130 (S=.01)
2	Kirthar (limestone)	616	24	2000 (S=.04)
3	Ghazij (claystone)	762	zero	Regional Seal
4	Laki (limestone & dolomite)	1,259	0.165	4 (S=0.00005)
5	Dunghan (dolomite and claystone)	-	-	-
6	Ranikot (Lakhra+Bara+Khadro)	1,799	0.68	14.97 (S=0.000007)
7	Pab (sandstone)	2,000 to onward	0.138	17.5 (S=0.000005)

The hydraulic conductivity of the unconsolidated deposits is about 19 m/day and the effective porosity is 0.25. Several hydrocarbon wells are producing gas in the Bhit concession area. The Ghazij (claystone) was found to be a cap rock (regional seal) over Pab sandstone - an enriched reservoir of hydrocarbon. The Ranikot formation is a prolific rock unit having good transmissivity and storativity to accumulate disposal waste of the producing gas Bhit concession.

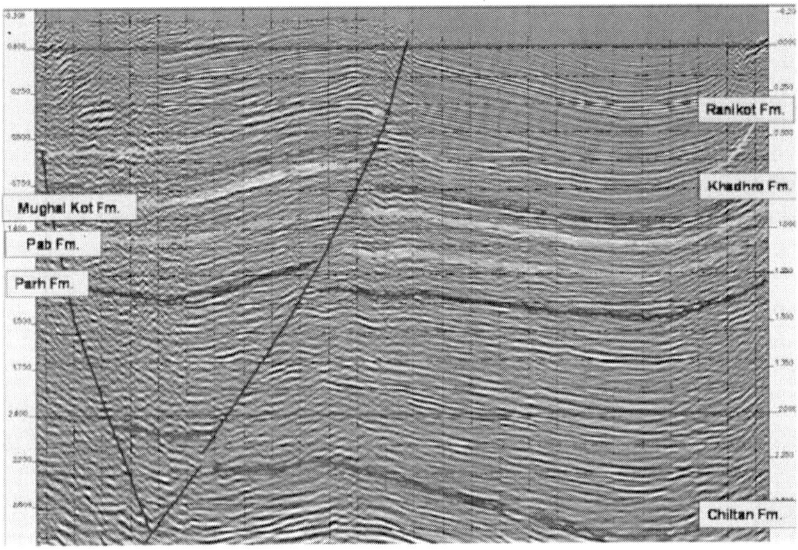

Figure 12. Interpretative seismic depth section indicating several geological formations

3.2.2. Model conceptualization and simulation

The groundwater flow system was treated as a multi-layered system. The upper layer aquifer is mainly unconfined and at depths where silt and clay horizons are present, the sand could probably cause partial confinement in some areas. The chances of transport contamination could take place much significantly through fracture zone of limestone and dolomitic formations. Seven aquifer layers were defined on the basis of physical characteristics of lithological formations (Table 1). The disposal of produced water in the injection well is set at 2,100 m depth in the Pab sandstone formation. The main source of recharge to groundwater is rainfall which is highly variable. During rains, different nullahs carry flows and infiltrations through the piedmonts and alluvial fans and cause subsequent lateral movement at depth. The recharge is higher in summer especially in the months of July and August. It may occur during rainy month of March in winter, a mean value for the limestone recharge may be taken as 200 mm/year based on the recharge data of rainfall. The main discharge components are groundwater extraction from water wells/dug wells, evapotranspiration, and spring discharge. The abstraction of groundwater becomes higher during months of little recharge to groundwater i.e. November to January, which may affect the storage of groundwater for a limited period.

The MODFLOW code and the MT3DMS code were used to solve the flow and transport equations. The model domain comprised of 40 x 30 grid network with total area of 1200 sq km (Figure 13). First, the model was run for steady state condition. The model took up groundwater extraction from the oil-field water wells (TW1 & TW2) and the community tube wells (TW3 & TW4). The discharge rate of the water wells was 0.0083 m^3/sec and discharge rate of deep seated injection well (DW) as supplied was 0.00152 m^3/sec. The injection of the produced wastewater from the injection well was considered during simulation. Once the flow model was completed and run was carried out, the contaminant transport model was set and simulated to evaluate the groundwater contamination and movement of plume.

3.2.3. Simulation of contaminant transport of injection well

The steady-state hydraulic heads were used as initial condition in MT3DMS option available in PM5 to simulate the dispersion of plume. The MT3DMS model simulates the processes i.e. advection, mechanical dispersion, retardation, decay and molecular diffusion related to the fate of contaminant. Initial concentration was set to zero in all the layers. In Advection package, 3rd order TVD scheme [12, 33] was selected. This method is considered as a good compromise between the standard finite difference and particle tracking approaches. In dispersion package, TRPT (Horizontal transverse dispersivity/Longitudinal dispersivity) was set to 0.3 for all the layers except in layer 3, where it was set to 0.1. In dispersion package, TRPV (Vertical transverse dispersivity/Longitudinal dispersivity) was set to 0.3 for all the layers except in layer 3. For this layer, it was set to 0.1. Longitudinal dispersivity was set to 10 m. There was no sorption selected in chemical reaction package. The injection well was set at layer 7 in the sink/source menu. The concentration of the injection well liquid was considered to be 100 ppm. The model was then simulated for 1-, 10-, and 30-year period for studying the behaviors produced wastewater in and around the injection well.

3.3. Results and discussion

The computations were carried out for three cases i.e. in case-I, Injection well was continuously discharged for one-year period, in case-II it well was simulated for 10-year period and in case-III for 30-year period. The hydraulic heads and drawdown were computed in all three cases. The velocity vectors prominent in layer-1 tend to move in the northeast direction towards Manchhar Lake. The groundwater flow had shown decline in confining layers like 4 and 7. The two tube wells of the oil-fields each discharging at the rate of 8.3 x 10^{-3} m^3/sec along with community wells were used for the study. The results obtained from the 3-D transport model are shown in Table 6 and transport of contaminant plumes in three simulation periods in Figures 14-16.

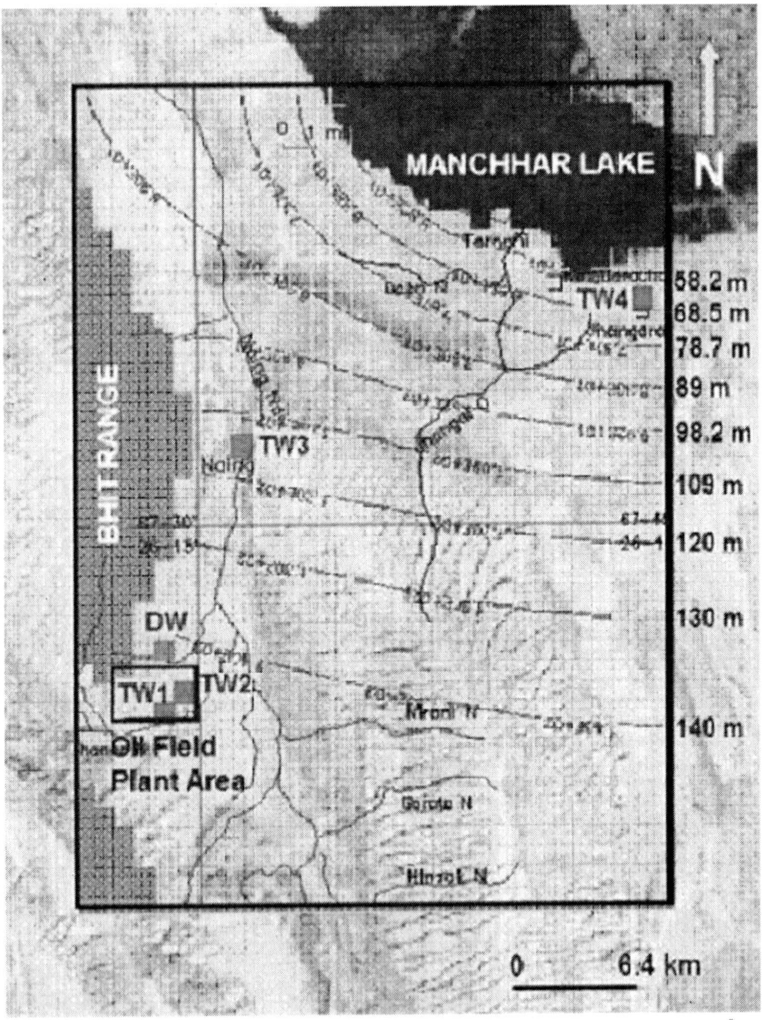

Figure 13. Manchhar Lake treated as constant head boundary; Bhit Range as impervious boundary. *DW* deep injection well, *TW* tube well

Table 6. Maximum concentration observed in different simulation periods (ppm)

Layer	1 Year	10 Years	30 Years
1	-	2.21×10^{-23}	1.96×10^{-20}
2	3.16×10^{-27}	3.88×10^{-21}	1.32×10^{-18}
3	3.14×10^{-21}	3.54×10^{-16}	4.01×10^{-14}
4	3.16×10^{-15}	3.18×10^{-11}	1.06×10^{-09}
5	9.58×10^{-10}	9.53×10^{-07}	2.45×10^{-05}
6	8.04×10^{-05}	7.95×10^{-03}	7.21×10^{-02}
7 Pab (Sandstone)	6.28×10^{-02}	0.626	1.861

After 30 years of simulation period, only traces of contamination were found in Ghazij Formation. Moreover, it is found that after 1-year period of simulation the produced wastewater will reach upward in layer-5 (Ranikot Formation) emerging from layer 7 (Pab sandstone) as shown in Figure 14. In this period, no contamination was found in layer 1 and 2. In 10-year simulation a plume of produced water moved from layer 7 to layer 5 (Figure 15). Only traces of contamination were found in layer 3 (Ghazi Formation). In Figure 6, plume of produced water contamination indicates movement from layer 7 to layer 4 after 30 years simulation. The layer 3 was found to be acting as a regional confining seal. In this layer, only traces of contamination were present. The movement of produced wastewater was found within a radius of 3 km at the bottom of injection well in the Pab sandstone. The upper aquifers in the alluvial deposit, Nari sandstone, and Kirthar limestone was remain safe from the effects of produced wastewater disposal from the deep seated injection well. The community wells tapping in the upper few tens of meters, naturally oozing springs and the Manchhar Lake located about 43 km from the injection well were also found to be safe from the effects of produced water injection even after contaminant transport simulation of 30-year period. The development of plume was significant in layer 7 and upward in the three cases (shown in Figures 14-16).

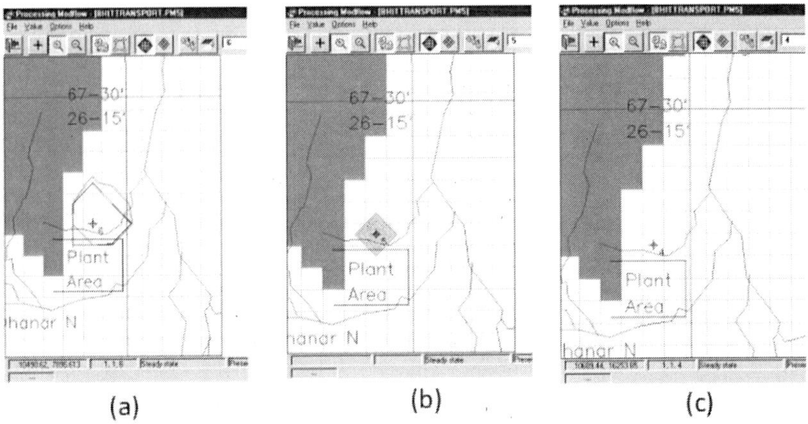

Figure 14. Upward movement of the plume from layer 7 to layer 6 (a); from layer 6 to layer 5 (b) and from layer 5 to layer 4 (c) after 1 year simulation period

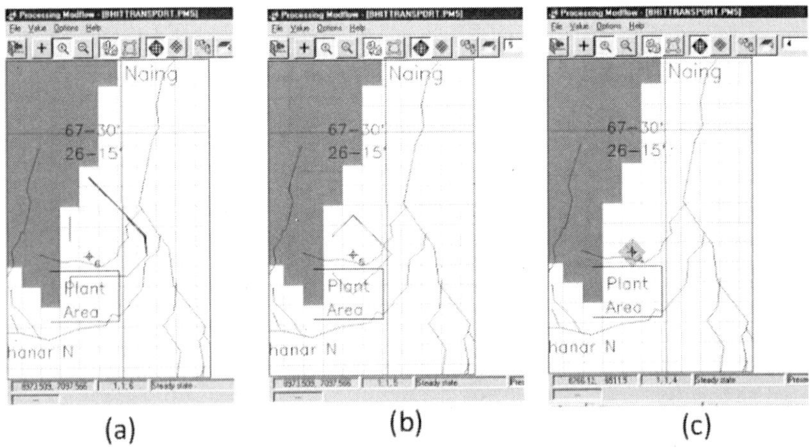

Figure 15. Upward movement of the plume from layer 7 to layer 6 (a); from layer 6 to layer 5 (b) and from layer 5 to layer 4 (c) after 10 year simulation period

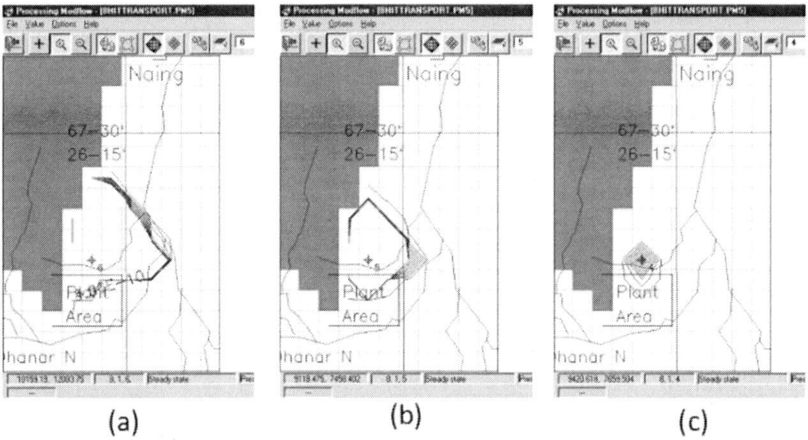

Figure 16. Upward movement of the plume from layer 7 to layer 6 (a); from layer 6 to layer 5 (b) and from layer 5 to layer 4 (c) after 30 year simulation period

4. CONCLUSIONS

The results of the three-dimensional groundwater modeling study highlighted the hydrogeological characteristics and features of the contaminant transport of the deep injected tube well in the Bhit oil-field area. The groundwater contaminant transport modeling technique has proved to be effective in simulating produced wastewater plume from the deep seated injection well. The study would provide base for evaluating risks of contaminants on long term basis in similar conditions in future. Risk of expansion of plume regionally does not exist as the disposal of wastewater is made in the deeper horizon well below the aquifers and also the quantity is quite limited.

Thorough understanding of surface hydrology, hydrogeological conditions and contaminant behavior in the aquifer system coupled with application of reliable modeling techniques could be helpful in dealing with water management issues related to contaminant hydrology.

REFERENCES

1. J. R Boulding, and J. S Ginn, Practical handbook of soil, vadose zone and groundwater contamination: assessment, prevention and remediation, CRC Press; 2004

2. D. A Walter, and J. P Masterson, Simulation of Advective Flow under Steady-State and Transient Recharge Conditions, Camp Edwards, Massachusetts Military Reservation, Cape Cod, Massachusetts, Water-Resources Investigations Report 034053USGS; 2003
3. M. P Anderson, and W. W Woessner, Applied Groundwater Modeling, Simulation of Flow and Advective Transport. Academic Press, Inc San Diego, California; 1992
4. M. G Mcdonald, A. W Harbaugh, ModFlow, A modular three-dimensional finite-difference ground-water flow model, U.S. Geological Survey Open-File Report 83875Chapter A1, Washington D.C.; 1988
5. C Zheng, and P. P Wang, A Modular Three-Dimensional Multispecies Transport Model, US Army Corps of Engineers, Washington, DC: 1999
6. E. P Poeter, and M. C Hill, UCODE, a computer modelling. *Computers & Geosciences* 199925457462
7. J Kim, and M. Y Corapcioglu, Modeling dissolution and volatilization of LNAPL sources migrating on the groundwater table, *Journal of Contaminant Hydrology* 2003651-2
8. Mott McDonald Pakistan (pvt) LtdA report on soil and groundwater Assessment of depots/ Installations/ airfields, Shell Pakistan Limited; 2000
9. A Ashraf, Z Ahmad, Regional groundwater flow modeling of upper Chaj Doab, Indus Basin. *Geophysical Journal International* 20081731724
10. El-A Kadi, Groundwater Modeling Services for Risk Assessment Red Hill Fuel Storage Facilities, NAVFAC Pacific, Oahu, Hawaii; 2007
11. E. L Periago, A. N Delgado, and F Diaz-fierros, F. Groundwater contamination due to cattle slurry: modeling infiltration on the basis of soil column experiments, *Water Research* 2000343
12. W Eric, E. W Strecker, and W Chu, Parameter Identification of a Ground-Water Contaminant Transport Model, *Groundwater* 1986241
13. S Jin, P Fallgren, J Cooper, J Morris, and M Urynowicz, Assessment of diesel contamination in groundwater using electromagnetic induction geophysical techniques, Journal of Environmental Science and Health 2008436
14. C. W Fetter, Applied hydrogeology (4th edition), Prentice Hall; 2000
15. KeyGroundwater Fate and Transport Evaluation Report: South Cavalcade Superfund Site, Houston, Texas, prepared by KEY Environmental, Inc. on behalf of Beazer East, Inc. for the U. S. Environmental Protection Agency; 1997
16. EPADetermining Soil Response Action Levels Based on Potential Contaminant Migration to Groundwater: A Compendium of Examples, EPA/540/2891989
17. ASTMStandard Guide for Risk-Based Corrective Action Applied at Petroleum Release Sites, American Society for Testing and Materials, ASTM E-173995Philadelphia, PA; 1995
18. C. J Newell, R. K Mcleod, and J. R Gonzales, BIOSCREEN: Natural Attenuation Decision Support System, User's Manual Version 1.3, EPA/600/R-96/087, National Risk Management Research Laboratory, Office of Research and Development, U. S. Environmental Protection Agency, Cincinnati, 45268, Ohio; 1996
19. W-H Chiang, and W Kinzelbach, Processing MODFLOW. A simulation system for modeling groundwater flow and pollution; 2001
20. D. W Pollock, Documentation of a computer program to compute and display path lines using results from the U.S Geological Survey modular three-dimensional finite-difference groundwater flow model: U.S Geological Survey, open-file report 89381Denver; 1989
21. Y. Z Zhou, Sampling frequency for monitoring the actual state of groundwater systems, *Jour of Hydrology* 1996180301318
22. R. W Plus, C. J Paul, Multi-layer sampling in conventional monitoring wells for improved estimation of vertical contaminant distribution and mass. *Jour of Contam. Hydrology* 19972585111
23. R. S Fisher, and P. T Goodmann, Characterizing groundwater quality in Kentucky: from site selection to published information. Proceedings of 2002 National Monitoring Conference, national water Quality Monitoring council, May 1923Madison, Wisconsin; 2002
24. A Zeru, Investigations numériques sur l'inversion des courbes de concentration issues d'un pompage pour la quantification de la pollution de l'eau souterraine / Numerical

investigations on the inversion of pumped concentrations for groundwater pollution quantification. PhD Thesis, Université Louis Pasteur (France); 2004192

25. S Bauer, M Bayer-raich, T Holder, C Kolesar, D Muller, and T Ptak, Quantification of groundwater contamination in an urban area using integral pumping tests *Jour of Contam. Hydrology* 200475183213
26. A Bockelmann, Z Zamfirescu, T Ptak, P Grathwohl, and G Teutsch, Quantification of mass fluxes and natural attenuation rates at an industrial site with a limiten monitoring network: a case study site. *Jour of Contam. Hydrology* 20036097121
27. NESPAKMaster Feasibility studies for flood management of hill-torrents of Pakistan, Supporting Vol-IV Sindh Province. National Engg. Services Pak. (Pvt) ltd. Lahore, Pakistan; 1998
28. http://enwikipedia.org/wiki/Manchar_Lake (accessed 10 September 2012
29. Z Ahmad, I Ahmad, and G Akhtar, A report on groundwater reserve estimation of the Ahmad Khan well field and its safe utilization for the 4000 TPD Luck Cement plant Pezu, D.I.Khan; 1994
30. N Ahmad, Groundwater resources of Pakistan, Ripon printing press; 1974
31. WAPDALower Indus Report. Physical Resources, groundwater, supl.6.1.6, Tube wells and Boreholes, Nara Command; 1965
32. WAPDARohri Hydropower Project. Feasibility study, HEPO 971989
33. A Harten, High resolution schemes for hyperbolic conservation laws. *Jour of Comput. Phys.* 198349357393

CHAPTER 2

Geochemical Characterization of Groundwater in a Volcanic System

Carmelo Bellia [1,†], Adrian H. Gallardo [2,†,*], Masaya Yasuhara [1,†] and Kohei Kazahaya [1,†]

[1]*National Institute of Advanced Industrial Science and Technology (AIST), Geological Survey of Japan, 1-1-1 Higashi, Tsukuba, Ibaraki 305-8561, Japan*

[2]*Argentina National Scientific and Technical Research Council (CONICET), FCFMyN, Department of Geology, San Luis National University, Ejercito de los Andes 950, San Luis 5700, Argentina*

ABSTRACT

A geochemical investigation was undertaken at Mt. Etna Volcano to better define groundwater characteristics of its aquifers. Results indicate that the Na–Mg ± Ca–HCO_3^- ± (SO_4^{2-} or Cl^-) type accounts for more than 80% of the groundwater composition in the volcano. The remaining 20% is characterized by elevated Ca^{2+}. Waters along coastal areas are enriched in SO_4^{2-} or Cl^-, mainly due to mixing with seawater and anthropogenic effects. The majority of the samples showed values between −4‰ to −9‰ for $\delta^{18}O$ and −19‰ to −53‰ for $\delta^{2}H$, suggesting that precipitation is the predominant source of recharge to the aquifers, especially in the west of the study area. The analysis of $\delta^{13}C$ and pCO_2shows values 1 to 3 times higher than those expected for waters in equilibrium with the atmosphere, suggesting a partial gas contribution from deep sources. The diffusion of gasses is likely to be controlled by tectonic structures in the volcano. The ascent of deep brines is also reflected in the CO_2 enrichment (up to 2.2 bars) and enriched $\delta^2H/\delta^{18}O$ compositions observed in the salt mounts of Paternò.

Keywords: groundwater; geochemistry; water quality; isotopes; Mt. Etna

1. INTRODUCTION

Mt. Etna is located on the east coast of Sicily (Italy). It is the tallest active volcano in Europe and one of the most active volcanoes in the world [1]. The volcanic deposits are the most important groundwater reservoir for the entire Sicily, as it is the only drinking water resource for over one million people who live at distances of more than 100 km [2]. The volcano rises over an important regional tectonic system, which causes the crust to break up into an intricate system of fractures and faults that together with geology, paleotopography and geometry of the sedimentary basement are the major factors governing the groundwater flow in the area.

Furthermore, seismic events and volcanic eruptions frequently reshape the morphology of the terrain, potentially altering flow conditions and modifying the groundwater composition. Agriculture and the development of urban and industrial centers such as Catania are largely dependent on the neighboring volcano. In this context, population growth and new economic activities continue to mount pressure on the environment and hence call for a more sustainable use of the region's groundwater resources. Geochemistry and isotope investigations have been used commonly on Etnean aquifers to study the hydrological processes. For instance [3] determined that groundwater concentrations of minor and trace elements stood out with respect to other Italian aquifers due to the major contribution of volcanic gases and hydrothermal fluids. Later, [4] determined the origin and effects of fluid-rock interaction within Mt Etna by analyzing B, O, H, and Sr concentrations. Oxygen and Cl isotopes were used by [5] to determine groundwater recharge and flow paths along the flanks of Mt. Etna. Findings showed that groundwaters beneath intensely cultivated areas were enriched in ^{18}O probably as a result of the evaporation of irrigation waters during summer. Finally, [6] investigated the isotopic signature of Etnean waters, and [7] used tritium activities to trace the age and movement of groundwater in the volcano in recent times.

The hydrology of Mt. Etna has been studied over a long period. This work further expands the current understanding by providing an updated snapshot in time of the geochemical and isotopic composition of Mt. Etna's groundwater.

2. STUDY AREA

2.1. Geological Background

Mt. Etna is located on the eastern coast of Sicily, Italy. The morphology of the volcano is shaped by four summit craters, a caldera of about 18 km perimeter and maximum depth of 1000 m called "Valle del Bove", and numerous side cones scattered along its flanks (Figure 1). Slopes are gentle (7°–8°) up to an elevation of 1800–2000 m above sea level (A.S.L.), increasing to 20°–25° at higher elevations. The volcanic edifice consists of a lower shield unit overlain by a stratovolcano. The shield rests discordantly on Miocene flysch deposits to the NW, and argillaceous Pleistocene sediments to the SE [8,9] and consists of plateau terraces of submarine lavas derived from fissural emissions generated at an early stage, approximately 500 ka. The basal unit was followed by pyroclastic material and sub-aerial tholeiitic lavas outcropping rather discontinuously in the southern sectors of the volcano approximately 300 ka [10]. Products changed in composition about 200 ka to transitional and later Na-alkaline tholeiites [11].

Currently, Mt. Etna releases about 11.66 kTons/day of CO_2 from the summit and as diffuse soil emanations from the upper flanks [12]. This quantity is much larger than other active volcanoes and corresponds to about 10%–15% of the CO_2produced by all the volcanoes on the planet [13].

The climate of the region follows the pattern of the Mediterranean region, with higher precipitations in autumn and winter. Most rainfall occurs to the east, reaching up to 1200 mm/year. To the south, the average precipitation decreases to about 440 mm/year, with a wide range of intermediate conditions in between. At heights above 2000 m precipitation tends to occur as snow.

2.2. Hydrogeological Setting

Mt. Etna hydrogeological settings are similar to other basaltic volcanoes: fissured and highly permeable lavas are interbedded with discontinuous layers of low permeability pyroclastics. According to [9,14], typical

Etnean aquifers can be described as unconfined and hosted by highly permeable volcanites. On the basis of structural, geological and geophysical data, three main hydrogeological basins were defined (Figure 1): (1) the eastern basin, with flow towards the Ionian Sea; (2) the southern and western basins flowing towards the Simeto River; and (3) the northern basin with flow to the Alcantara River [14,15]. Groundwater recharge preferentially occurs at high elevations where rainfall infiltrates through the unsaturated sediments and then follows a radial pattern towards the peripheral sedimentary terrains. Groundwater flow originating at lower heights, where the volcanic cover is much thinner, is mainly controlled by the shape of the impermeable substrate. According to [16], Etna's volcanites generally have a high intrinsic permeability (2.5×10^{-7} to 2.9×10^{-6} cm^2). In contrast, the permeability of the basement sediments would average 10^{-10} cm^2 [17].

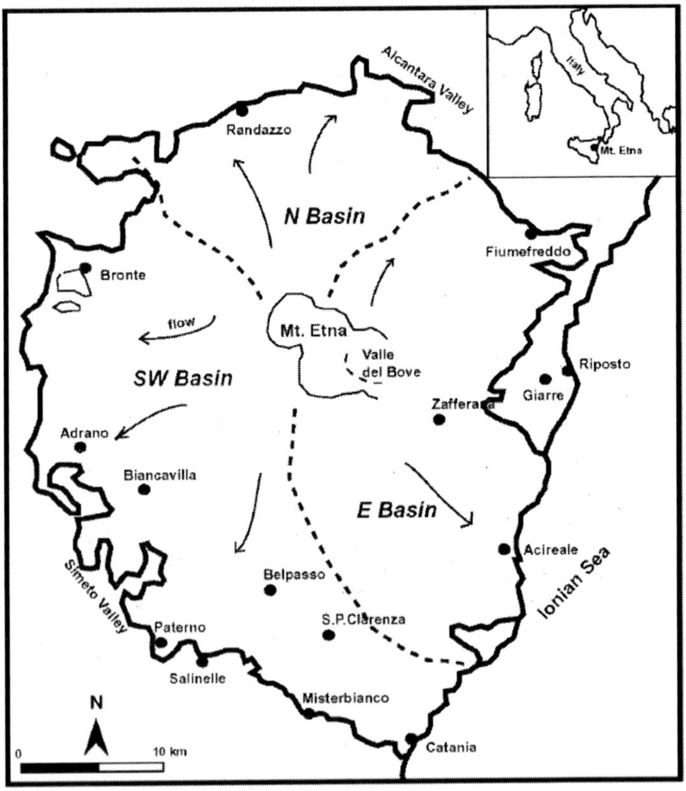

Figure 1. Outline of the investigation area and hydrogeological basins within Mt. Etna.

The absence of a developed hydrographic system on the surface suggests an important circulation of groundwater, highlighted by the presence of hundreds of springs and wells of significant flow rates. Yields over 80 L/s have been described by several authors since the 70s [16,18].

Recent hydrogeochemical studies (e.g., [19,20]) indicated that groundwater in Mt. Etna has a general composition of bicarbonate type, with a few samples of chloride-sulphate type. The relative abundance of major elements in solution is generally (Na, Mg) > Ca > K for cations, whilst HCO_3^- always prevails over other anions [20].

Findings from [21,22] using $\delta^{18}O$ and $\delta^{2}H$ data suggested that Etnean groundwaters are meteoric in origin. Nevertheless, more recent work from [5] showed that the isotopic imprint of groundwater in Mt. Etna might reflect several sources such as evaporation from the Mediterranean Sea to the east, moisture from the Atlantic Ocean on the lower northern flanks, and volcanic vapor affecting precipitation on the upper regions of the cone.

3. SAMPLING AND ANALYTICAL METHODS

An extensive sampling campaign was undertaken in early 2014 at Mt. Etna to determine the chemical characteristics of groundwaters in the volcano. Samples were collected during three stages from 46 boreholes, 14 springs, 2 surface water points, and 6 locations within the so-called "mud volcanoes" (Figure 2). The bores are all being used for water supply or are connected to storage tanks. As a first measure, piped water-supply and taps were disinfected with a 10% hypochlorite solution and water run for a few minutes before sampling commenced. Samples were then directly collected from the source, field-filtered at 0.45 μm, and stored in 500 mL plastic containers for the analysis of dissolved inorganic elements. Bottles were cooled to 4° and dispatched for analysis within 48 h. A number of field blanks and duplicates were also sent to the laboratory for quality control. Standard parameters (pH, EC, temperature) were measured *in situ* using a Hydrolab Qanta probe, while a colorimetric titration kit was employed to calculate the alkalinity (HCO_3^-) content in waters. Chemical concentrations were determined at the laboratories of the Geological Survey of Japan in Tsukuba. Major cations were analyzed by an inductively coupled argon plasma atomic spectrophometer (ICP-AES), and anion determinations were carried out by ion-chromatography. Additional samples were collected for the analysis of carbon-13 ($\delta^{13}C$), and treated in the field with $HgCl_2$ to prevent biological fractionation. These samples were stored in glass bottles and analyzed by a mass spectrometer. Results were reported as

‰ deviations (per mil) from the PeeDee Belemnite standard. Oxygen and hydrogen isotopes were also measured by an isotope ratio mass spectrometer. Measurements were referenced to the VSMOW international standard, and reported in the conventional delta notation.

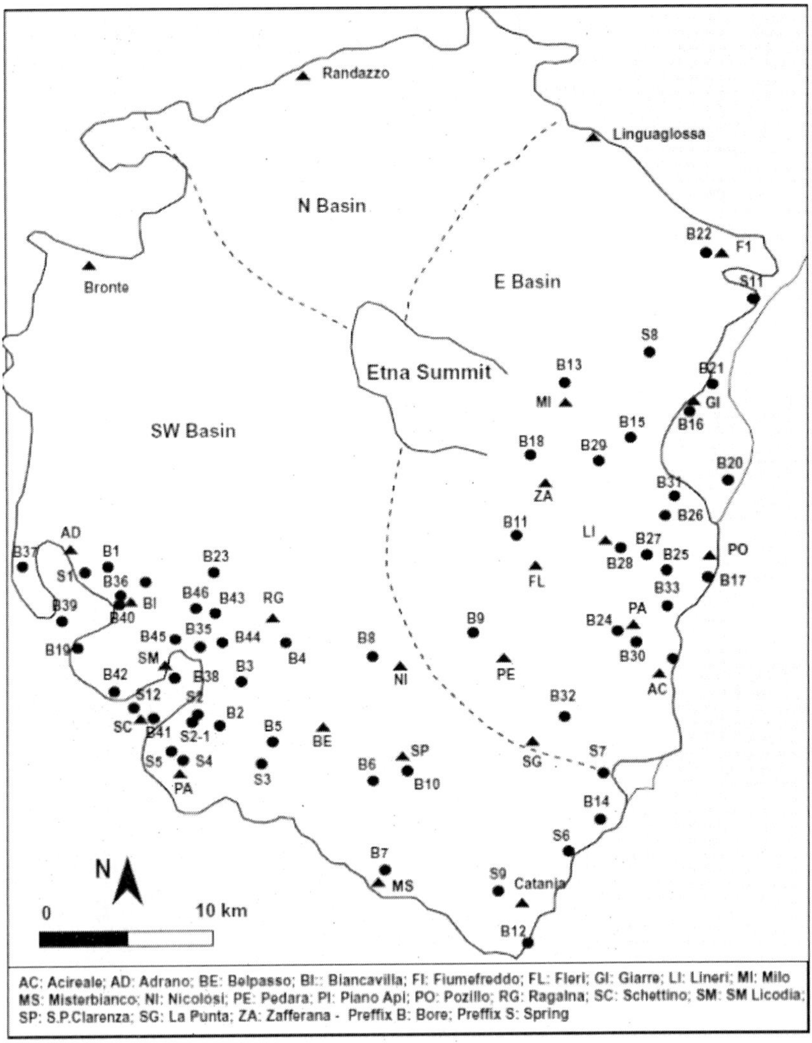

Figure 2. Location of the sampling sites in Mt. Etna.

4. RESULTS AND DISCUSSION

4.1. Groundwater Composition

Table 1 summarizes the groundwater composition of Mt. Etna groundwaters found in this study.

With the exception of SO_4^{2-}, ion concentrations increase towards the west, where values are usually two to three times higher than neighboring regions (Figure 3).

There is a general inverse relationship between water mineralization and elevation. The lowest groundwater contents were recorded on the north, and towards the cone summit in the eastern basin. In contrast, maximum concentrations were measured on the southernmost flank of the volcano (western basin). Results also indicate that Etnean waters have low temperatures, with an average of 17 °C. The highest water temperatures were recorded in intense tectonically fractured areas such as Paternò, Biancavilla, and Adrano in the West, and between Pozzillo and Zafferana in the East. This is consistent with [23], who argued that in an active volcanic system such as Mt. Etna, the transfer of deep gases (and the associated heat) toward the surface occurs principally along zones of high permeability in the crust.

Furthermore, high-temperature volatiles released from the magma could interact with descending meteoric waters causing a general increase of the temperature in the rest of the aquifer. This is in line with [24], who suggested a main magmatic signature linked to degassing of an enriched mantle beneath Mt. Etna. [25] characterized Etna's magma as CO_2-rich while [26], hypothesized that the asthenosphere beneath the volcano rises to a depth that permits the continuous escape of CO_2 from the mantle. Present findings of CO_2 partial pressure °pCO_2), also suggest that Etnean waters might interact with CO_2 of deep origin, arguably magmatic. Measured pCO_2 varies between 5×10^{-4} and 2.2 bars and more than 90% of the samples exhibit values 1 to 3 orders of magnitude higher than those expected for waters in equilibrium with the atmosphere ($pCO_2 = 10^{-3.6}$ bar). These positive emission anomalies could be attributed to the release of CO_2 by fresh magma that intruded into the volcano plumbing system [27,28]. In contrast, the anomalous high pCO_2 values (1.04 bars) in sample S4 could be ascribed to local waters rapidly charged with CO_2 and dissolved elements from the surrounding soil matrix. This would also explain the high conductance values among the sampled springs. Maximum pCO_2 values (up to 2.15 bars) were recorded at the "Salinella" mounts near Paternò, on the southern fringes of Etna.

Table 1. Average major and trace elements composition of waters sampled in the Mt. Etna.

Sample	Basin	Temp. °C	EC µS/cm	pH	HCO_3 meq/L	F meq/L	Cl meq/L	NO_3 meq/L	PO_4 meq/L	SO_4 meq/L	Na meq/L	K meq/L	Ca meq/L	Mg meq/L	TDS mg/L	Hardness °fH	PCO_2 at 20° bars
B11	E	9.1	219	7.35	2.62	0.02	0.29	0.02	0.02	0.30	1.35	0.21	0.69	1.38	149	10.3	0.007
B12	E	18	1138	7.77	8.16	0.04	3.28	0.92	-	1.70	6.07	0.65	3.50	4.57	736.6	40.3	0.009
B13	E	11.4	289	7.92	1.26	0.01	1.49	0.30	-	0.57	1.33	0.25	0.64	1.87	252.3	12.6	0.001
B14	E	14.7	1123	8.29	3.84	0.03	6.15	1.48	-	2.99	7.73	0.52	1.91	5.95	997.9	39.2	0.001
B15	E	10.6	282	8.37	2.02	0.01	0.96	0.09	0.02	0.79	2.21	0.24	0.57	1.13	191.1	8.5	0.001
B16	E	-	300	0.03	3.00	-	-	1.43	2.75	-	0.34	0.33	1.1	-	-	-	-
B17	E	15.8	1329	6.85	8.86	0.03	4.25	0.73	-	3.76	8.69	0.77	3.73	5.53	912.9	46.2	0.079
B18	E	5.6	424	7.82	4.58	0.03	0.96	0.01	-	1.68	3.79	0.35	1.60	1.46	307.5	15.3	0.004
B20	E	14.1	298	8.39	2.70	0.01	0.51	0.24	0.05	0.52	1.12	0.33	1.92	0.61	203.7	12.6	0.001
B21	E	16.4	967	7.32	8.24	0.06	2.47	0.46	-	1.91	5.07	0.64	2.15	7.67	723.4	49.0	0.025
B22	E	13.9	373	7.87	3.66	0.02	0.70	0.08	-	0.65	2.46	0.28	1.20	1.30	224.7	12.5	0.003
B24	E	17.4	544	7.8	3.90	0.03	1.37	0.31	-	0.69	2.99	0.26	0.70	1.75	268.2	-	0.004
B25	E	16	1056	6.58	7.08	0.02	1.72	0.61	-	2.35	4.15	0.57	2.42	4.60	574.1	-	0.118
B26	E	17.5	1570	6.57	10.67	0.03	2.74	0.28	0.02	3.86	7.07	0.85	3.45	6.75	838.6	-	0.181
B27	E	17	718	6.58	5.03	0.02	1.03	0.85	-	1.42	2.62	0.40	1.78	3.10	410.7	-	0.084
B28	E	16	886	6.4	6.44	0.02	1.35	0.21	-	1.62	3.47	0.54	2.09	3.87	458.8	-	0.162
B29	E	26	944	7.33	8.52	0.01	1.14	0.16	-	0.95	5.15	0.46	1.88	3.59	449.4	-	0.025
B30	E	18.8	588	7.8	3.33	0.03	1.63	0.50	-	0.79	3.08	0.25	0.76	1.90	303.1	-	0.003
B31	E	29.8	1527	7	8.89	0.05	2.86	0.41	-	4.15	5.43	0.76	3.66	6.80	827	-	0.056
B32	E	17.8	996	7.25	6.93	0.03	2.39	0.19	-	1.57	4.82	0.47	1.45	4.64	522.7	-	0.025
B33	E	21.3	1180	6.8	8.54	0.03	1.77	0.61	0.02	2.52	4.35	0.52	2.78	5.71	647.5	-	0.085
S6	E	17.6	2300	7.95	4.37	0.00	17.3	0.77	-	3.59	18.62	0.82	3.74	4.65	1693.5	41.9	0.003
S7	E	16.8	-	7.64	1.52	0.03	2.97	1.77	-	2.78	3.25	0.81	1.44	5.03	638.1	32.3	0.002
S8	E	14.9	330	8.15	1.95	0.02	0.77	0.18	-	1.14	2.14	0.23	0.73	1.34	210.8	10.4	0.001
S9	E	15.5	981	8.44	6.14	0.04	2.88	1.25	0.05	2.52	5.49	0.50	2.68	5.29	729.9	39.8	0.001
S10	E	15.3	676	7.74	5.00	0.04	2.46	0.41	-	0.75	3.77	0.41	2.23	2.39	441.7	23.1	0.006

Sample	Basin	Temp. °C	EC µS/cm	pH	HCO_3 meq/L	F meq/L	Cl meq/L	NO_3 meq/L	PO_4 meq/L	SO_4 meq/L	Na meq/L	K meq/L	Ca meq/L	Mg meq/L	TDS mg/L	Hardness °fH	PCO_2 at 20° bars
S11	E	14.3	443	7.79	3.50	0.02	0.86	0.33	-	0.96	2.61	0.30	1.31	1.85	283.6	15.8	0.004
S12	E	19.7	1670	7.5	13.10	0.04	2.84	1.01	-	3.25	7.51	0.66	2.79	8.90	935.2	-	0.026
R1	River E	11.2	490	8.73	5.72	0.03	0.78	0.10	-	1.03	2.11	0.19	1.84	4.61	395.2	32.2	0.001
B1	W	7	1268	7.98	17.51	0.06	3.56	0.10	0.05	1.10	10.04	0.66	6.75	2.84	869.3	47.9	0.012
B2	W	15.9	1335	6.46	16.34	0.02	3.07	0.02	-	0.18	8.81	0.69	3.85	9.48	939.2	66.6	0.355
B3	W	15.1	1004	6.8	12.80	0.03	2.56	0.09	-	1.49	5.10	0.77	2.72	6.46	679.7	45.9	0.128
B4	W	8.8	832	6.2	12.18	0.03	0.91	0.08	-	0.87	4.12	0.48	0.64	6.82	486.9	37.3	0.502
B5	W	15.4	1441	6.71	17.71	0.04	2.17	0.21	-	0.56	6.82	0.78	6.09	6.07	852.2	60.8	0.218
B6	W	15	1153	7.11	12.36	0.04	2.86	0.29	-	0.93	5.65	0.57	5.09	4.83	746.7	49.5	0.061
B7	W	16.7	1013	7.49	7.94	0.03	2.90	0.75	-	1.46	4.89	0.54	3.28	3.58	624.3	34.3	0.016
B8	W	18.1	989	6.48	8.40	0.03	2.50	0.07	0.02	1.46	4.66	0.66	3.44	3.53	574.2	34.8	0.176
B9	W	13.9	384	8.24	2.68	0.04	0.66	0.08	-	0.68	2.26	0.23	0.81	0.94	182.3	8.8	0.001
B10	W	16.4	1951	6.84	23.85	0.06	4.65	0.17	0.02	0.72	6.83	0.59	1.21	18.04	1192	96.2	0.217
B19	W	17.1	1093.6	7.71	7.60	0.01	9.68	2.20	0.02	6.63	13.19	0.82	7.84	3.84	1488.4	58.3	0.009
B23	W	12.1	924	7.42	13.72	0.03	0.90	0.03	-	0.56	4.87	0.54	4.70	3.78	571.2	43.3	0.033
B34	W	18.7	1582	6.51	17.49	0.05	1.87	0.01	-	0.45	6.50	0.52	2.62	10.68	827.3	-	0.341
B35	W	14.9	1182	6.06	12.43	0.04	1.27	0.09	-	0.94	4.54	0.54	2.41	7.68	638.3	-	0.683
B36	W	23	2051	7.1	17.87	0.03	3.63	1.51	-	2.09	10.24	0.95	2.40	10.94	1114.8	-	0.090
B37	W	16.3	1800	7.41	13.33	0.04	4.40	0.93	-	3.09	9.94	0.56	2.02	8.87	995.6	-	0.033
B38	W	14.5	1194	6.51	11.84	0.03	1.44	0.19	0.03	1.15	4.40	0.53	2.41	7.62	648.3	-	0.231
B39	W	17.6	1766	6.98	14.72	0.04	3.16	0.97	-	2.62	8.30	0.68	2.55	9.65	976.8	-	0.097
B40	W	19	2560	7.3	21.90	0.09	5.32	4.80	-	4.42	11.72	0.91	5.83	14.72	1784.3	-	0.069
B41	W	18	1825	7.07	13.64	0.04	3.34	1.36	-	4.19	8.43	0.68	2.95	9.80	1060.1	-	0.073
B42	W	20	1860	6.8	14.57	0.04	3.51	0.97	-	3.15	8.64	0.61	3.17	9.42	1020.3	-	0.146
B43	W	16.4	1228	6.08	14.70	0.04	1.48	0.02	-	0.53	5.29	0.50	2.65	8.78	708.4	-	0.772
B44	W	20	1315	6.18	14.77	0.04	1.47	0.00	-	0.53	5.32	0.51	2.64	8.83	708.9	-	0.616
B45	W	16.2	1212	6.28	12.02	0.03	1.38	0.25	0.02	1.03	4.51	0.53	2.35	7.57	645.4	-	0.398

Sample	Basin	Temp.	EC	pH	HCO_3	F	Cl	NO_3	PO_4	SO_4	Na	K	Ca	Mg	TDS	Hardness	PCO_2 at 20°
		°C	μS/cm		meq/L	meq/L	meq/L	meq/L	meq/L	meq/L	meq/L	meq/L	meq/L	meq/L	mg/L	°fH	bars
B46	W	14.6	1250	5.93	13.72	0.04	1.22	-	-	0.63	4.73	0.55	2.51	8.33	662.3	-	1.017
S1	W	14.2	1264	8.48	10.82	0.04	3.05	0.61	0.06	2.70	8.46	0.66	5.13	3.50	822.8	43.1	0.002
S2	W	15	1169	6.85	15.47	0.03	1.75	0.07	-	0.33	7.24	0.62	1.82	9.82	778	58.2	0.138
S2-1	W	13.7	1215	7.08	16.73	0.05	2.23	-	-	0.20	8.64	0.67	3.24	10.22	902.5	67.3	0.086
S2-2	W	15	1240	7	15.85	0.04	2.06	0.02	-	0.25	6.81	0.53	5.30	4.64	717.4	49.7	0.100
S3	W	19.8	1734	6.96	22.34	0.05	1.98	0.06	-	1.09	7.83	0.37	0.86	12.64	866.8	67.5	0.158
S4	W	19.2	1851	6.15	22.56	0.03	2.15	0.04	-	0.69	8.49	0.49	0.48	13.91	917.4	71.9	1.039
S5	W	16	1505	8.33	15.61	0.04	3.85	0.32	-	1.74	9.19	0.66	5.22	4.53	881	48.7	0.005
SM1	Salinella	10.2	67,600	6.1	47.2	1.03	1343.4	-	-	-	1266.2	16.6	14.1	51.7	81,393.9	328.9	2.2
SM2	Salinella	36.5	110,600	6.1	26.0	-	1234.7	1.3	-	14.0	1158.1	19.0	17.0	96.1	75,158.2	565.5	1.3
SM2a	Salinella	12.6	71,700	6.1	28.4	2.03	1286.1	-	-	-	1186.9	19.6	16.3	88.9	77,382.8	525.6	1.4
SM2b	Salinella	11.8	71,600	6.3	35.4	2.36	1317.9	-	-	8.1	1215.7	19.5	17.6	88.5	80,007.4	530.4	1.2
SM2c	Salinella	16.4	72,100	6.3	28.2	1.39	1289.1	-	-	11.6	1186.4	16.5	19.3	76.7	77,444.8	479.6	0.9
SM3	Salinella	18.8	60,100	6.2	30.8	0.91	1138.8	-	-	10.2	1037.3	16.9	33.3	54.5	68,910.1	438.5	1.3
R	River W	9.5	1098	8.54	6.04	0.01	2.91	0.30	-	6.76	8.26	0.53	5.49	4.90	1198.8	51.9	0.001
M1	Seawater	10.0	39,800	7.8	3.3	1.07	626	-	-	58.6	546	12.9	69.8	45.7	40.308	577	0.003

In here, the source of the fluids would be associated with a hydrothermal system enriched in CO_2 with temperatures between 100 °C to 150 °C that extends between Paternò and the central part of the Etna [29].

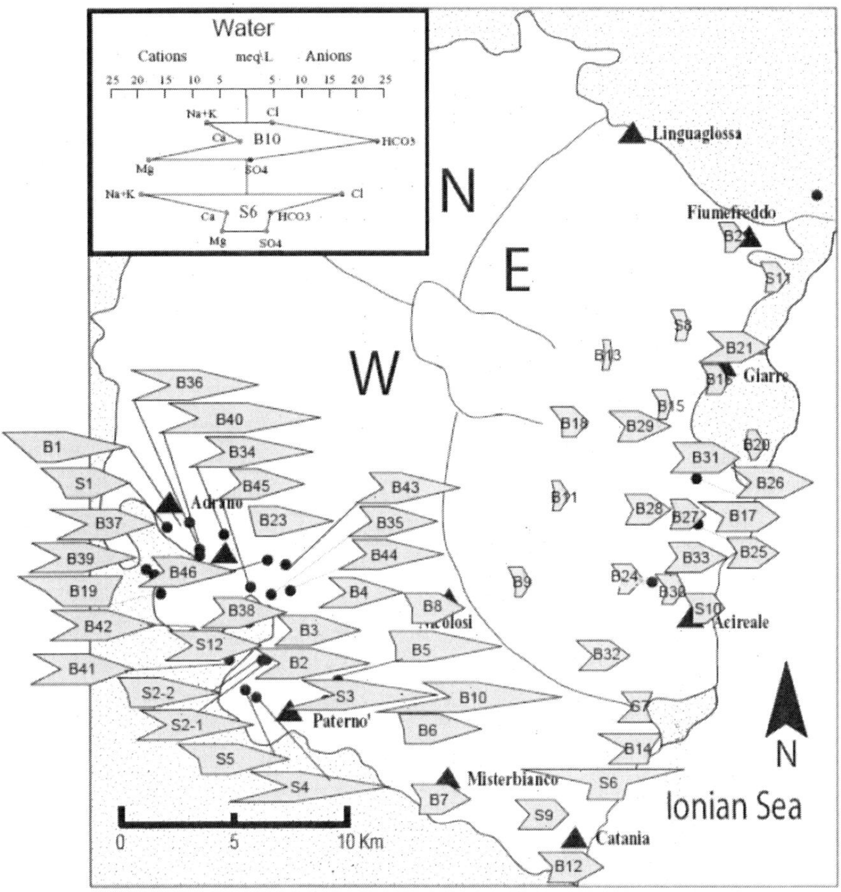

Figure 3. Stiff diagrams displaying average ion concentrations in the sampled waters.

Direct inputs of deep CO_2 are a key factor in determining the chemistry of Etnean waters. In effect, the interaction between CO_2 and infiltrating rain water lowers the pH to values below 4 [30]. These low-pH waters become highly reactive resulting in chemical weathering of the host basaltic rocks. As a result of this process, HCO_3^- (along with H_2CO_3 and CO_3^{2-}) is gradually generated, whilst Mg, Ca, K, and Na are released into solution. The general positive relationship between HCO_3^- and these cations (especially Na^+ and Mg^{2+}) supports the idea that dissolution from acidic waters is as a major mechanism for the mineralization of Etnean aquifers (Figure 4). Bicarbonate concentrations are normally higher in the western basin, with values up to 23.8 meq/L around the towns of Paternò, S.P. Clarenza, and Biancavilla. In contrast, groundwater in the eastern basin shows a more limited enrichment in HCO_3^-, with concentration values approximately 1/3 (~5 meq/L) of those observed in the west. This is in line with [31], who argued that chemical weathering in Etna is not spatially uniform since the presence of CO_2 in soils is more abundant in particular areas of the volcano, such as the south-western flank.

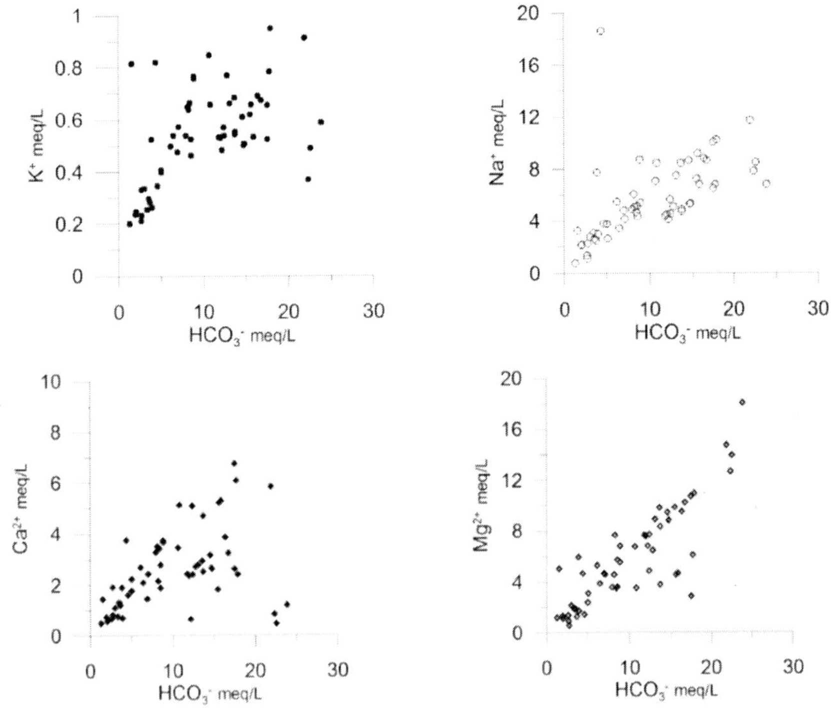

Figure 4. HCO_3^- against major cations in groundwater.

Furthermore, HCO_3^- concentrations appear to be influenced by the ground surface elevation. There is a negative trend between HCO_3^- and elevation, with the lowest concentrations towards the volcano's summit.

As discussed, more restricted hydrological circuits mean that CO_2-enriched waters near the top of the edifice have fewer opportunities to react with the host basaltic rocks. Lower dissolution rates translate into lower HCO_3^- concentrations and a general decrease in the total dissolved solids content (TDS) of the waters.

Electrical conductivity (EC) from 1000 to 2000 μS/cm, and TDS values between 700 mg/L and 1400 mg/L were found to be representative for groundwater in the southernmost flank of Mt. Etna. Exceptional salinities detected at S6 near Catania (EC = 2300 μS/cm; TDS = 1694 mg/L) are attributed to partial mixing with seawater in proximities to the Ionian coast. Longer residence times favoring basalt leaching would be an additional factor explaining the higher salinity in the SW sector of Etna [32]. Conversely, the eastern flank is characterized by EC values between 500 and 1000 μS/cm, and TDS from 300 to 1100 mg/L. This lower TDS might be ascribed to water circulation in more transmissive sediments that facilitate the flow of meteoric recharge and reduce the transit time underground. Elevation would be another factor controlling the TDS distribution in groundwater. Dissolved solids in samples above 600m A.S.L. are usually below 500 mg/L, which can be attributed to the limited water-rock interaction at high elevations. As expected, groundwater at higher elevations receives direct recharge from precipitation, which is unable to interact with the host rocks for long enough to produce major changes in its chemical composition.

Major cations (Ca^{2+}, Mg^{2+}, Na^+, K^+) show a similar distribution to TDS. Calcium concentrations average 2.7 meq/L. Higher values were recorded south of Adrano (7–7.8 meq/L), and between Belpasso and Camporotondo (5–7 meq/L) towards the western margin of the volcano. The lowest concentrations were measured south of Paternò and north of Ragalna, on the southwest (<1.5 meq/L). Again, groundwater in the western basin reflects more extensive water-rock interaction due to longer transit and residence times. This may be explained by the rainfall distribution, as maximum precipitation (*i.e.*, groundwater recharge) occurs on the eastern flank of the volcano, contrary to the western flank that receives lower rainfall and consequently shows prolonged interactions between groundwater and the host rock [33].

In particular, Ca^{2+} and K^+ have a similar distribution, with sharp concentration variations around the region of Paternò. These changes occur over short distances and reflect different aquifers. Higher salinities can be related to contact with alkaline brines discharged by the Paternò mud volcanoes (Salinelle), although groundwater mineralization through permeable faults can still exert some influence.

Magnesium reaches maximum concentrations in the western basin at S.P Clarenza, and over a vast area up to Catania. There exists a close relationship between Mg^{2+} and HCO_3^-. This suggests that dissolution caused by CO_2-enriched waters on the host rocks is common to both species. In particular, the source of Mg^{2+} can be explained by the leaching of olivines and pyroxenes from the basaltic rocks in the substrate.

Chloride concentrations range from 0.3 to 17 meq/L. In principle, the abundance of Cl^- might be attributed to alteration of the volcanic rocks and to the interaction of groundwater with volcanic gases [3]. As in the case of Na^+, maximum Cl^-contents (~17.3 meq/L) were measured at S6, near Catania, likely due to mixing with seawater. Figure 5 shows that waters near the coast exhibit a Na/Cl ratio closer to the 1:1 line, unlike samples from the western basin that are partially depleted of Cl^-. Seawater intrusion and mixing between shallow groundwater and deep brines could be the reason for these patterns. This is also coincident with [9], who considered both mixing and water-rock interaction to be responsible for the increased salinity of groundwater in Mt. Etna. It is worthy to note that although evaporates do not crop out in the area, their presence beneath the volcanic cover has been hypothesized both from geological and hydrochemical data [21]. These deposits could thus be another contributing factor for Cl^- in groundwater.

Maximum concentrations of NO_3^- and SO_4^{2-} were observed in the valleys that exist at low elevations in the volcano (e.g., south of Adrano). High concentrations are also visible within the stretch of land south of Giarre to Fiumefreddo, suggesting the leakage of fertilizers into the aquifers. The availability of water resources and the quality of the soils in the lower flanks of the volcano have favored the agricultural exploitation of the region since ancient times. Under conditions of high oxygen, part of the ammonium—sulphate fertilizers applied on the ground would be rapidly converted to NO_3^-, which is not sorbed by the negatively charged soil colloids and moves readily to the water table [34]. At higher elevations (e.g., Valle del Bove), the SO_4^{2-} inputs into groundwater could be controlled by the ascent of volatiles from vapor-dominated systems such as fumaroles rather than anthropogenic effects.

In summary, groundwater in the area of study can be grouped into four main types: (1) Mg–Na ± Ca–HCO_3^- (with rare Cl^- or SO_4^{2-}) (54%); (2) Na–Mg ± Ca–HCO_3^- ± (SO_4^{2-} or Cl^-) (28%); (3) Na–Ca–Mg–HCO_3^- (12%); and (4) Na–Mg ± Ca–Cl (with HCO_3^- or SO_4^{2-}) (6%). The differences between the first and second group are minor and mainly associated with variations in the Na–Mg and SO_4^{2-}–Cl^- content (with relative concentrations normally higher for the second type). This implies that types "(1)" and "(2)" explain 82% of the waters in the region. Waters of the third group are associated with elevated Ca^{2+} concentrations, while the last type is characterized by high salinity. Samples enriched in Cl^-

could be the result of mixing with seawater or solute diffusion from marine clay aquitards.

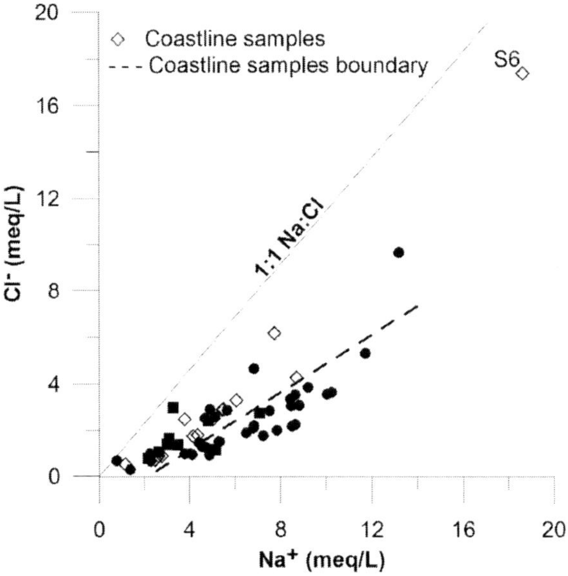

Figure 5. Relation of samples against the Na:Cl ratio.

Additionally, carbonate waters are generally predominant in the western basin, probably due to the interaction with hydrothermal CO_2. In contrast, the Cl^-–SO_4^{2-} type is mainly found along the coast of the eastern basin due to saline influxes. Furthermore, the eastern coast is the area most densely inhabited and some anthropogenic effects are already reflected in the more elevated NO_3^- contents, likely derived from agricultural and urban wastewater (e.g., samples B14, S7, S9, S1).

4.2. Isotopic Signature

The isotopic signature of the sampled waters was also used to determine the recharge areas and circulation pathways in Mt. Etna (Table 2). Values for groundwater and spring samples fall between −4‰ to −9‰ for $\delta^{18}O$ and −19‰ to −53‰ for δ^2H. This is in close correlation with the Global Meteoric line ($\delta^2H = 8 \times \delta^{18}O + 10‰$, [35]) and the eastern Mediterranean local meteoric water line (EMMWL) defined by [36]: $\delta^2H = 8 \times \delta^{18}O + 22‰$, suggesting a predominant meteoric origin for the collected samples (Figure 6). Similar trends had been reported by [4], who postulated that most waters in Mt. Etna originated as local precipitation infiltrated at an elevation between 1100 and 1900 m. Minor variations in the $\delta^{18}O$ of the samples also suggest limited effects of rock interaction on the original water composition.

Table 2. Typical isotopic composition of Etnean waters.

Sample No.	Sample Type	Basin	DIC $\delta^{13}C$ (PDB ‰)	$\delta^{18}O$ (‰ VSMOW)	$\delta^{2}H$ (‰ VSMOW)
B1	Bore	W	−0.1	−9.1	−51.1
B2	Bore	W	−1.6	−8.6	−45.3
B3	Bore	W	−2.4	−8.6	−47.0
B4	Bore	W	−1.3	−7.8	−41.2
B5	Bore	W	−1.6	−8.6	−45.3
B6	Bore	W	−2.4	−8.6	−47.0
B7	Bore	W	−1.3	−7.8	−41.2
B8	Bore	W	+0.2	−7.4	−38.8
B9	Bore	W	−1.0	−6.6	−38.9
B10	Bore	W	−1.2	−6.7	−28.6
B19	Bore	W	−10.3	−6.8	−36.7
B23	Bore	W	−7.9	−6.1	−29.0
S1	Spring	W	−1.8	−8.6	−50.2
S2	Spring	W	−1.8	−8.6	−50.2
S2-1	Spring	W	−1.2	−8.7	−46.3
S2-2	Spring	W	−0.6	−8.7	−46.6
S3	Spring	W	−0.1	−7.4	−40.1
S4	Spring	W	−0.7	−8.6	−46.6
S5	Spring	W	−0.1	−7.3	−40.1
R	River	W	−7.5	−8.2	−44.4
B11	Bore	E	−1.0	−7.5	−37.4
B12	Bore	E	−4.6	−6.7	−39.3
B13	Bore	E	−8.6	−7.2	−38.9
B14	Bore	E	−1.5	−7.1	−38.6
B15	Bore	E	−2.0	−7.4	−37.5
B16	Bore	E	-	−6.3	−33.0
B17	Bore	E	−1.8	−7.0	−39.1
B18	Bore	E	−4.6	−6.7	−37.7
B20	Bore	E	−9.4	−4.2	−19.2
B21	Bore	E	−14.5	−6.0	−34.3
B22	Bore	E	−6.9	−7.3	−46.7
B24	Bore	E	−7.7	−7.1	−31.5
B27	Bore	E	−1.8	−6.9	−39.1
B28	Bore	E	−5.0	−7.8	−42.2
B29	Bore	E	−5.8	−7.5	−41.2
B30	Bore	E	−9.4	−4.1	−19.2
B31	Bore	E	−14.5	−5.9	−34.3
B32	Bore	E	−6.9	−7.3	−46.7
B33	Bore	E	+0.8	−9.1	−53.3
S6	Spring	E	−11.2	−6.2	−33.7
S7	Spring	E	−11.2	−6.2	−33.7

Sample No.	Sample Type	Basin	DIC δ¹³C (PDB ‰)	δ¹⁸O (‰ VSMOW)	δ²H (‰ VSMOW)
S8	Spring	E	−11.8	−6.7	−37.9
S9	Spring	E	−11.4	−6.7	−38.1
S10	Spring	E	−6.1	−6.9	−36.5
S11	Spring	E	−9.1	−7.1	−35.6
S12	Spring	E	−1.2	-	-
R1	River	E	−5.4	−8.0	−44.9
SM1	Salt mount	Paternò	-	+8.8	−14.5
SM2	Salt mount	Paternò	-	+10.4	−12.0
SM3	Salt mount	Paternò	-	+7.5	−13.6
Sm2a	Salt mount	Paternò	+5.3	+9.78	−18.2
Sm2b	Salt mount	Paternò	+3.0	+10.2	−21.5
Sm2c	Salt mount	Paternò	+1.4	+10.1	−19.8
M1	Seawater	-	-	+1.1	+1.3

Adifferent trend is observed for waters collected in the salt mount brines. The "positive" isotopic ratio of the samples (*i.e.*, $\delta^{18}O$ up to about +10‰, and $\delta^{2}H$ ranging from −21.5‰ to a maximum of −12‰) suggests that these waters could be mixed with hydrothermal fluids of deep origin following a more prolonged interaction with the host rocks.

Figure 6 also evidences the contrast between the mainly meteoric groundwater against Cl-enriched fluids from the salt mounts. As postulated by [4], the anomalous enrichment of Cl^- might be related to the proximity of the sampled bodies to areas of intense magmatic outgassing, where Cl-rich gases are likely to interact with shallow aquifers. Considering that the ratio $Cl/\delta^{18}O$ in the salt mounts is higher than seawater (sample M1), the direct mixing between these two fluids would be improbable.

Additional information may be inferred by assessing the geographical distribution of the stable isotope ratios in the volcano. Waters from the southern flanks of Mt. Etna are relatively homogeneous and generally fit with the EMMWL. In contrast, waters in the eastern sector deviate from the meteoric line suggesting that more complex processes could take place there. These processes might be associated to heterogeneous recharge and irregular circulation patterns, especially on the alluvial lowlands of the eastern basin, where surface runoff and upward flows could also affect the chemical signature of the waters. These observations are consistent with [4,15]), who described considerably more diversity for waters in the eastern flanks of the volcano.

The isotopic composition of the waters might also be influenced by the terrain elevation, although the relationship is generally limited and not always distinguished (Figure 7).

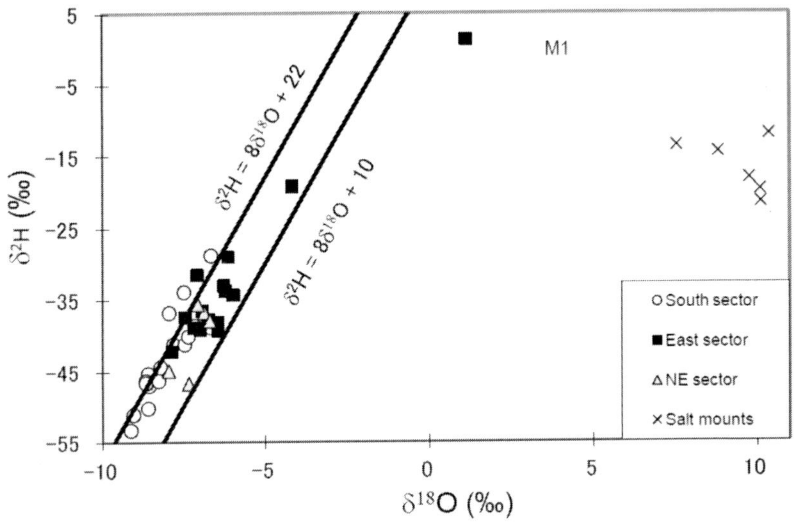

Figure 6. Relation between oxygen and hydrogen isotopic composition in Etnean groundwater.

In effect, samples with a lighter isotope composition often coincide with higher elevations towards the cone summit, but the groundwater character still varies considerably, likely in response to additional underlying factors. As clouds rise up the volcano, the heavy isotopes are depleted and the residual precipitation gets isotopically lighter [37]. The most depleted samples would locate in the southern basin where the δ^2H composition approaches −55‰.

These differences between the isotopic compositions in the south and other regions of the volcano could be another indication of variations in meteoric inputs and the heterogeneity of the hydrological circuits within Etna.

A second trend suggests that waters become lighter from East to West, which is coincident with the main wind direction and the geographical distribution of precipitations. In effect, rainfall amounts peak on the eastern flanks of the Etna, mainly in relation to cooling sea breezes and clouds from the neighbor Mediterranean Sea [8]. The water vapor from the sea and the subsequent clouds and precipitation, are characterized by a high ^2H excess which in turn, is reflected in the composition of the waters [38]. In its migration across the volcano, falling precipitation undergoes fractionation and becomes increasingly lighter.

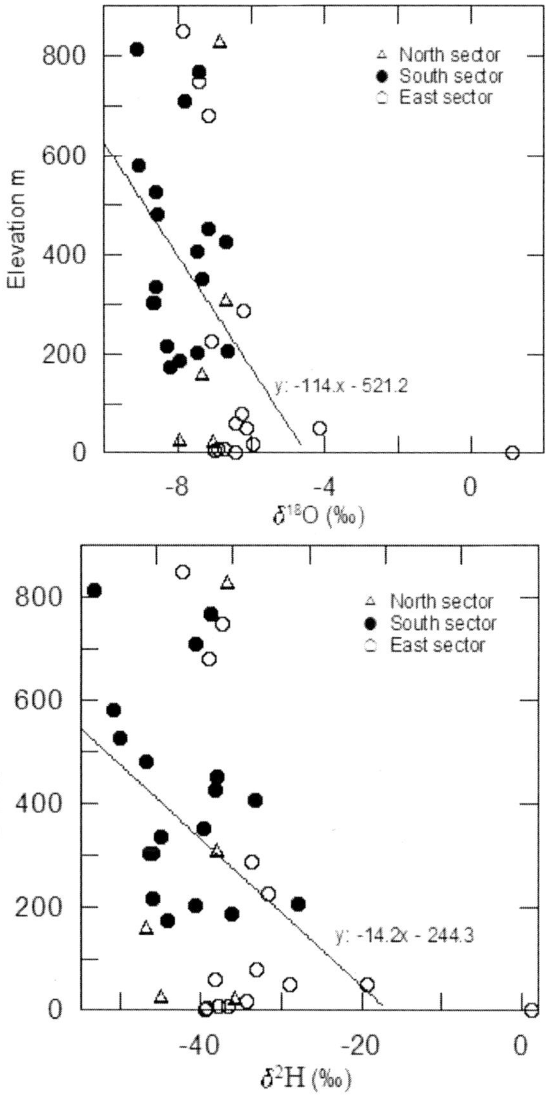

Figure 7. Stable isotopes in groundwater against elevation in the volcano.

Values for $\delta^{13}C$ vary from as low as −14‰ in proximities to the town of Giarre (B31), up to about 5‰ in waters nearby the mud volcanoes of Paternò. Measured values plot above the line of pCO_2 in the atmosphere and largely outside the typical range of groundwater suggesting a contribution of external CO_2 (Figure 8). Given that groundwater in Etna does not come in contact with outcrops or superficial carbonate rocks [14], the prevailing pCO_2 source would be at depth.

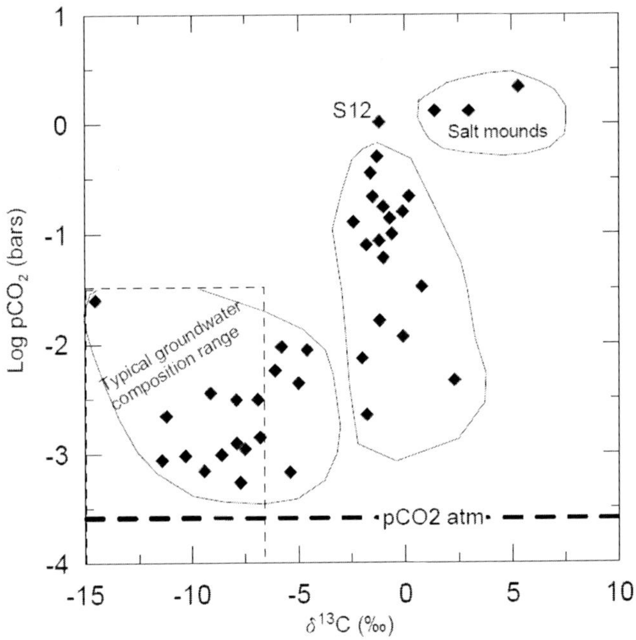

Figure 8. Partial pressure of CO_2 *vs.* $\delta^{13}C$ showing the possible contribution of magmatic gasses. Typical groundwater composition within the dashed area.

Under low HCO_3^- concentrations, $\delta^{13}C$ values range broadly from −2‰ to −15‰ (Figure 9). An increase in HCO_3^- (>10 meq/L) is coincident with $\delta^{13}C$ in the range of −2.4‰ and +2.3‰, with most samples clustering around −1.5‰. In such conditions, the equilibrium between HCO_3^- and CO_2 causes the isotopic composition of the latter species to become more positive. Therefore, the Etnean magmatic CO_2 becomes isotopically heavier [39]. This pattern suggests some additional inputs of CO_2, possibly related to hydrothermal fluids that stripped the gasses from a magmatic reservoir and transported them into the shallow aquifers. In such conditions, the rise of CO_2 would lower the pH of the circulating waters and result in higher concentration of dissolved HCO_3^- after weathering the host rocks. Thus, many waters in the studied area might be partially influenced by this secondary CO_2 despite their overall meteoric origin.

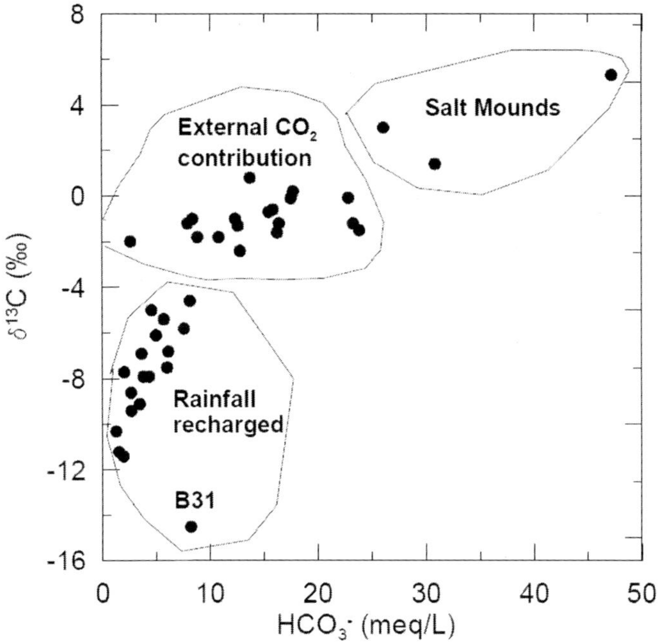

Figure 9. Change in groundwater composition in relation to the HCO_3^- and $\delta^{13}C$ concentrations.

In short, two major groundwater groups can be discriminated on the basis of $\delta^{13}C$: (1) waters with compositions below −4.6‰, mainly recharged by direct percolation of rainfall; (2) waters with $\delta^{13}C > -2.4‰$ in a stretch of land between Adrano and Misterbianco in the south, and around Pozillo to the east, that include a component of external dissolved gas. This trend suggests that the diffusion of CO_2 gasses in Mt. Etna is unevenly distributed, and essentially controlled by the main tectonic structures of the volcano. During their ascent to the surface, these gases interact in different ways with shallow water-bearing strata changing their concentrations as they cross the aquifers [23].

It is worthy to note that a more particular isotopic signature was recorded at the "Salinelle di Paternò". In these mud volcanoes, the $\delta^{13}C$ compositions ranged from 1.4‰ to 5.3‰ while the pCO_2 values varied from 1.3 to 2.2 bars. The enrichment in CO_2 and the high "positives" stable isotopic ratio observed ($\delta^2H \sim -12‰$ and $\delta^{18}O \sim 10.4‰$) seem to indicate a very deep origin of these fluids.

5. SUMMARY AND CONCLUSIONS

A new survey dataset of major ion concentrations and isotope ratios was used to update the knowledge of geochemical characteristics of groundwaters and to better understand the flow system around Mt. Etna. The Etnean groundwaters possess a marked bicarbonate-alkaline chemistry, which is consistent with an abundance of dissolved CO_2 gas and the composition of the volcanic host rocks. Chloride-sulphate and nitrate dominated waters can locally prevail along the Ionian Sea coast, largely due to urban contamination and the leakage of agricultural fertilizers. Other distinctive characteristics of the sampled waters include low temperatures, high conductance, and elevated hardness. The salinity of the waters decreases with elevation due to the proximity to recharge areas and shorter travel paths. Oxygen-deuterium isotopes showed that waters are essentially recharged by infiltrating rainfall. Especially to the south, most of the samples display a good correlation with the eastern Mediterranean meteoric water line. To the east, collected samples deviate from the meteoric line, suggesting more heterogeneous circulation paths, and variable degrees of interaction between meteoric waters and the aquifer rocks.

Furthermore, the isotope composition is influenced by the provenance of wet air masses from the Mediterranean Sea. In this regard, waters become isotopically lighter to the west, following the distribution of precipitation on the volcano. Similarly, the liquid-vapor fractionation of waters results in lighter waters along with an increase in elevation or in proximity to the cone summit.

The analysis of $\delta^{13}C$ indicates that at least a proportion of the waters, mainly in the southern region of the volcano, would be affected by external CO_2 contributions, possibly of hydrothermal origin. This is also supported by pCO_2 values 1 to 3 orders of magnitude higher than those expected for waters in equilibrium with the atmosphere. Furthermore, a high enrichment in CO_2 along with high positive values for $\delta^2H/\delta^{18}O$ suggest that waters in the salt mounts around Paternò would be influenced by brines originating at depth within the system.

ACKNOWLEDGMENTS

The authors are deeply grateful to Gaetano Punzi, current Director of the Regional Geological Office for Land Reclamation, Sicily, for his help throughout the field work. This paper would never have been achieved without him. Many thanks also to Glenn Harrington, Adjunct Supervisor in the National Centre for Groundwater Research and Training, Flinders University, and Principal Hydrogeologist at Innovative Groundwater Solutions Pty Ltd, Australia, for his comments and assistance to improve the manuscript. The authors also wish to thank John Luczaj, and two anonymous reviewers for their suggestions and useful discussions. Special thanks to the AIST, Geological Survey of Japan for allowing us to use its facilities and equipment for the chemical analyses.

REFERENCES

1. Gerlach, T.M. Etna's greenhouse pump. *Nature* **1991**, *315*, 352–353.
2. D'Alessandro, W.; Bellomo, S.; Bonfanti, P.; Brusca, L.; Longo, M. Salinity variations in the water resources fed by the Etnean volcanic aquifers (Sicily, Italy): Natural *vs.* anthropogenic causes. *Environ. Monit. Assess.* **2011**, *173*, 431–446.
3. Giammanco, S.; Ottaviani, M.; Valenza, M.; Veschetti, E.; Principio, E.; Giammanco, G.; Pignato, S. Major and trace elements geochemistry in the ground waters of a volcanic area: Mount Etna (Sicily, Italy). *Water Res.* **1998**, *32*, 19–30.
4. Pennisi, M.; Leeman, W.P.; Tonarini, S.; Pennisi, A.; Nabelek, P. Boron, Sr, O, and H isotope geochemistry of groundwaters from Mt. Etna (Sicily)—Hydrologic implications. *GCA* **2000**, *64*, 961–974.
5. D'Alessandro, W.D.; Federico, C.; Longo, M.; Parello, F. Oxygen isotope composition of natural waters in the Mt Etna area. *J. Hydrol.* **2004**, *296*, 282–299.
6. Liotta, M.; Grassa, F.; D'Alessandro, W.; Favara, R.; Gagliano Candela, E.; Pisciotta, A.; Scaletta, C. Isotopic composition of precipitation and groundwater in Sicily, Italy. *Appl. Geochem.* **2013**, *34*, 199–206.
7. Catalano, R.; Imme, G.; Mangano, G.; Morelli, D.; Giammanco, S. Natural tritium determination in groundwater on Mt Etna (Sicily, Italy). *J. Radioanal. Nucl. Chem.* **2014**, *299*, 861–866.
8. Chester, D.K.; Duncan, A.M.; Guest, J.E.; Kilbum, C.R.J. *Mount Etna: The Anatomy of a Volcano*; Stanford University Press: Stanford, CA, USA, 1985.
9. Aiuppa, A.; Bellomo, S.; Brusca, L.; D'Alessandro, W.; Federico, C. Natural and anthropogenic factors affecting groundwater quality of an active volcano (Mt. Etna, Italy). *Appl. Geochem.* **2003**, *18*, 863–882.
10. Clocchiatti, R.; Schiano, P.; Ottolini, L.; Bottazzi, P. Earlier alkaline and transitional magmatic pulsation of Mt. Etna volcano. *Earth Planet. Sci. Lett.* **1998**, *132*, 25–41.
11. Tanguy, J.C.; Condomines, M.; Kieffer, G. Evolution of Mount Etna magma: Constraints on the present feeding system and eruptive mechanism. *J. Volcanol. Geotherm. Res.* **1997**, *75*, 221–250.
12. Pering, T.D.; Tamburello, G.; Aiuppa, A.; McGonigle, A.J.S. The First Record of a High Time Resolution Carbon Dioxide Flux for the North-East Crater of Mount Etna. In Proceedings of the IAVCEI 2013 Scientific Assembly, Kagoshima, Japan, 20–24 July 2013.
13. McGee, K.A.; Delgado, H.; Cardenas Gonzales, L.; Venegas Mendoza, J.J.; Gerlach, T.M. High CO_2 emission rates at Popocatepetl volcano, Mexico. In Proceedings of the Abstracts AGU Fall Meeting, Baltimore, MD, USA, 29 May–2 June 1995.
14. Ogniben, L. Lineamenti idrogeologici dell'Etna. *Riv. Min. Sicil.* **1966**, *100–102*, 151–174. (In Italian).
15. Ferrara, V. Valutazione della vulnerabilità degli acquiferi. In *Carta della vulnerabilità all'inquinamento dell'acquifero vulcanico dell'Etna*; Civita, M., Ed.; SELCA: Firenze, Italy, 1990. (In Italian)

16. Aureli, A. Idrogeologia del fianco occidentale etneo. In Proceedings of the 2nd International Congress on Underground Waters, Palermo, Italy, 28 April–1 May 1973; pp. 425–487. (In Italian).
17. Schilirò, F. Proposta metodologica per una zonazione geologicotecnica del centro abitato di Maletto. *Geol. Tec.* **1988**, *3*, 32–53. (In Italian).
18. Ferrara, V. Idrogeologia del versante orientale dell'Etna. In Proceedings of the 3rd international symptom on groundwaters, Palermo, Italy, 1–5 November 1975; pp. 91–134. (In Italian).
19. Aiuppa, A.; Allard, P.; D'Alessandro, W.; Giammanco, S.; Parello, F.; Valenza, M. Review of magmatic gas leakage at Mount Etna (Sicily, Italy): Relationships with the volcano-tectonic structures, the hydrological pattern and the eruptive activity. In *Etna Volcano Laboratory*; Calvary, S., Bonaccorso, A., Coltelli, M., Del Negro, C., Falsaperla, S., Eds.; Geophysical Monography Series AGU: Washington, DC, USA, 2004; Volume 143.
20. Kozlowska, B.; Morelli, D.; Walencik, A.; Dorda, J.; Altamore, I.; Chieffalo, V.; Giammanco, S.; Imme, G.; Zipper, W. Radioactivity in waters of Mt. Etna (Italy). *Radiat. Meas.* **2009**, *44*, 384–389.
21. Anza, S.; Dongarra, G.; Giammanco, S.; Gottini, V.; Hauser, S.; Valenza, M. Geochimica dei fluidi dell'Etna. *Miner. Petrogr. Acta* **1989**, *32*, 231–251. (In Italian).
22. Allard, P. Endogenous magma degassing and storage at Mount Etna. *Geophys. Res. Lett.* **1997**, *24*, 2219–2222.
23. D'Alessandro, W.D.; de Gregorio, S.; Dongarra, G.; Gurrieri, S.; Parello, F.; Parisi, B. Chemical and isotopic characterization of the gases of Mount Etna (Italy). *J. Volcanol. Geotherm. Res.* **1997**, *78*, 65–76.
24. Nakai, S.; Wakita, H.; Nuccio, P.M.; Italiano, F. MORB-type neon in an enriched mantle beneath Etna, Sicily. *Earth Planet. Sci. Lett.* **1997**, *153*, 57–66.
25. Spilliaert, N.; Allard, P.; Métrich, N.; Sobolev, A.V. Melt inclusion record of the conditions of ascent, degassing, and extrusion of volatile-rich alkali basalt during the powerful 2002 flank eruption of Mount Etna (Italy). *J. Geophys. Res.***2006**, *111*.
26. Hirn, A.; Nicolich, R.; Gallart, J.; Laigle, M.; Cernobori, L. ETNASEIS Scientific Group Roots of Etna volcano in faults of great earthquakes. *Earth Planet. Sci. Lett.* **1997**, *148*, 171–191.
27. Giammanco, S.; Bonfanti, P. Cluster analysis of soil CO_2 data from Mt. Etna (Italy) reveals volcanic influences on temporal and spatial patterns of degassing. *Bull. Volcanol.* **2009**, *71*, 201–218.
28. Camarda, M.; de Gregorio, S.; Gurrieri, S. Magma-ascent processes during 2005–2009 at Mt. Etna inferred by soil CO_2emissions in peripheral areas of the volcano. *Chem. Geol.* **2012**, *330–331*, 218–227.
29. Giammanco, S.; Neri, M. Rapporto sull'attività parossistica della Salinella dello Stadio di Paternò. In *UF Vulcanologia e Geochimica*; INGV Sezione de Catania: Sicily, Italy, 2005. (In Italian)
30. Anza, S.; Badalamenti, B.; Giammanco, S.; Gurrieri, S.; Nuccio, P.M.; Valenza, M. Preliminary study on emanation of CO_2 from soils in some areas of Mount Etna (Sicily). *Acta Vulcanol.* **1993**, *3*, 189–193.
31. Giammanco, S.; Gurrieri, S.; Valenza, M. Soil CO_2 degassing on Mt. Etna (Sicily) during the period 1989–1993: Discrimination between climatic and volcanic influences. *Bull. Volcanol.* **1995**, *57*, 52–60.
32. Brusca, L.; Aiuppa, A.; D'Alessandro, W.; Parello, F.; Allard, P.; Michel, A. Geochemical mapping of magmatic gas-water-rock interactions in the aquifer of Mount Etna volcano. *J. Volcanol. Geotherm. Res.* **2001**, *108*, 199–218.
33. D'Alessandro, W.; Federico, C.; Aiuppa, A.; Longo, M.; Parello, F.; Allard, P.; Jean-Baptiste, P. Groundwater circulation at Mt Etna: Evidenses from ^{18}O, ^{2}H and ^{3}H contents. In Proceedings of the 10th International Symposium on Water-Rock Interaction, Villasimius, Italy, 10–15 June 2001.
34. Follet, R.F. Fate and Transport of Nutrients: Nitrogen. In *Agricultural Research Service, Soil-Plant-Nutrient Research Unit*; Working Paper No. 7; United States Department of Agriculture: Fort Collins, CO, USA, 1995; p. 33.

35. Craig, H. Isotopic variations in meteoric waters. *Science* **1961**, *133*, 1702–1703.
36. Gat, J.R.; Carmi, H. Evolution of the isotopic composition of atmospheric waters in the Mediterranean Sea area. *J. Geophys. Res.* **1970**, *75*, 3039–3040.
37. Mazor, M. *Chemical and Isotopic Groundwater Hydrology*; Marcel Dekker Publisher: New York, NY, USA, 1991.
38. Gat, J.R.; Klein, B.; Kushnir, Y.; Roether, W.; Wernli, H.; Yam, R.; Shemesh, A. Isotope composition of air moisture over the Mediterranean sea: An index of the air—Sea interaction pattern. *Tellus. Ser.* **2003**, *B55*, 953–965.
39. Caracausi, A.; Italiano, F.; Paonita, A.; Rizzo, A.; Nuccio, P.M. Evidence of deep magma degassing and ascent by geochemistry of peripheral gas emissions at Mount Etna (Italy): Assessment of the magmatic reservoir pressure. *J. Geophys. Res.* **2003**, *108*.

CHAPTER 3

Climate Change and Its Influence on the Karst Groundwater Recharge in the Jinci Spring Region, Northern China

Zhenxing Jia 1,2, Hongfei Zang 3, Xiuqing Zheng 1,* and Yongxin Xu 1,4

1 College ofWater Resources Science and Engineering, Taiyuan University of Technology, Taiyuan 030024, China

2 Shanxi Institute of Water Resources, Taiyuan 030001, China

3 School ofWater Conservancy, North China University of Water Resources and Electric Power, Zhengzhou 450046, China

4 Department of Earth Sciences, University of the Western Cape, Cape Town 7535, South Africa

ABSTRACT

Due to climate change and human activities over the last fifty years, the spring flow volume of karst groundwater has sharply diminished in China. Climate change is one of the critical factors that initiates a series of karst hydrogeologic and water ecological environmental problems, because the precipitation shows a decreasing trend while the temperature shows an increasing trend. The Jinci Spring is one of the largest, most famous springs in northern China. This study employed data from the Taiyuan Meteorological Station and ten precipitation stations in and around the Jinci Spring region as well as the runoff data gathered from two hydrological monitoring stations during 1960–2012. The sliding average method and the Mann-Kendall test were used to analyze the variation tendency of precipitation, temperature, and land evaporation in this area. Finally, the following were calculated: the varying pattern of the karst groundwater recharge amount and the response of the recharge amount to precipitation, land evaporation, and river runoff by quantitative analysis. The results indicated that the precipitation and land evaporation amount decreased at first and then subsequently increased. Likewise, the variation trend of the karst groundwater recharge amount in the spring region was roughly consistent with the precipitation variation pattern. In contrast, the temperature displayed an increasing trend. The climate change resulted in a reduction of the karst groundwater recharge amount, and it had the greatest influence in the 1990s, which caused the karst groundwater recharge amount to decrease 26.75 mm as compared to that of the 1960s (about 39.68% lower than that of the 1960s). The Jinci Spring had zero flow during this period. The reduction in precipitation was one of main factors that caused the cutoff of the Jinci Spring.

Keywords: karst groundwater; climate change; trend test; abrupt change test

1. INTRODUCTION

The karst groundwater in the Jinci Spring region is endowed with rich groundwater resources, stabilized dynamic state, and good water quality. Additionally, it has become the water supply source for urban water supply and industrial water. Due to both climate change and human activity, the Jinci Spring has been cutoff, and the karst groundwater level in the spring source area has declined, which has greatly influenced the urban water supply security and water ecological environment. This was a notable event marking a water environment degradation in Taiyuan over the last 50 years. The recharge of karst groundwater in the Jinci Spring region is primarily affected by local climate change; thus, the influence of climate change on the recharge of groundwater in the spring region has particular significance to the optimization of utilization, the implementation of scientific scheduling, and the management of regional water resources. In general, water management is a strong key to use water resources in a correct way [1].

Since the 21st century, countries in Europe and America have begun to research the quantitative calculation methods of groundwater recharge on different temporal-spatial scales [2,3]. The water balance model includes surface runoff as well as groundwater recharge, and it also takes into consideration the influence of climate change and human activities on groundwater recharge, while simultaneously researching the influence of climate change on water resources [4,5]. Doveri, M. et al. utilized water isotopes as environmental tracers for the conceptual understanding of groundwater flow [6]. Cervi, F. et al. studied 12 springs from the northern Apennines of Italy by means of a comprehensive hydrogeological investigation to unravel recharge processes taking place in a highly fractured slab of flysch rock hosting the corresponding aquifers [7]. At present, the development of a groundwater recharge model and a sensitivity analysis of groundwater to climate change are still in the research and exploration stages in China [8]. Based on precipitation data obtained from over 600 precipitation stations and the annual average temperature data obtained for over 100 meteorological stations in northern China, Weitai Wang et al. analyzed the change tendency and the change relationship among the sunspot quantity, the precipitation, and climate change as well as estimated the influence of climate change on regional karst groundwater recharge in 2012 [9].

There has been considerable research focused on the Jinci Spring; however, much of the study has been concentrated on analyzing the cause of the spring flow exhaustion, analyzing the influence of human activities on the spring flow, and analyzing the conditions of the Jinci Spring reflow [10,11]. In contrast, there has been considerably less research conducted on the influence of climate change on karst groundwater recharge in the spring region [12]. For example, only Qinghai Guo et al. analyzed the guiding significance of the spring flow process in the context of climate change [13].

In this paper, the sliding average method and the Mann-Kendall test method have been adopted to analyze the annual variations of climatic factors [14], such as precipitation, temperature, and land evaporation [15,16]. The purpose of this study is to reveal the influence of atmospheric precipitation and Fenhe River leakage on karst groundwater recharge in the spring region and to provide a scientific basis for the reasonable exploitation and utilization of karst groundwater resources in the spring region and the establishment of karst groundwater management and protection policies [17,18].

2. RESEARCH AREA OVERVIEW

The Jinci Spring is located 25 km southwest of Taiyuan, Shanxi Province, and it is one of the largest karst springs in northern China.

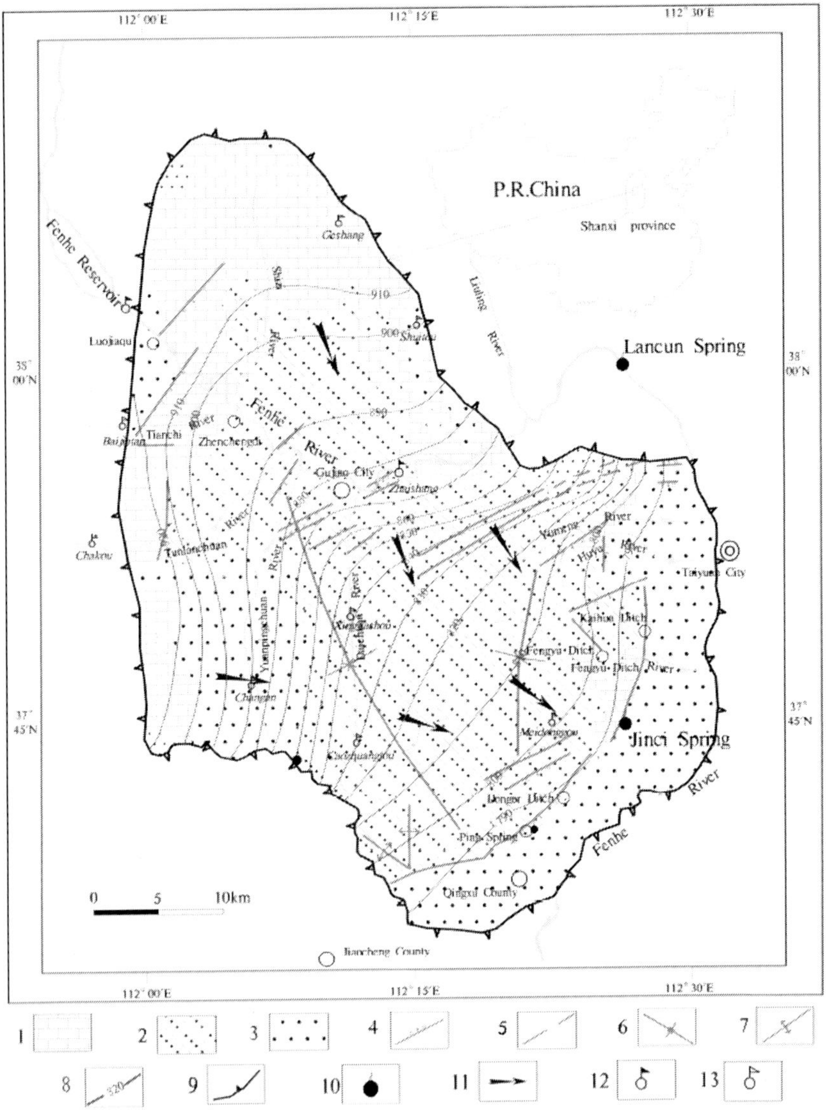

Figure 1. Simplified hydrogeological map of the Jinci Spring region. 1. Cambrian and Ordovician carbonate rocks; 2. Carboniferous-Permian clastic rocks; 3. Quaternary loose sediments; 4. fault structure; 5. buried fault; 6. syncline structure; 7. anticline structure; 8. groundwater contour lines; 9. boundary of spring region; 10. spring; 11. groundwater flow direction; 12. hydrologic station; 13. precipitation station.

The spring region is 2030 km^2 (Figure 1). The formation of the Jinci Spring was primarily a result of the karst groundwater of Taiyuan Xishan becoming blocked by the stratum of quaternary aquitard on the eastern side of the Taiyuan-Jiaocheng fault zone when there was basin movement. As a result, it overflowed through the ground and formed the spring. It is a classic piedmont fault overflow spring.

The Jinci Spring region has a typical temperate semi-arid continental monsoon climate, and it is typically characterized by droughts and windy conditions, a high precipitation concentration, intensive evaporation, four distinct seasons, a large temperature difference between day and night, and a short frost-free season. The multi-year (1960–2012) average precipitation of this region was 457.8 mm. The inter-annual, annual, and geographical distributions are extremely uneven. The precipitation of the highest flow year is over three times greater than that of the lowest flow year, and 60% of the annual precipitation is concentrated in the flood season, which extends from June to September. The multi-year average evaporation amount in this region is 1871.8 mm (an observed value of 20 cm from an observation vessel), and the multi-year average temperature is 10 °C.

The Jinci Spring region is part of the Fenhe River system of the Yellow River basin. The river system within the area is relatively developed, and the Fenhe River is the largest river in the area; it runs through this area from west to east and then turns south after it reaches the mountainous outlet in Lancun. The Tianchi River, Shizi River, Tunlanchuan, Yuanpinchuan, Dachuanhe, Liulin River and others are the first-grade tributaries of the Fenhe River, but they are seasonal rivers.

The spring region is composed of two parts, mountain area and plain area. It is bound by the Taiyuan-Jiaocheng fault. The northwestern part of the fault is a middle-low mountain area, and the terrain is higher in the northwest than in the southeast. The area is about 1771 km^2, and the elevation is 900–1500 m. The eastern fault zone is a basin plain area, with an area of 259 km^2 and an elevation of 780–850 m. The mountainous area and the basin plain area connect directly with terrain change.

The main aquifer of the Jinci Spring karst groundwater is the limestone of the upper and lower Majiagou Formation and the Fengfeng Formation of middle Ordovician, and the secondary is the dolomite and dolomitic limestone of the lower Ordovician and Cambrian system. The recharge source of the spring karst groundwater system primarily originates from three sources: the first is the precipitation penetrating recharge in the carbonatite bare field to the north of the Fenhe River, which is the main recharge of the Jinci Spring; the second is the seepage recharge in the section of the carbonatite river that reaches Fenhe (the seepage reach of the Fenhe River in this spring mainly includes the Luojiaqu-Zhenchendi seepage section (the length is about 22 km), the Gujiao–Zhaishang seepage section (the length is about 5 km), and the Zhaishang–Shaoshi seepage section (the length is about 16 km)); and the third is the leaking recharge from the groundwater of the overlying formation.

The karst groundwater in the spring region migrates from northwest to southeast. The Malan syncline is located in the middle area of this spring region, which displays in the northwest~southeast direction and extends for 43 km. Because the Ordovician limestone is buried, the karst development is weak and has poor permeability in the synclinal axis. The region in the west Malan syncline and to the south of the Fenhe River is a stagnant zone, and the karst groundwater basin that is located on the east side of this syncline is in the strong runoff zone. In addition, a water-abundance zone with strong karstification and water conductivity runs along the Taiyuan-Jiaocheng fault zone.

Except for the natural discharge points at the Jinci Spring and the Ping Spring, the karst groundwater in this spring discharges a small amount of water into the Taiyuan Basin as subsurface flow. In addition, the exploitation of karst groundwater is another source of groundwater discharge [19].

The Jinci Spring flow has exhibited a reduction tendency year by year since the 1960s, especially in the late 1980s. It declined from 1.69 m^3/s in the 1960s to 1.13 m^3/s in the 1970s, to 0.46 m^3/s in the 1980s, and to 0.15 m^3/s in the 1990s. The Jinci Spring was cutoff on 30 April 1994. The groundwater level of the entire Jinci Spring sharply declined from 1994 to 2009. The water level at the spring vent of the Jinci Spring declined to its lowest point in history (774.94 m), which is 27.65 m lower than the elevation at the spring vent of the Jinci Spring (802.59 m). However, it has exhibited a rising tendency from 2010 to 2012 and has recovered from 777.54 m in 2010 to 788.21 m in 2012 (Table A1 and Figure 2).

The average annual precipitation of the Jinci Spring region in the 1960s, 70s, 80s and 90s are 507.7 mm, 456.1 mm, 424.1 mm and 396.7 mm respectively, which shows a descending trend. According to the precipitation and discharge data of the 1960s and 1990s, the precipitation decreased by 21.9%, and the average spring flow decreased by 91.8%, from 1.735 m^3/s to 0.142 m^3/s. According to this result, precipitation is not the main factor of the spring flow attenuation. The average annual precipitation from 2000 to 2012 is 493.9 mm, which is 7.9% higher than that from 1960 to 2012 (458 mm). Therefore, the change of precipitation has a small contribution to the change of the karst water level in the Jinci Spring region. The reason why the spring dried up is mainly attributed to groundwater over-abstraction.

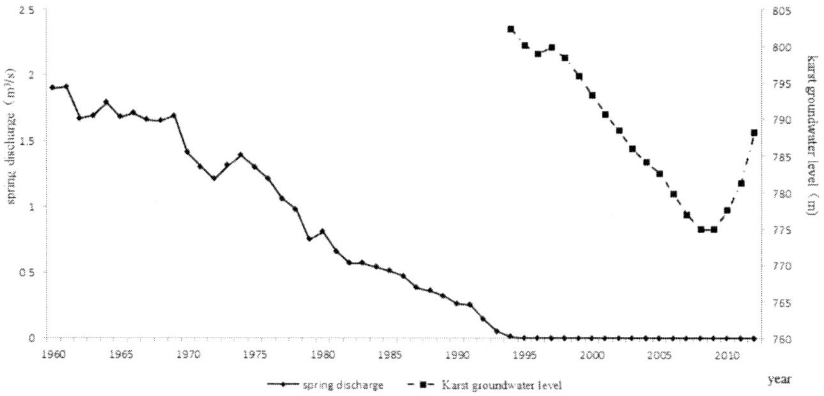

Figure 2. The change trend chart of Jinci Spring discharge and the karst groundwater level of the spring vent.

The main reasons for the rise of the groundwater level in recent years are as follows: first, by means of utilizing the Yellow River surface water, the local government closed part of the groundwater exploitation wells to reduce the amount of groundwater exploitation; second, the high water consumption enterprises in this region have been shut down or moved; third, the implementation of the Fenhe River resuming flow engineering has achieved continuous flow of the Fenhe River for many years, which increases the recharge of groundwater; fourth, through the implementation of regional ecological environment comprehensive management, vegetation coverage has been raised, and water resources are conserved. After taking the above measures, the karst groundwater level in the spring area has risen remarkably in recent years.

3. RESEARCH METHOD

3.1. Time-Series Test

To analyze climate change characteristics in this spring and the influence of climate on the karst groundwater recharge, the Mann-Kendall non-parametric test method was employed to conduct a trend analysis and abrupt change test [20].

3.1.1. Trend Test

The Mann-Kendall rank correlation test (referred to as the M-K trend test) is a non-parametric statistical test, which can accurately verify the sequence variation trend [21].

(1) The sample sequence $(x_1, x_2, ..., x_n)$ has no trend.

(2) The statistic is calculated, and the calculation formula is:

$$S = \sum_{i=1}^{n-1} \sum_{j=i+1}^{n} \operatorname{sgn}(x_j - x_i) \tag{1}$$

(3) S is normal distribution, and its average value is 0. The variance is $Var(S) = n(n - 1)(2n + 5)/18$, and the calculation formula of the standard normal statistic is:

$$Z = \begin{cases} (S-1)/\sqrt{Var(S)} & S > 0 \\ 0 & S = 0 \\ (S+1)/\sqrt{Var(S)} & S < 0 \end{cases} \tag{2}$$

(4) After testing, if the statistic is $|Z| \geq Z1 - \alpha/2$, then the original hypothesis is refused, which means the sequence has an obvious rising or declining trend.

(5) The tilt rate (quantization of monotonous trend) is calculated, and the calculation formula is:

$$\beta = median\left[\frac{X_i - X_j}{i - j}\right], \forall j < i \tag{3}$$

where $1 < j < i < n$, and the median is the median function. When $\beta > 0$, the sequence exhibits a rising trend; when $\beta<0$, the sequence exhibits a declining trend [22].

3.1.2. Abrupt Change Test

(1) The one time sequence is assumed to be $Xi = (x_1, x_2..., x_n)$, and a one-order sequence mi is constructed to show the accumulative quantity of a sample when $xi > xj$ $(1 \leq j \leq i)$. One statistic d_k is defined as:

$$d_k = \sum_{i=1}^{k} m_i,\ 2 \leq k \leq n \tag{4}$$

(2) Under the assumption that the original sequence is stochastic independence, the average value and variance of dk are, respectively:

$$E(d_k) = k(k - 1) / 4 \tag{5}$$

$$\mathrm{Var}\ (d_k) = k(k-1)(2k+5) / 72 \tag{6}$$

(3) d_k is standardized, and then:

$$u(d_k) = \frac{dk - E(d_k)}{\sqrt{\mathrm{var}(d_k)}} \tag{7}$$

where $u(d_k)$ is a standard distribution, and its probability can be achieved through calculation or looking it up in a table. Assuming one significance level to be a_0, when $a_1 > a_0$, the original hypothesis is accepted; however, when $a_1 < a_0$, the original hypothesis is refused, which means that this sequence exhibits a strong rising or declining trend. All the $u(d_k)(1 \le k \le n)$ will chart a curve UF_k [23].

(4) The time sequence x_k is ranked in reverse and is then calculated according to formula (7), and at the mean time:

$$\begin{cases} UB_k = -UF_k \\ k' = n + 1 - k \end{cases} \qquad (k, k' = 1, 2, \cdots, n) \tag{8}$$

The variation trend of sequence x_k can be further analyzed through analyzing the statistical series UF_k and UB_k, and the mutation time and area can be explicitly known. If $UF_k > 0$, then the sequence exhibits a rising trend, and if $UF_k < 0$, then the sequence exhibits a declining trend. If they are higher or lower than the critical lines, then the rising or declining trend is remarkable. If there is any point of intersection between the two curves of UF_k and UB_k, and the point of intersection is between critical lines, the time corresponding to the point of intersection is the mutation starting time [24].

3.2. Groundwater Recharge Amount Calculation

The main influencing factors of climate change on the karst groundwater recharge amount in the spring region are precipitation, land evaporation and river runoff [25]. The change of the groundwater recharge amount due to precipitation, land evaporation, and river runoff can be analyzed based on a quantitative calculation of karst groundwater recharge amount in the spring region. The water balance method is a general calculation method used for the hydrological cycle elements. This research uses the Fenhe Reservoir hydrologic station as the basin import and the Zhaishang hydrologic station as the basin export to determine the scope of the study area, which is 1587 km^2.

The length of the seepage zone in the Fenhe River from the Fenhe Reservoir to Zhaishang in the Jinci Spring is 27 km, and the length of the seepage zone in the Fenhe River from Zhaishang to Shaoshi is 16 km. The seepage zone from the Fenhe Reservoir to Zhaishang occupies 62.8% of the total seepage zone length. The exposed limestone area in this spring region is 391 km^2, which is mainly located to the north of this spring and to the north of the Fenhe River channel. The exposed limestone area in the basin controlled by the Zhaishang hydrologic station is 317 km^2, which occupies 81.1% of the total exposed area in this spring region. Thus, the research on the recharge of precipitation and the Fenhe River seepage of the zone from the Fenhe Reservoir to Zhaishang to the karst groundwater in the Jinci Spring region is highly important to the research on the karst groundwater recharge in the entire spring region.

The water balance in the basin from the Fenhe Reservoir to Zhaishang can be expressed as the following formula [26]:

$$Q_1 + P = Q_2 + E + R \tag{9}$$

where Q_1 is the observed runoff amount in the Fenhe Reservoir hydrologic station; P is the basin precipitation; Q_2 is the observed runoff amount in the Zhaishang hydrologic station; E is the land evaporation amount; and R is the recharge amount of precipitation and river channel seepage on the groundwater in this spring region.

4. RESULTS AND DISCUSSION

The data used in this research includes meteorological data and runoff data from two hydrologic stations located at the Fenhe Reservoir and Zhaishang from 1960 to 2012; the station location map is shown in Figure 1.

4.1. Climate Change Characteristic Analysis

The precipitation infiltration recharge is the most important recharge source in the Jinci Spring, and its main means of recharge include the direct precipitation infiltration recharge of the exposed carbonatite area, the indirect precipitation infiltration recharge of the overlay area, and the seepage recharge of the surface runoff yield formed by the precipitation in the carbonatite river channel. The precipitation will be transformed in three ways after landing on the ground: to form surface runoff, to infiltrate underground and then form groundwater, and to land-evaporate via vegetation and ground surface. In these three ways, the slope surface has a short lag time that allows the precipitation to quickly import into the channel and flow away, and the land evaporation and underground infiltration demonstrate the reciprocal relationship under the transduction function of all kinds of energy, including water and vapor in an unsaturated zone. Additionally, the temperature controls the land evaporation capacity to some extent. Thus, the precipitation amount and the temperature have important influences on the infiltration of karst groundwater in this area.

4.1.1. Precipitation Variation

The precipitation data in the research uses the annual precipitation data from ten precipitation stations in the Jinci Spring region, and the areal precipitation in the spring region can be obtained through the Thiessen polygons method. The annual precipitation in the spring varies from 217.87 to 676.63 mm, and the multi-year average value is 457.85 mm. The precipitation in the spring region exhibits greater fluctuation at different times. For example, the 1960s and 1970s have a relatively larger fluctuation, the 1980s and 1990s have relatively less fluctuation, and the precipitation amount since 2000 is a significant increase as compared to the 1990s (Table A2 and Figure 3a). The most obvious precipitation decline took place during the 1990s, and it is 111 mm less than that in the 1960s (as shown in Table 1).

According to the sliding average method and the linear tendency estimate method, the number of years is 53, and by setting k = 5, the 5a sliding average value of the annual precipitation in the Jinci Spring region can be obtained, and the sliding curve can be charted (Figure 3a).

The following conclusions were made from the analysis. The 5a sliding average precipitation from 2007 to 2011 is the largest (549.87 mm), whereas the value from 1997 to 2001 is the lowest (352.77 mm). The annual precipitation in the Jinci Spring region generally exhibits an initial decrease and then subsequent increase trend. The variation amplitude of the precipitation in the Jinci Spring region shows volatility. The annual distribution is uneven. Additionally, the average precipitation is 457.85 mm, the maximum precipitation occurred in 1964 (676.63 mm), and the minimum occurred in 1972 (217.87 mm); their quantitative difference is 458.76 mm.

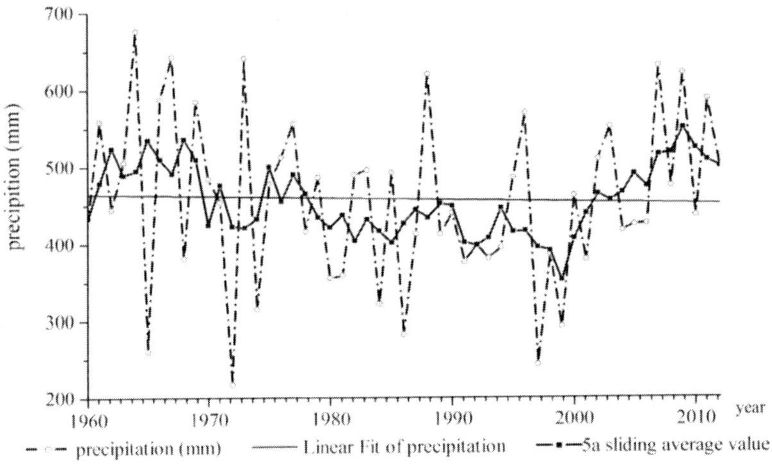

(a)

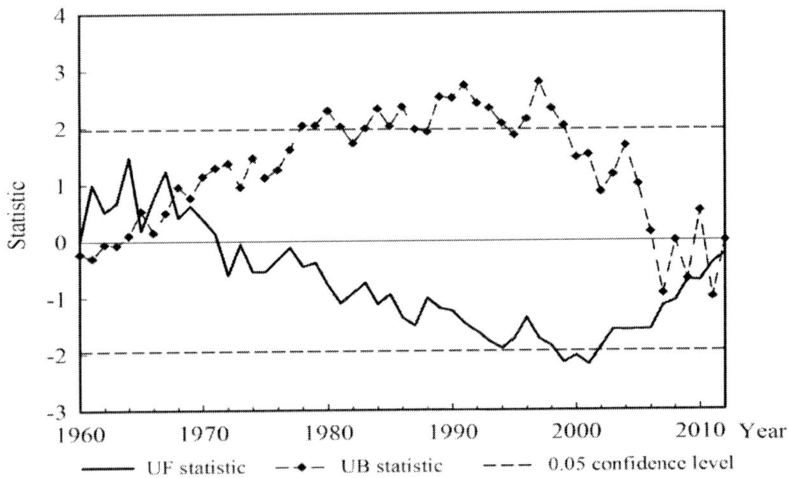

(b)

Figure 3. Annual precipitation variation trend and mutation analysis in the Jinci Spring region: (**a**) annual precipitation variation trend; (**b**) annual precipitation mutation analysis.

Table 1. Annual average precipitation, temperature, and evaporation for each time period in the Jinci Spring region: Statistical Table.

Time	1960s	1970s	1980s	1990s	2000–2012	Variation Degree between the 1960s and after 2000(%)
Average precipitation/mm	507.7	456.1	424.1	396.7	493.9	−2.73
Average temperature/°C	9.5	9.62	9.71	10.38	10.98	15.63
Average evaporation/mm	383.90	363.13	350.3	339.47	400.33	4.28
Average karst groundwater recharge/mm	67.40	63.47	65.98	40.66	59.53	−11.68

The statistic Z, β, and significance level of the annual precipitation in the Jinci Spring region can be obtained by the Mann-Kendall method to conduct a trend test, and the test results are, respectively, −0.22, −0.41, and 0.05. These results reveal that the precipitation test value Z of the study area is less than 0, but it does not pass the significance test level of 0.05, which means the precipitation in the Jinci Spring region from 1960 to 2012 generally shows a declining trend, and the multi-year average declining degree of precipitation is 0.41 mm/a [27].

According to the M-K abrupt change test result of the annual precipitation time variation (Figure 3b), the precipitation in the spring region before 1971 exhibited a rising trend, whereas a declining trend has been evident since 1972. It passes the 0.05 significance test ($U_{0.05}$ = 1.96). According to the intersecting point of the UF_k and UB_k curves, it can be determined that the mutation points of the annual precipitation existed in 1967 and 2009 in the Jinci Spring region [28].

4.1.2. Temperature Variation

The annual average temperature and the 5a sliding average temperature time sequence from 1960 to 2012 in the Jinci Spring region are shown in Figure 4a. According to Figure 4a, the two curves generally exhibit a rising trend over time.

The annual average temperature in the Jinci Spring region varies from 8.7 to 11.8 °C, and the multi-year average value is 10.09 °C over the last fifty years. As for the annual average temperature, the highest occurred in 2006 (11.8 °C), and the lowest occurred in 1967 (8.7 °C). The average temperature the Jinci Spring region increased over time (as shown according to decade in Table 1). The difference between the 1960s, 1970s, and 1980s is not very large, but the average temperature in the 1990s increased sharply to 0.88 °C (about 9.3%) higher than that of the 1960s, and in the 2000s it was 1.48 °C (about 15.63%) higher than that of the 1960s. This conclusion is in accordance with the conclusion found in literature [29]. According to the statistical temperature data, the 5a sliding average temperature from 1967 to 1971 in the Jinci Spring region is the lowest (9.06 °C), and the value from 2006 to 2010 is the highest (11.8 °C); the difference between them is 2.74 °C.

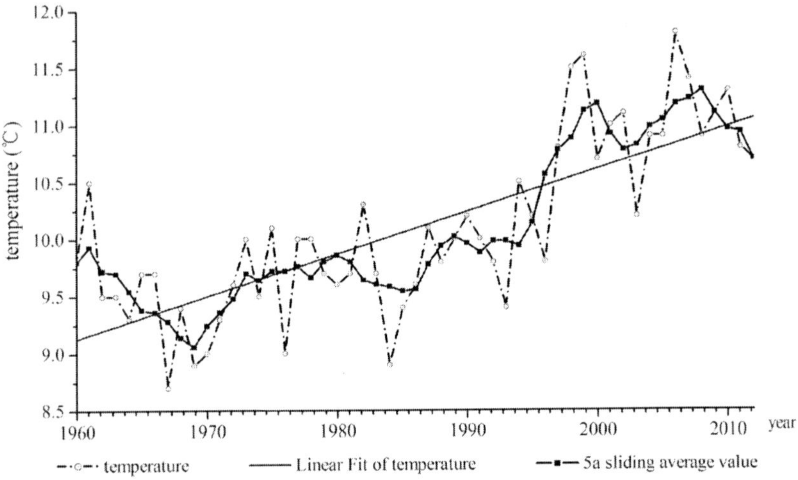

(a)

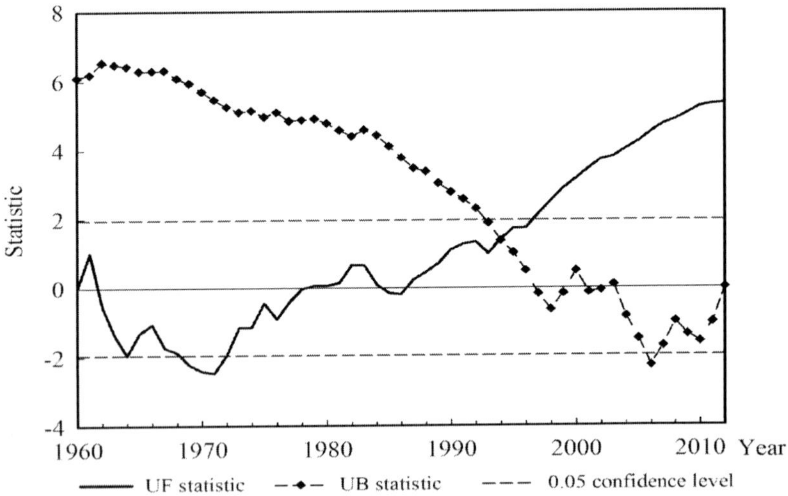

(b)

Figure 4. Annual average temperature variation trend and mutation analysis in the Jinci Spring region: (**a**) average temperature variation trend; (**b**) annual average temperature mutation analysis.

The multi-year average temperature trend test result, as tested by the Mann-Kendall method, is shown in Table 2. It can be seen that the statistic |Z| of the multi-year average temperature in the Jinci Spring region is larger than $Z_{1-\alpha/2}$, which means that the temperature in the entire spring region refused the null hypothesis, and it has a monotonous trend with statistical meaning. The β value of the multi-year average temperature in the table is 0.04, which is larger than 0; it also indicates that the temperature has exhibited a rising trend in the Jinci Spring over the last five decades [30].

Table 2. The trend and abrupt change test of the annual average precipitation, temperature, land evaporation, and karst groundwater recharge.

Element	Multi-Year Average Value/mm	M–K Trend Test			Mutation Point (Year)
		Z	β	Trend	
Precipitation	457.85	−0.22	−0.41	Declining	1967, 2009
Temperature	10.09	6.10	0.04	Rising	1994
Land evaporation	369.29	0.41	0.15	Rising	1964, 2007
Karst groundwater recharge	59.37	−1.04	−0.25	Declining	1964, 2007

The M-K abrupt change test of the annual temperature time variation (Figure 4b) shows that the average temperature in the spring region from 1962 to 1978 exhibited a declining trend; however, the average temperature has been on a rising trend since 1979 (in addition to 1985 and 1986), and it is still rising today. This rising trend passes the 0.05 significance test ($U_{0.05}$ = 1.96), and the mutation occurred in 1994 (the intersection of the UF_k–UB_k curve in the figure).

4.1.3. Evaporation Variation

According to the annual precipitation and annual average temperature data from 1960 to 2012, the evaporation in the spring region is calculated based on the Turc land evaporation empirical formula [31]:

$$Q_z = P \div \sqrt{[0.9 + P^2 \div (300 + 25 \times T + 0.05 \times T^3)^2]} \tag{10}$$

where Q_z is the land evaporation amount (mm), P is the precipitation (mm), and T is the temperature (°C).

Curves representing the annual evaporation and 5a sliding average evaporation time sequence in the spring region from 1960 to 2012 are shown in Figure 5a. The two curves generally show an initial decreasing and then subsequent increasing trend over time, which is similar to the precipitation variation trend (Figure 5a).

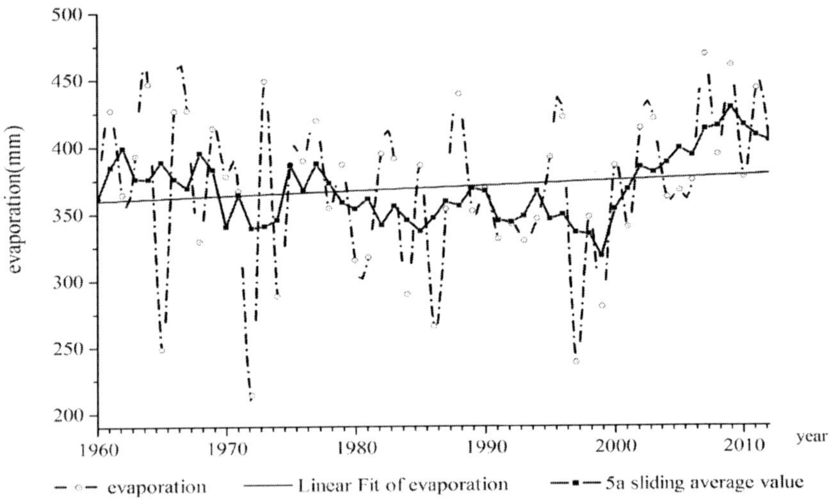

(a)

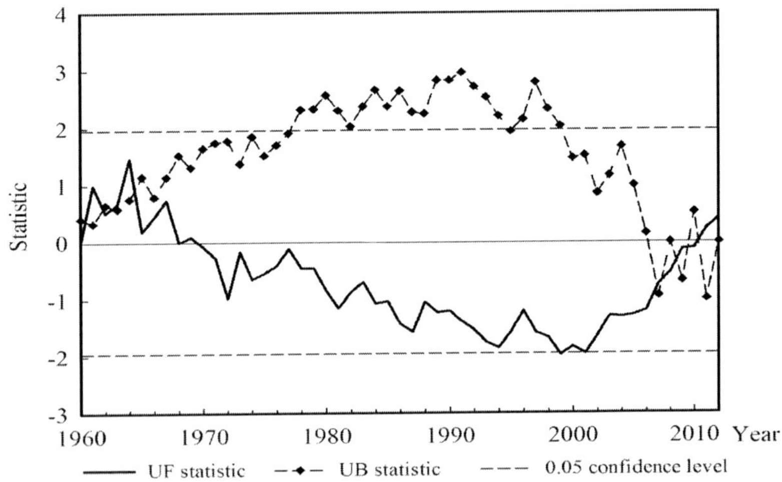

(b)

Figure 5. Annual evaporation variation trend and mutation analysis in the Jinci Spring region: (**a**) annual evaporation variation trend; (**b**) annual evaporation mutation analysis.

The annual land evaporation in the Jinci Spring region varies from 213.74 to 467.9 mm, and the multi-year average value is 369.29 mm. The lowest was 213.74 mm in 1972, and the highest was 467.9 mm in 2007; the difference between them is 254.15 mm.

According to Table 1, the evaporation at each time period decreased first and then subsequently increased over time. As compared to the 1960s, the 1970s decreased 20.77 mm (about 5.4%), the 1980s decreased 33.6 mm (about 8.75%), the 1990s decreased 44.43 mm (about 11.57%), and 2000–2012 increased 16.43 mm (about 4.28%). As for the 5a sliding average evaporation, the value from 2007 to 2011 is the highest (428.01 mm), and the value from 1997 to 2001 is the lowest (317.85 mm); the difference between them is 110.15 mm.

The M-K mutation test of the annual land evaporation variation over time (Figure 5b) has shown that the land evaporation in the spring region from 1960 to 1969 exhibited a rising trend. From 1970 to 2010, it exhibited a declining trend, and 2011 and 2012 demonstrated a rising trend. According to the intersecting point of the UF_k and UB_k curves, it can be determined that the mutation existed in 1964 and 2007 [32].

4.2. Analysis of the Influence of Climate Change on Karst Groundwater Recharge in the Spring Region

The recharge amount of the Jinci Spring region from the precipitation and river channel seepage in the Fenhe Reservoir-Zhaishang zone can be obtained according to Formula (9).

The annual recharge amount in the Fenhe Reservoir-Zhaishang zone of the Jinci Spring region varies from 29.32 to 140.28 mm (about 11,500–54,900 thousand m^3), and the multi-year average value is 59.37 mm (about 23,200 thousand m^3).

The karst groundwater recharge amount in the study region has a larger fluctuation at different times. The 1960s, 1970s, and 1980s are relatively higher, whereas the 1990s is relatively lower, and since 2000 the recharge amount has slightly increased (Figure 6a). As compared to the 1960s, the recharge amount of the 1980s decreased 1.42 mm (about 2.11%), that of the 1990s decreased 26.75 mm (about 39.68%), and that of 2000–2012 decreased 8.07 mm (about 11.97%).

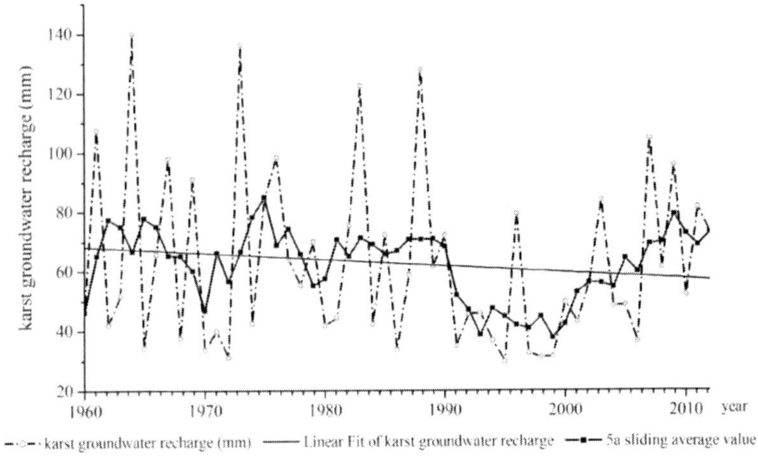

(a)

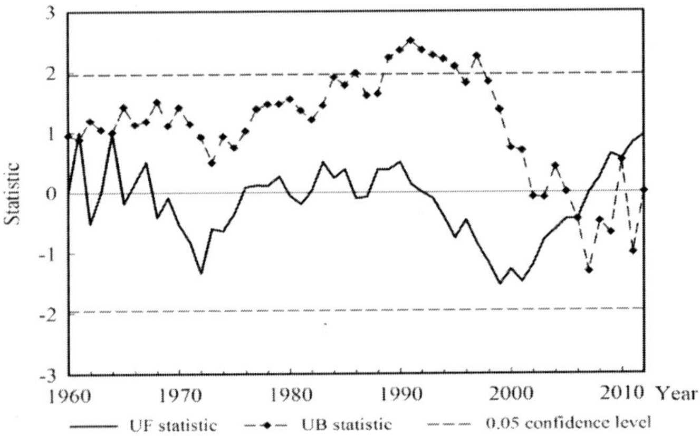

(b)

Figure 6. Annual karst groundwater recharge amount variation trend and mutation analysis in the Fenhe Reservoir-Zhaishang zone of the Jinci Spring region: (**a**) annual karst groundwater recharge amount variation trend; (**b**) annual karst groundwater recharge amount variation trend mutation analysis.

The curves of the annual recharge amount and 5a sliding average recharge amount time sequence in the study area from 1960 to 2012 are shown in Figure 6a. The following conclusions were drawn from the analysis. The highest 5a sliding average recharge amount was from 1973 to 1977 at 84.98 mm (converted into 33,200 thousand m^3), and the lowest value was from 1997 to 2001 at 37.96 mm (about 14,850 thousand m^3). Similar to the precipitation variation trend, the karst groundwater recharge amount in the spring region generally exhibited an initial declining and then subsequent increasing trend [33].

The Mann-Kendall method was used to conduct the trend test, and the test result shows that the statistic Z, β, and the significance level of the annual recharge amount in the study region are, respectively, −1.04, −0.25, and 0.05. The result reveals that the karst groundwater recharge amount test value Z of the study region is less than 0, which means that the karst groundwater recharge amount in the study region from 1960 to 2012 generally exhibits a declining trend. The M-K mutation test shows that the karst groundwater recharge amount exhibited a rising trend before 1967, and then a declining trend predominated; however, it has been rising once again since 2007. According to the intersecting point of the UF_k and UB_k curves, it can be determined that the mutation existed in 1964 and 2007 (Figure 6b) [34].

5. CONCLUSIONS

This study employed the M-K non-parametric test method to analyze the variation trend and mutation characteristic of the annual precipitation, land evaporation, temperature, and karst groundwater recharge in the Jinci Spring region. The main conclusions are as follows:

1) By comparing the relationship between the change in precipitation and the discharge of the spring, the change in precipitation has a small contribution to the change in karst water level in the Jinci Spring region. The reason why the spring dried up is mainly attributed to groundwater over-abstraction.

2) A total of 53 years of statistics reveal that the annual precipitation of the Jinci Spring region generally follows an initial decreasing and then subsequent increasing trend, and the most obvious declining trend occurred in 1990s, which was 111 mm less than that of the 1960s. Likewise, the variation trend of land evaporation is similar to that of precipitation; however, the annual average temperature has maintained an increasing trend.

3) The variation process of the karst groundwater recharge in the Jinci Spring region is generally consistent with that of the precipitation. As compared to the climate of the 1960s, all the climate changes at the other time periods resulted in a reduction in the recharge amount. The climatic change had the strongest influence in the 1990s, which

caused the karst groundwater recharge amount in the spring region to be reduced by 26.75 mm, about 39.68% lower than that of the 1960s. The reduction of precipitation is one of main reasons underlying the cutoff of the Jinci Spring.

4) The results of the Mann-Kendall abrupt change test revealed that the karst groundwater recharge amount in the Fenhe Reservoir-Zhaishang zone of the Jinci Spring region generally exhibits a declining tendency. The karst groundwater recharge amount exhibited a rising trend before 1967. After that, a declining trend took place; however, a rising trend has been occurring since 2007. The mutations existed in 1964 and 2007.

ACKNOWLEDGEMENTS

The authors would like to acknowledge the support of the National Natural Science Foundation of China (Grant No. 41572239). Furthermore, we gratefully acknowledged the Water Resources Department of Shanxi Province for providing the historical data, and we thank local partners, Xiaopeng Xing, Luxiu Li, and Wenzhong Zhang, for their support and assistance in the field.

APPENDIX A

Table A1. The data of spring discharge and karst groundwater levels.

Year	Spring Discharge (m^3/s)	Karst Groundwater Level (m)
1960	1.9	
1961	1.91	
1962	1.67	
1963	1.69	
1964	1.79	
1965	1.68	
1966	1.71	
1967	1.66	
1968	1.65	
1969	1.69	
1970	1.41	
1971	1.3	
1972	1.21	
1973	1.31	
1974	1.39	
1975	1.3	
1976	1.21	

1977	1.06	
1978	0.98	
1979	0.75	
1980	0.81	
1981	0.66	
1982	0.57	
1983	0.57	
1984	0.54	
1985	0.51	
1986	0.47	
1987	0.38	
1988	0.36	
1989	0.32	
1990	0.26	
1991	0.25	
1992	0.14	
1993	0.05	
1994	0.01	802.32
1995	0	800.11
1996	0	798.97
1997	0	799.79
1998	0	798.37
1999	0	795.94
2000	0	793.3
2001	0	790.69
2002	0	788.42
2003	0	785.96
2004	0	784.07
2005	0	782.59
2006	0	779.86
2007	0	777
2008	0	774.94
2009	0	774.94
2010	0	777.54
2011	0	781.26
2012	0	788.21

Table A2. The data of precipitation, temperature and observed runoff amount.

Year	The Precipitation in the Jinci Spring Area (mm)	Average Temperature (°C)	The Observed Runoff Amount in the Fenhe Reservoir Hydrologic Station (m^3/s)	The Observed Runoff Amount in the Zhaishang Hydrologic Station (m^3/s)
1960	433.59	9.8	6.81	8.09
1961	558.89	10.5	6.98	8.2
1962	444.83	9.5	12.3	14.2
1963	506.34	9.5	10.6	13.7
1964	676.63	9.3	11.0	15.5
1965	260.63	9.7	19.9	19.3
1966	588.16	9.7	5.07	9.89
1967	642.10	8.7	27.2	33.1
1968	381.16	9.4	18.2	20.4
1969	584.83	8.9	14.3	18.3
1970	484.44	9	14.9	18.6
1971	453.94	9.3	8.54	10.9
1972	217.87	9.6	8.06	7.71
1973	641.07	10	5.55	8.38
1974	315.23	9.5	14.8	14
1975	477.61	10.1	5.34	5.68
1976	512.10	9	5.43	6.68
1977	555.91	10	9.95	13.6
1978	416.21	10	13.9	14.2
1979	486.16	9.7	13.6	15.1
1980	355.13	9.6	13.5	13.4
1981	358.19	9.7	6.97	6.78
1982	489.41	10.3	7.72	8.7
1983	495.87	9.7	8.65	7.79
1984	321.51	8.9	9.22	8.7
1985	492.36	9.4	4.89	6.62
1986	282.86	9.6	8.49	8.67
1987	413.17	10.1	3.78	3.84
1988	619.87	9.8	8.52	11.2
1989	412.33	10	9.54	9.48
1990	439.46	10.2	9.34	9.29
1991	376.33	10	8.43	8.95
1992	397.77	9.8	5.24	5.75
1993	381.11	9.4	7.73	8.02
1994	395.11	10.5	5.56	6.19
1995	485.76	10.2	12.5	15.7
1996	569.90	9.8	21.7	24.7
1997	243.46	10.8	9.21	8.87
1998	385.76	11.5	8.30	8.8
1999	292.43	11.6	7.35	7.41
2000	462.59	10.7	3.89	4.05
2001	379.63	11	3.8	4.09
2002	509.67	11.1	3.04	3.7
2003	551.73	10.2	5.46	5.77
2004	416.67	10.9	5.84	4.98
2005	425.05	10.9	4.02	3.32
2006	425.79	11.8	3.12	2.99
2007	630.25	11.4	2.36	2.67
2008	474.58	10.9	7.16	6.63
2009	621.03	11.1	6.13	7.07
2010	436.44	11.3	6.51	7.14
2011	587.04	10.8	7.37	8.52
2012	499.89	10.7	6.64	7.48

REFERENCES

1. Tazioli, A.; Mattioli, A.; Nanni, T.; Vivalda, P.M. Natural Hazard Analysis in the Aspio Equipped Basin. In *Engineering Geology for Society and Territory–Volume 3: River Basins, Reservoir Sedimentation and Water Resources*; Springer International Publishing: Cham, Switzerland, 2015; pp. 431–435.

2. Goderniaux, P.; Brouyere, S.; Fowler, H.J.; Blenkinsop, S.; Therrien, R.; Orban, P.; Dassargues, A. Large scale surface-subsurface hydrological model to assess climate change impacts on groundwater reserves. *J. Hydrol.* **2009**, *373*, 122–138.
3. Kundzewicz, Z.W.; Doll, P. Will groundwater ease freshwater stress under climate change. *Hydrol. Sci. J.* **2009**, *54*, 665–675.
4. Liu, C.Z.; Xie, Z.H. Recent advances in research on sensitivity of groundwater to climatic changes. *J. China Hydrol.***2007**, *27*, 1–6. (In Chinese).
5. Holman, I.P.; Tascone, D.; Hess, T.M. A comparison of stochastic and deterministic downscaling methods for modelling potential groundwater recharge under climate change in East Anglia, UK: Implications for groundwater resource management. *Hydrogeol. J.* **2009**, *17*, 1629–1641.
6. Doveri, M.; Mussi, M. Water Isotopes as Environmental Tracers for Conceptual Understanding of Groundwater Flow: An Application for Fractured Aquifer Systems in the "Scansano-Magliano in Toscana" Area (Southern Tuscany, Italy). *Water* **2014**, *6*, 2255–2277.
7. Cervi, F.; Corsini, A.; Doveri, M.; Mussi, M.; Ronchetti, F.; Tazioli, A. Characterizing the Recharge of Fractured Aquifers: A Case Study in a Flysch Rock Mass of theNorthern Apennines (Italy). In *Engineering Geology for Society and Territory—Volume 3: River Basins, Reservoir Sedimentation and Water Resources*; Springer International Publishing: Cham, Switzerland, 2015; pp. 563–567.
8. Jackson, C.R.; Meister, R.; Prudhomme, C. Modelling the effects of climate change and its uncertainty on UK Chalk groundwater resources from an ensemble of global climate model projections. *J. Hydrol.* **2011**, *399*, 12–28.
9. Wang, W.T.; Liang, Y.P.; Wang, Z.H.; Zhao, C.H. Characteristics of climate change in northern China and its effect on groundwater in karst areas. *Hydrolgeol. Eng. Geol.* **2012**, *39*, 6–10. (In Chinese).
10. Huang, H.L. The relationship between the cutoff of Jinci Spring and the conservation of groundwater resource. *Coal. Geol. China* **2003**, *15*, 26–28. (In Chinese).
11. Long, Y.Q.; Li, W.; Li, Y.G. Tank model of Jinci karst groundwater system under the on the condition of simultaneous draining in minging. *Carsologica Sin.* **2011**, *30*, 27–33. (In Chinese).
12. Sun, C.Z.; Wang, J.S.; Lin, X.Y. Research on the Shanxi Jinci Spring's recovery after the use of water from the yellow river as municipal water supply. *Carsol. Sin.* **2001**, *20*, 11–16. (In Chinese).
13. Guo, Q.H.; Wang, Y.X.; Ma, T. The flow variation processes of large karstic spring in Shanxi in the recent 50 years and its guiding significance on global climatic change. *Sci. China. Ser. D* **2005**, *35*, 167–175. (In Chinese).
14. Green, T.R.; Taniguchi, M.; Kooi, H.; Gurdak, J.J.; Allen, D.M.; Hiscock, K.M.; Treidel, H.; Aureli, A. Beneath the surface of global change: Impacts of climate change on groundwater. *J. Hydrol.* **2011**, *405*, 532–560.
15. Bloomfield, J.P.; Williams, R.J.; Gooddy, D.C.; Cape, J.N.; Guha, P. Impacts of climate change on the fate and behaviour of pesticides in surface and groundwater—A UK perspective. *Sci. Total Environ.* **2006**, *369*, 163–177.
16. Ali, R.; McFarlane, D.; Varma, S.; Dawes, W.; Emelyanova, I.; Hodgson, G.; Charles, S. Potential climate change impacts on groundwater resources of south–western Australia. *J. Hydrol.* **2012**, *475*, 456–472.
17. Barron, O.; Froend, R.; Hodgson, G.; Ali, R.; Dawes, W.; Davies, P.; McFarlane, D. Projected risks to groundwater–dependent terrestrial vegetation caused by changing climate and groundwater abstraction in the Central Perth Basin, Western Australia. *Hydrol. Process.* **2014**, *28*, 5513–5529.
18. Waibel, M.S.; Gannett, M.W.; Chang, H.; Hulbe, C.L. Spatial variability of the response to climate change in regional groundwater systems—Examples from simulations in the Deschutes Basin, Oregon. *J. Hydrol.* **2013**, *486*, 187–201.
19. Yadav, R.; Tripathi, S.K.; Pranuthi, G.; Dubey, S.K. Trend analysis by Mann–Kendall test for precipitation and temperature for thirteen districts of Uttarakhand. *J. Agrometeorol.* **2014**, *16*, 164–171.

20. Hamed, K.H. Trend detection in hydrologic data: The Mann–Kendall trend test under the scaling hypothesis. *J. Hydrol.* **2008**, *349*, 350–363.
21. Lacombe, G.; McCartney, M.; Forkuor, G. Drying climate in Ghana over the period 1960–2005: Evidence from the resampling-based Mann–Kendall test at local and regional levels. *Hydrol. Sci. J.* **2012**, *57*, 1594–1609.
22. Keilholz, P.; Disse, M.; Halik, U. Effects of Land Use and Climate Change on Groundwater and Ecosystems at the Middle Reaches of the Tarim River Using the MIKE SHE Integrated Hydrological Model. *Water* **2015**, *7*, 3040–3056.
23. Liu, Y.L.; Zhai, X.L.; Zheng, A.Q. Analysis of precipitation variation trend in Guanzhong Basin based on the Mann–Kendall. *Yellow River* **2012**, *34*, 28–30. (In Chinese).
24. Hamed, K.H. Exact Distribution of the Mann–Kendall Trend Statistic for Persistent Data. *J. Hydrol.* **2009**, *365*, 86–94.
25. Gain, A.K.; Giupponi, C.; Renaud, F.G. Climate Change Adaptation and Vulnerability Assessment of Water Resources Systems in Developing Countries: A Generalized Framework and a Feasibility Study in Bangladesh. *Water***2012**, *4*, 345–366.
26. Fathelrahman, E.; Davies, A.; Davies, S.; Pritchett, J. Assessing Climate Change Impacts on Water Resources and Colorado Agriculture Using an Equilibrium Displacement Mathematical Programming Model. *Water* **2014**, *6*, 1745–1770.
27. Baruffi, F.; Cisotto, A.; Cimolino, A.; Ferri, M.; Monego, M.; Norbiato, D.; Cappelletto, M.; Bisaglia, M.; Pretner, A.; Galli, A. Climate change impact assessment on Veneto and Friuli plain groundwater. Part I: An integrated modeling approach for hazard scenario construction. *Sci. Total. Environ.* **2012**, *440*, 154–166.
28. Samper, J.; Li, Y.M.; Pisani, B. An evaluation of climate change impacts on groundwater flow in the Plana de La Galera and Tortosa alluvial aquifers (Spain). *Environ. Earth. Sci.* **2015**, *73*, 2595–2608.
29. Tang, G.; Ding, Y.; Wang, S.; Ren, G.; Liu, H.; Zhang, L. Comparative analysis of the time series of surface air temperature over China for the last 100 years. *Adv. Clim. Chang. Res.* **2009**, *5*, 71–78. (In Chinese).
30. Pasini, S.; Torresan, S.; Rizzi, J.; Zabeo, A.; Critto, A.; Marcomini, A. Climate change impact assessment in Veneto and Friuli Plain groundwater. Part II: A spatially resolved regional risk assessment. *Sci. Total. Environ.* **2012**, *440*, 219–235.
31. Turc, L. The water balance of soils—Relation between precipitation evaporation and flow. *Ann. Agron.* **1954**, *5*, 491–569.
32. Ribeiro, L.; Kretschmer, N.; Nascimento, J.; Buxo, A.; Rotting, T.; Soto, G.; Senoret, M.; Oyarzun, J.; Maturana, H.; Oyarzun, R. Evaluating piezometric trends using the Mann–Kendall test on the alluvial aquifers of the Elqui River basin, Chile. *Hydrol. Sci. J.* **2015**, *60*, 1840–1852.
33. Stefanova, A.; Hesse, C.; Krysanova, V. Combined Impacts of Medium Term Socio-Economic Changes and Climate Change on Water Resources in a Managed Mediterranean Catchment. *Water* **2015**, *7*, 1538–1567.
34. Ligaray, M.; Kim, H.; Sthiannopkao, S.; Lee, S.; Cho, K.H.; Kim, J.H. Assessment on Hydrologic Response by Climate Change in the Chao Phraya River Basin, Thailand. *Water* **2015**, *7*, 6892–6909.

CHAPTER 4

Groundwater Level Changes Due to Extreme Weather—An Evaluation Tool for Sustainable Water Management

Jadwiga R. Ziolkowska [1,*] and Reuben Reyes [2]

[1]*Department of Geography and Environmental Sustainability, The University of Oklahoma, 100 East Boyd St. SEC 650, Norman, OK 73019-1081, USA*

[2]*Oklahoma Climatological Survey, The University of Oklahoma, 120 David L. Boren Blvd., Suite 2900, Norman, OK 73072, USA*

ABSTRACT

In the past decade, extreme and exceptional droughts have significantly impacted many economic sectors in the US, especially in California, Oklahoma, and Texas. The record drought of 2011–2014 affected almost 90% of Texas areas and 95% of Oklahoma state areas. In 2011 alone, around $1.6 billion in agricultural production were lost as a result of drought in Oklahoma, and $7.6 billion in Texas. The agricultural sectors in Oklahoma and Texas rely mainly on groundwater resources from the non-replenishable Ogallala Aquifer in Panhandle and other aquifers around the states. The exceptional droughts of 2011–2014 not only caused meteorologically induced water scarcity (due to low precipitation), but also prompted farmers to overuse groundwater to maintain the imperiled production. Comprehensive studies on groundwater levels, and thus the actual water availability/scarcity across all aquifers in Oklahoma and Texas are still limited. Existing studies are mainly focused on a small number of selected sites or aquifers over a short time span of well monitoring, which does not allow for a holistic geospatial and temporal evaluation of groundwater level variations. This paper aims at addressing those issues with the proposed geospatial groundwater visualization model to assess availability of groundwater resources for agricultural, industrial, and municipal uses both in Oklahoma and Texas in the time frame of 2003–2014. The model is an evaluation tool that can be used by decision-makers for designing sustainable water management practices and by teachers and researchers for educational purposes.

Keywords: drought; geospatial analysis; groundwater; US; visualization; water management

1. INTRODUCTION

Over the past centuries, extreme and exceptional drought events have significantly affected both surface water and groundwater resources [1,2,3]. While low surface water levels might be an immediate indicator of drought, changes to groundwater levels indicate long-term water scarcity. At the same time, while it is relatively straightforward to monitor and assess surface water changes, measuring variations in groundwater resources (aquifers) is very challenging and time consuming, which—in many cases—leads to missing or scattered data sets. However, while surface water resources can be replenished by adequate precipitation, groundwater aquifers might be non-replenishable once exploited (like, e.g., the Ogallala Aquifer containing fossil water, especially in the southern parts of the High Plains) [4] or it may take years for an aquifer to replenish naturally, subject to geological formations in different regions. While these conditions make groundwater resources extremely valuable, groundwater monitoring and measurements have been inconsistent geographically and over time.

This research aims at developing a model that can serve as an evaluation tool for variations in groundwater levels in two states, Oklahoma and Texas that have been mostly affected by the recent drought events. While the model presents a user-friendly way to analyze geographical and temporal variations in water availability in Oklahoma and Texas aquifers, it also sets a ground for our work on an integrated drought indicator to serve as a predictor of drought in the future. For the last decade, Texas and Oklahoma have been exposed to exceptional multiyear droughts, ranking number 1 and number 2 among the driest US states in 2010–2011, respectively (Figure 1). This trend has been recorded by the National Oceanic and Atmospheric Administration [5], while drought conditions can undoubtedly be called cyclical events recurring more or less frequently in sinusoidal cycles.

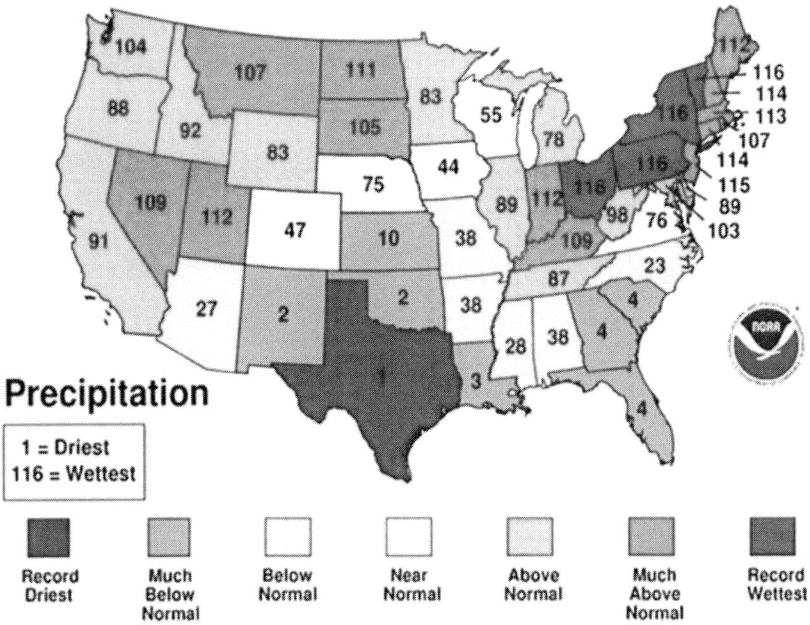

Figure 1. US statewide precipitation ranks (October 2010–September 2011). Source: National Oceanic and Atmospheric Administration [5].

Figure 2 shows a temporal overlap of drought events in both states, which is plausible due to their geographical proximity. In 2011, almost 65% of Oklahoma areas were in exceptional drought, ~95% in extreme drought, and 100% in severe drought. In 2013, even though the area in exceptional drought decreased to 40%, ~98% of the state areas suffered from extreme drought, and 100% from severe drought. Oklahoma also experienced severe and moderate droughts in 2006 and 2009; however, their magnitude was not comparable to the 2011–2014 droughts. A similar situation occurred in Texas, with ~85% of areas in exceptional drought, ~98% in extreme drought, and 100% in severe drought in fall 2011. The 2013 drought in Texas was less severe than in Oklahoma;

however, Texas experienced more severe droughts on record in 2006 and 2009.

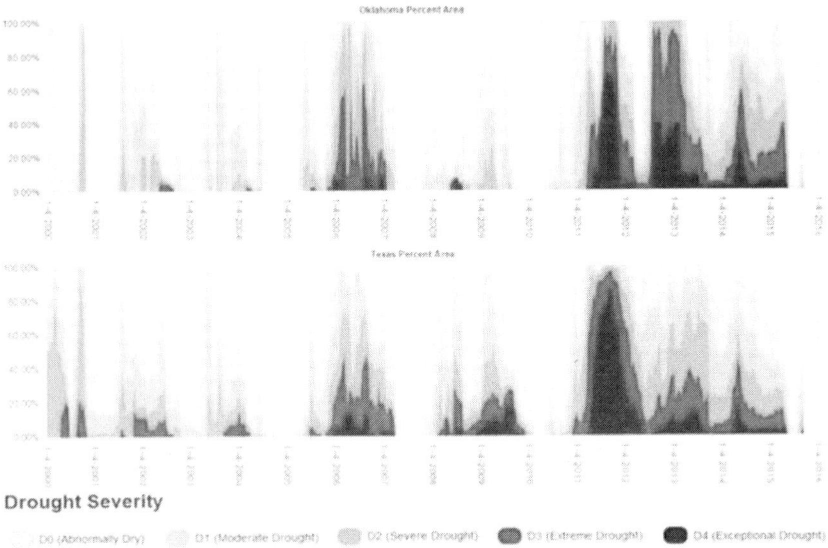

Figure 2. Drought index in Oklahoma and Texas in 2000–2016. Source: Drought Monitor [6], NDMC [7]. Legend: This figure shows categorical US Drought Monitor statistics that describe the percent of the area in a certain drought category, based on precipitation. It excludes areas that are better or worse. The statistics will add up to 100 percent for a given week.

Although droughts have a significant impact on all economic sectors, including municipal water provision, industrial operations, and thermoelectric power generation [8], the agricultural sector is particularly affected. This can be explained with the fact that agriculture is the most intensive user of water for irrigation among all economic sectors both in Oklahoma and Texas. The 2011 drought caused considerable economic losses in agricultural production of nearly $1.6 billion in Oklahoma [9] and $7.6 billion in Texas [10,11], while the entire Texas economy suffered economic losses of around $17 billion [12].

Even though the presented numbers refer to only a one-year drought event, they represent the severity of potential economic impacts as a result of extreme drought conditions, which could potentially recur both in the agricultural sector as well as in other sectors in the future. As the 2011 drought was a record drought, we use this example as a representative data trend for the geospatial analysis of changing groundwater levels in Oklahoma and Texas in 2003–2014 presented in this paper.

The relevance of conservative groundwater use stems not only from the instant meteorological water scarcity, but also from long-term predictions regarding future water availability. According to the Texas Water Development Board (TWDB) [13], Texas' groundwater supplies are expected to fall by 30% in the next 50 years. Similarly, according to the Oklahoma Water Resources Board (OWRB) [14], approximately one-third of the river basins in Oklahoma exhibit poor surface water quality, which may make them unavailable for reliable supply, while bedrock groundwater depletions are anticipated to occur in 34 basins.

In the face of decreasing water availability and an increasing demand for water due to growing population (33% increase in the next 50 years) [14], a question arises about the actual levels of available groundwater resources and possible mitigation/adaptation measures to the current and anticipated droughts in the future. Due to the relevance and urgency of the problem, a multitude of studies emphasized the importance of groundwater evaluation and modelling approaches for sustainable water management [15,16,17,18,19,20,21]. Also, some studies applied GIS-based models to address recharge volumes of aquifers from the hydrogeology perspective [22] or groundwater potential zones [23,24]. However, research on geospatial and temporal variability of groundwater resources and a holistic analysis of groundwater changes is still very limited [25,26]. Missing or inconsistent temporal groundwater monitoring records have been a restraining factor for comprehensive research in this field; however, it is common across most US states due to fluctuating human and monetary resources. Figure 3 displays an example in which for the first two wells data on well levels were available only until 2000, while the third well was monitored throughout 2012. In our study we found this common trend both in Oklahoma and Texas.

Furthermore, a multitude of studies analyzed volumetric changes of groundwater resources in different parts of the world to quantify effects of drought. For instance, Konikow [28] evaluated long-term cumulative depletion volumes in 40 separate aquifers in the US by means of calibrated groundwater models, analytical approaches, and volumetric budget analyses. Bhanja et al. [29] analyzed spatio-temporal groundwater storage anomalies in India to prove strong seasonality, with annual maxima observed during the monsoon season and minima during pre-monsoon season. Sakakibara et al. [30] determined comprehensive groundwater recharge processes in a catchment with large seasonal hydrological variations by means of field surveys in the Wangkuai Reservoir watershed in North China, while Zhang et al. [31] estimated groundwater storage indirectly from daily streamflow based on hydraulic groundwater theory. Similarly, Manna et al. [32] used the Chloride Mass Balance method to obtain long-term recharge values in an upland sandstone aquifer of southern California. Studies are also known that investigated impacts of crop rotations on groundwater storage and recharge in agricultural watersheds [33]. Several studies also used GRACE (Gravity Recovery and Climate Experiment) satellite data [34,35,36] or a combination of satellite data and in-situ observations [37,38], as well as geospatial software tools [39].

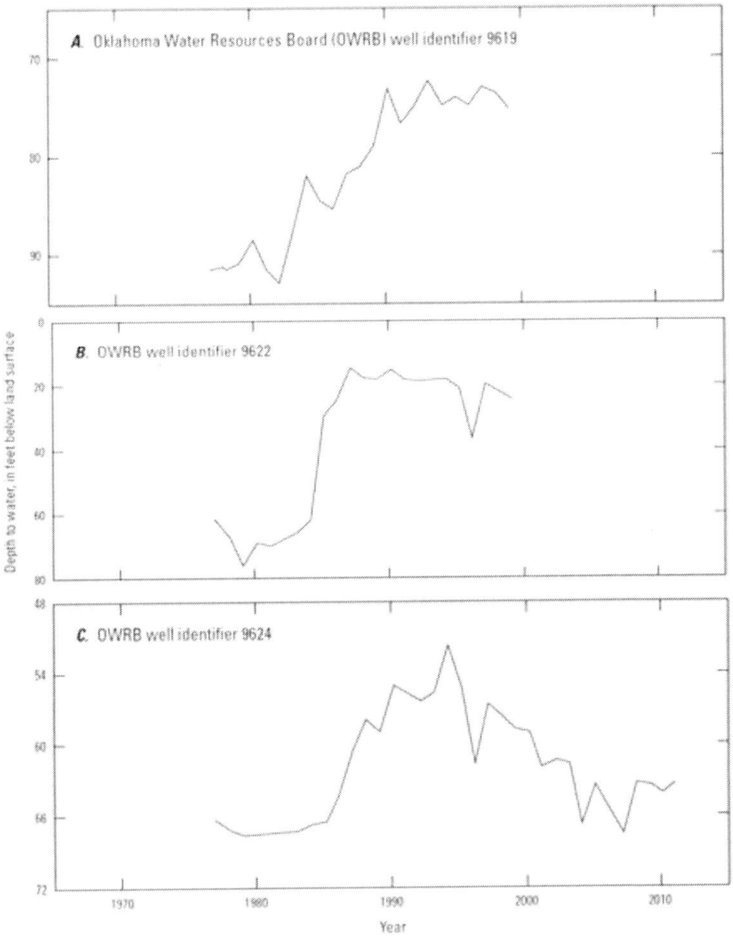

Figure 3. Water levels from Oklahoma Water Resources Board (OWRB) Mass Measurement Program wells. Source: Mashburn et al. [27]. (**A**) Well 9619, (**B**) Well 9622, (**C**) Well 9624.

The research presented with this paper aims at extending literature in the field by proposing a new model for monitoring groundwater level changes based on a method combination of normalized in-situ averages (rather than absolute water volumes) and the Palmer Drought Severity Index (PDSI) values. As opposed to GRACE measuring groundwater levels based on gravity at a large geographical scale, this model provides a detailed picture of groundwater level changes at each individual well, thus allowing for higher granularity and geospatial resolution of the results.

The proposed geospatial and temporal model for Oklahoma and Texas is used as a case study to explain the benefits of this methodological procedure. The selection of those regions was determined by continued drought events in both states (as described above) and data availability. While Oklahoma has the highest quality weather monitoring network in the world (the Oklahoma Mesonet) [40], Texas has a more comprehensive database with regular groundwater measurements. Those variations in data availability, and statistical calculations conducted for this study, allowed for inference analyses providing a holistic and interactive model of geospatial and temporal groundwater level changes in 2003–2014.

It needs to be emphasized that this paper does not address the groundwater exploitation, total storage, water balances, or policy issues related to aquifer exploitation. Even though those are relevant issues, this research is focused on evaluating spatio-temporal variations in groundwater well levels and developing a tool that can be used for decision-making regarding sustainable water management.

2. RESEARCH OBJECTIVES

The main goal of this paper is to analyze changes in groundwater well levels in 2003–2014 and determine potential long-term indicators of drought in Oklahoma and Texas. To achieve this goal we developed a geospatial and temporal visualization model with the aim to: (a) identify regional differences in groundwater level changes (geospatial dimension); (b) evaluate changes in groundwater wells over time (2003–2014) (temporal dimension); and (c) emphasize practical benefits of the model by pinpointing potential impacts of those changes in the future as well as potential locations of concern for decision-makers in their efforts to design sustainable water management programs.

As sampling and monitoring of groundwater well levels across US states and regions is often irregular and inconsistent (in some cases it occurs once per year only), the proposed model visualizing normalized changes in groundwater levels over time can also be used for mitigating data paucity. Analyzing past and current groundwater availability is essential for determining water scarcity in the short, mid, and long term. By developing an evaluation tool this research can help decision-makers, water managers, and scientists to recognize geospatial and temporal patterns in groundwater level changes. It can further be used to predict anticipated regional groundwater changes in the future, as well as potential impacts of those changes, thus providing a tool for emergency planning, especially in times of extreme and exceptional droughts.

3. METHODS AND DATA

To evaluate changes in groundwater well levels in different regions of Oklahoma and Texas over time, we developed a four-dimensional (4D) visualization model based on the Ternary Visual Shape Logic introduced for the first time by Ziolkowska and Reyes [41] (Figure 4).

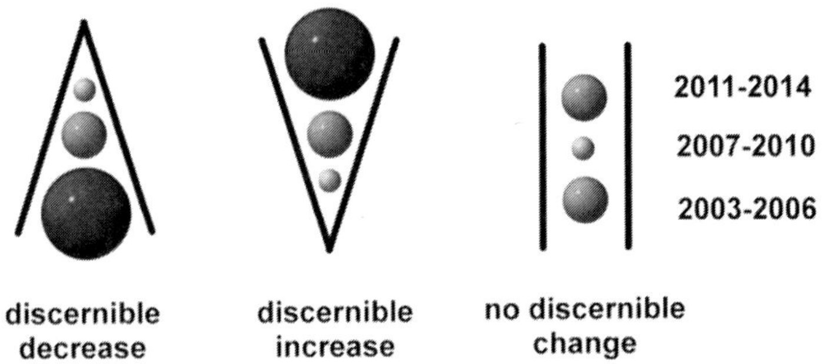

Figure 4. Ternary Visual Shape Logic diagram for the analyzed groundwater changes in 2003–2014. Source: Ziolkowska and Reyes [41].

The model offers a visual representation of temporal changes in groundwater wells. It uses normalized in-situ averages that are further validated with the Palmer Drought Severity Index values. This research does not address absolute water volume changes and balances, as commonly addressed in the literature. Rather, it proposes a more holistic approach of monitoring groundwater level changes even in situations when consistent in-situ measurements are compromised or scattered over time, as is the case with the presented case studies for Oklahoma and Texas. In both states, there is a significant variability in groundwater measurements at the respective wells across the state over the analyzed time span of 12 years. Accordingly, some wells were monitored more frequently (and thus consistent data sets are available for the entire time series), while other wells were monitored less frequently (which resulted in scattered data sets during that time period). For that reason, a comprehensive analysis of water balances is not possible to generate statistically robust results. To mitigate this problem, and provide a comprehensive pattern of groundwater level changes over time, this paper estimates a statistical trend of temporal changes at each well included in the analysis. To validate the correctness of this approach and the robustness of the groundwater model, the trend results have been compared and correlated with the Palmer Drought Severity Index values.

The model was developed for the time span of 12 years (2003–2014), with three-year sections: 2003–2006, 2007–2010, and 2011–2014, which facilitated a temporal analysis of the trends in groundwater levels and drought severity. The time period for this research was selected based on data availability. Data measurements of groundwater well levels before 2002 were very sparse, while the year 2003 was a breakthrough point in terms of the number of monitored wells across Oklahoma and Texas. We attribute this change to technological development and new monitoring instruments that allowed for more frequent and precise groundwater level measurements. Thus, including any analyses before the year 2003 would compromise the robustness of the results, and thus was forgone for the purpose of this study.

The presented model was developed through large data processing, and encompasses four major variables: latitude represented on the x-axis, longitude on the y-axis, time on the z-axis, and the extent of groundwater scarcity (groundwater well levels) as color-shaded spheres. While latitude and longitude allow for displaying geographical dimension, the time variable represents temporal dimension of groundwater level changes.

The model utilizes the US Geological Survey (USGS) water database containing 20,162 wells in Oklahoma and a database from Texas Water Development Board on ~140,000 monitored water wells in Texas. Due to data disparities and missing data points for all wells, we 'cleaned' the data set by determining selection criteria for each well to be included in the model, as follows: (a) wells sampled at least once in each of the three time sections: 2003–2006, 2007–2010, 2011–2014; and (b) wells with at least five samples minimum in the entire time frame. In this way, we specified 7211 water wells in Texas and 390 water wells in Oklahoma. The data was further normalized in order to provide a data-based benchmark and comparison basis for all groundwater data entries. Normalization of the groundwater well levels in both states was conducted for each well separately in two steps. First, minimum and maximum values in the entire raw data set were determined and averaged. In order to provide a more detailed distribution and precise analysis, the sampled values that fell between the minimum and the average value were averaged, thus generating the low-average value. Similarly, the sampled values that were observed between the maximum and the average value were averaged, thus generating the high-average value. This specification provided five hard values (minimum, low-average, average, high-average, and maximum) and categories to constitute a histogram with four bins that were further used to determine distribution of groundwater well levels over time (Figure 5 and Figure 6).

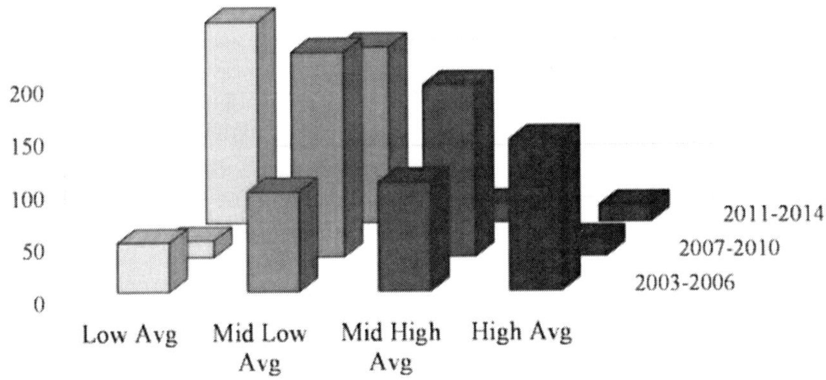

Figure 5. Distribution and categorization of normalized groundwater levels in Oklahoma in 2003–2014. Source: Authors' calculations and visualization.

Second, the raw data set was analyzed for temporal changes by averaging the values for the three time sections: 2003–2006, 2007–2010, and 2011–2014, respectively and assigning them to the corresponding bins. In this way, four values were generated that are represented with four categories of spheres of different colors and different sizes: small cyan spheres (representing the lowest groundwater levels), small medium light blue spheres, big medium blue spheres, and big dark blue spheres (representing the highest groundwater levels). The spheres closest to the ground surface display the years 2003–2006, the spheres second in height from the ground surface display the years 2007–2010, while the spheres highest from the surface represent the most recent years 2011–2014.

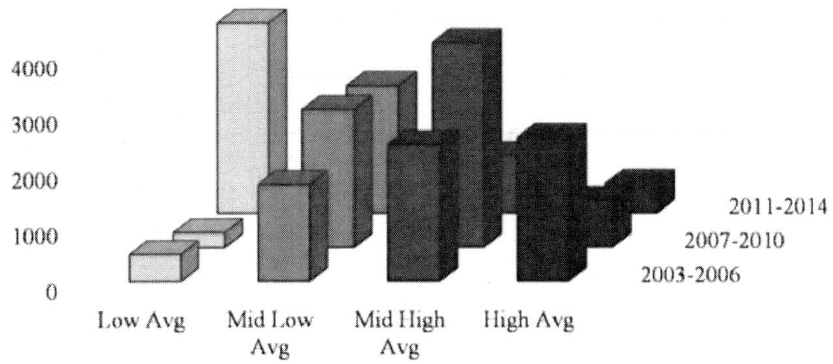

Figure 6. Distribution and categorization of normalized groundwater levels in Texas in 2003–2014. Source: Authors' calculations and visualization.

In a next step, we used the C++ language to develop computer code and generate KML (Keyhole Markup Language) data files that allow model users to interact with the results both in Google Earth or Cesium and Google Maps (or other platforms) [41]. The model shows three categories of regional trends and variations in groundwater well depletion over the analyzed time span: increasing, decreasing, and inconclusive trends (no change in water levels) in different regions of Oklahoma and Texas. The trends indicate regions of current water scarcity and potential limited water availability in the future.

Due to its interactive feature, the model allows for an analysis of groundwater level changes from a close and a broader perspective in all counties in Oklahoma and Texas over the analyzed time span 2003–2014. The model can be used as a decision-support tool to instantly recognize groundwater scarcity and derive potential socio-economic implications. The strength of the model stems from its ability to visually represent changes in all analyzed groundwater wells in all regions in two states, emphasizing well level gains and losses in different time frames. The large number of raw data is beneficial for a visual representation as discussed with this paper. However, it is overwhelming (due to the large number of data points) for a one-dimensional representation of time series.

To validate the model results and the robustness of our methodological approach, we applied the Palmer Drought Severity Index (PDSI) [42] that has been used extensively in the US and acknowledged as one of the most comprehensive indicators of drought. It incorporates water supply, water demand, temperature, and precipitation. Even though there is a multitude of drought indicators, and PDSI is not the universal measure to evaluate drought [43], we use this indicator due to its reliable and most comprehensive methodology as of today [44]. For validation purposes and geospatial visualization, we represented the Palmer Drought Severity Index as a gradient indicator for the analyzed regions. To prepare the basis for correlation analysis, the PDSI raw data were assigned to the respective drought categories for the three time spans (2003–2006, 2007–2010, and 2011–2014) (Figure 7 and Figure 8) in the same way as for groundwater levels (Figure 5 and Figure 6).

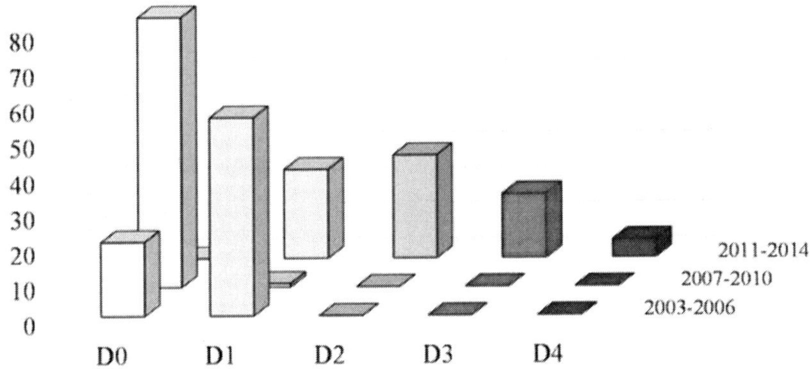

Figure 7. Categorization of averaged drought severity in Oklahoma in 2003–2014. Source: Authors' calculations and visualization.

Data from the Drought Monitor was evaluated for each week in the entire analyzed time period of 2003–2014. A total of 206,544 samples for all counties in Texas and Oklahoma were used for the analysis. The samples represent drought indexes [D0—abnormally dry, D1—moderate drought, D2—severe drought, D3—extreme drought, and D4—exceptional drought] reflecting the percent of the county in each of the drought conditions, respectively (compare with Figure 2). It needs to be mentioned that the presented categorization of groundwater levels (Figure 5 and Figure 6) and drought severity (Figure 7 and Figure 8) has low regional granularity, as it refers to the entire state of Oklahoma and Texas, respectively. The following geospatial distribution offers a more accurate picture of given conditions at a very detailed scale of specific groundwater wells or at the county level. Moreover, a follow-up analysis resonating from this research will allow us to evaluate these correlations for each climate region in both states.

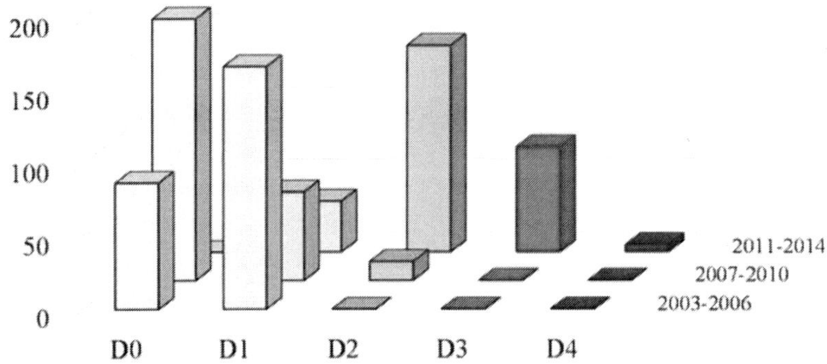

Figure 8. Categorization of averaged drought severity in Texas in 2003–2014. Source: Authors' calculations and visualization.

4. RESULTS AND DISCUSSION

The model allows for the estimation of the extent of water shortages as a result of droughts in all three analyzed time spans of 2003–2006, 2007–2010, and 2011–2014. It also provides a basis for revealing potential lingering uncertainty resulting from water scarcity and socio-economic implications of groundwater shortages in the future.

The groundwater well levels are represented with the blue-colored spheres (Figure 9) that are clickable and bring up a balloon information window with statistical information (including groundwater levels) in addition to a link to the original well data. Representing all time periods simultaneously is recommended in a model view to interactively navigate/fly through the wells and analyze time series changes over time (Figure 10).

The model shows that groundwater levels in most of the wells have been decreasing since 2003, reaching the peak in the number of wells with the lowest water levels in 2011–2014—the period of exceptional and extreme droughts in Oklahoma and Texas. Only a minority of the wells were inconclusive about the direction of change in groundwater well levels over time. As there is no clear pattern to these inconclusive data points, they can be explained either as outliers or deviating from the trend due to geological formations, rock type, and porosity in the given geographic location. Moreover, there are only a few wells with increasing water levels in both states in that time period. While there is a clear pattern of decreasing groundwater well levels in general, differences also exist from well to well over time. A comprehensive analysis of each well in different time periods can be viewed in the interactive model accessible at: http://www.hitechmex.org/OK_TX/index.html.

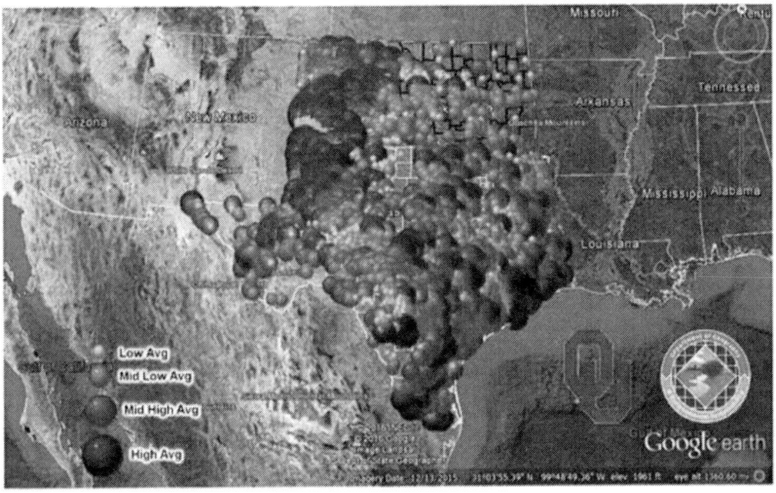

Figure 9. Screenshot of groundwater well levels in 2011–2014 in Oklahoma and Texas. Source: Authors' calculations and visualization.

Figure 10. Screenshot of groundwater wells in a Texas region in 2003–2014. Source: Authors' calculations and visualization.

To validate the model results visually, the Palmer Drought Severity Index was applied at this stage as a correlation variable for each county in Oklahoma and Texas. The results of the Palmer Drought Severity Index representation visibly show drought severity in 2011–2014 and weather variations in the three analyzed time spans (compare with Figure 2). By overlapping the PDSI gradient map with the interactive 4D groundwater well model a distinct visual correlation of those two variables unfolds. Figure 11 represents the PDSI gradient in Oklahoma and Texas and also a selection of wells with the lowest water levels in the analyzed time period. It clearly shows that in 2003–2006 the Palmer Drought Severity Index gradient indicated mostly abnormally dry conditions in both states, while the groundwater model shows a low number of wells with the lowest water levels. In 2007–2010, the Palmer Drought Severity Index showed a change in drought severity to moderate drought in some regions, while the number of wells with the lowest water levels either did not change in some regions or increased in regions shadowed with the light orange color (moderate drought according to PDSI). A statistical breakthrough is visualized in the last time period of 2011–2014, with PDSI indicating exceptional and extreme drought in many Oklahoma and Texas regions. At the same time, thc number of wells with the lowest water levels spiked markedly. The correlation of the groundwater well model with the independently calculated Palmer Drought Severity Index is a strong indicator of the methodological robustness and validity of our model. While this model is a visual representation of groundwater level changes over the time span of 12 years, it was not the goal of this study to present differences in aquifer storage coefficients across monitored regions or the relationship between water extraction and changes in groundwater levels.

The visual correlation was further tested statistically according to the formula:

$$r = \frac{1}{n-1}\sum\left(\frac{x-\overline{x}}{S_x}\right)\left(\frac{y-\overline{y}}{S_y}\right) \quad (1)$$

where *r*—correlation factor; *n*—number of observations; *x*—groundwater well levels; *y*—Palmer drought severity; $\bar{x}$—mean of groundwater level values; $\bar{y}$—mean of drought severity values; Sx—standard deviation of groundwater level values; Sy—standard deviation of drought severity values.

Accordingly, the correlation indicator for Oklahoma amounted to 0.75, while it was 0.95 for Texas, which indicates very strong correlations between the groundwater levels and severity of drought, and thus validates our hypothesis.

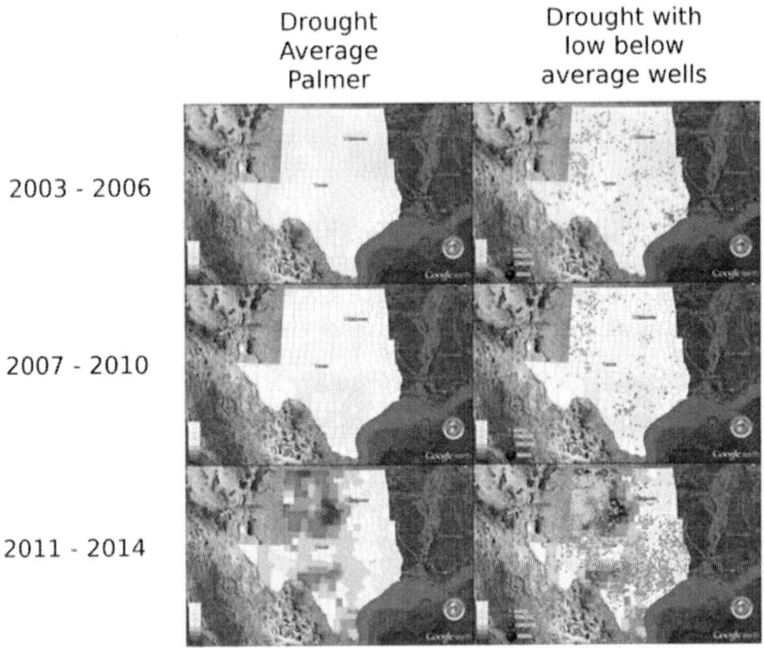

Figure 11. Palmer Drought Severity Index (PDSI) and groundwater wells with the highest water scarcity in 2003–2014. Source: Authors' calculations and visualization.

The visual model is an interactive tool that can be used for decision-making processes and designing sustainable water management measures. The interactive feature of the model allows the user to learn by navigating though the respective wells, clicking on the respective spheres to retrieve detailed statistical information about temporal variability of groundwater levels at the respective wells and in different regions as a whole, as well as context switching to display temporal correlations between the groundwater level changes and the Palmer Drought Severity Index. However, the model results are only a first step in deriving conclusions on potential mid- and long-term socio-economic impact as a result of drought and decreasing groundwater levels. As accentuated by the model results, the most impacted regions are in the Oklahoma and Texas Panhandle, which are the main agricultural production areas applying water from the Ogallala Aquifer in the southern parts of the High Plains. As discussed at the beginning of this article, the Ogallala Aquifer is non-replenishable, as it contains fossil water. Thus, low groundwater levels displayed by the model are an indicator of severe aquifer depletion (as a combined effect attributed to excessive water withdrawals and insufficient aquifer replenishment), which can lead to unexpected repercussions for the agricultural sector itself, as well as for other economic sectors (municipalities, industry sector, tourist facilities, etc.). Those changes can also further impact the production output, water prices, social welfare, and finally the economic growth in both states. Another question that arises in this context is the extent of drought and its impacts on employment levels and wages in the agricultural sector and beyond. Consequently, the length of drought will determine the scope of the enumerated socio-economic impacts in the mid and long term.

5. CONCLUSIONS

The results of this study show a correlation between decreasing groundwater levels and the Palmer Drought Severity Index. Both in Oklahoma and Texas, a significant decrease in groundwater levels has been found in 2001–2014 that correlates with the biggest drought on record. While regional differences have been detected in different regions in both states over time, the general trend of changing groundwater levels corresponds with the recorded regional and state-wide drought severity.

The results were generated and visualized with a geospatial model that allows for an interactive analysis of those trends and patterns in different geographic locations and regions in both states, and over time. In this way, past and current groundwater availability (and water scarcity) at the state, regional, and local level was evaluated holistically with a consistent methodology. The results and the model can be used as a decision-making support tool to assess potential regional groundwater changes in the future, and to analyze potential impacts of those changes. Finally, the outcomes of this research can be useful for long-term planning of emergency situations related to water scarcity.

Furthermore, the results of this study generated a methodological basis for future drought predictions. The positive correlation between the groundwater levels (measured by OWRB, TWDB, and USGS less frequently) and the Palmer Drought Severity Index (measured frequently on a weekly basis) provides a benchmark for determining future groundwater levels (aquifer depletion) even in situations of missing or inconsistent data from in-situ groundwater measurements. Accordingly, in such cases, the Palmer Drought Severity Index could be used as a predictor of groundwater well levels based on statistical inference.

The results and the interactive model can help decision makers with discussions on potential restrictions to water use to be placed on different sectors or users (e.g., farmers, municipalities, industries), as well as with sustainable water allocation. **Model access**: http://www.hitechmex.org/OK_TX/index.html.

ACKNOWLEDGMENTS

This material is based on work supported by the National Science Foundation under Grant No. IOA-1301789 and Oklahoma Climatological Survey (Oklahoma Mesonet). The authors thank anonymous reviewers for their valuable comments that helped us improve this paper.

REFERENCES

1. Lange, B.; Holman, I.; Bloomfield, J.P. A framework for a joint hydro-meteorological-social analysis of drought. *Sci. Total Environ.* **2017**, *578*, 297–306.
2. Ge, S.; Gorelick, S.M. Hydrology, Floods and Droughts. Groundwater and Surface Water. In *Reference Module in Earth Systems and Environmental Sciences, from Encyclopedia of Atmospheric Sciences*, 2nd ed.; Elsevier: Amsterdam, The Netherlands, 2015; pp. 209–216.
3. Fendekova, M.; Demeterova, B.; Slivova, V.; Macura, V.; Fendek, M.; Machlica, A.; Gregor, M.; Jalcovikova, M. Surface and groundwater drought evaluation with respect to aquatic habitat quality applied in Torysa river catchment, Slovakia. *Ecohydrol. Hydrobiol.* **2011**, *11*, 49–61.
4. Steward, D.R.; Allen, A.J. Peak groundwater depletion in the High Plains Aquifer, projections from 1930 to 2110. *Agric. Water Manag.* **2016**, *170*, 36–48.
5. NOAA (National Oceanic and Atmospheric Administration). National Temperature and Precipitation Maps, 2016. Available online: https://www.ncdc.noaa.gov/monitoring-content/sotc/national/statewidepcpnrank/statewidepcpnrank-201010-201109.gif (accessed on 19 July 2016).
6. Drought Monitor. *US Drought Monitor for Oklahoma*; The National Drought Mitigation Center: Lincoln, NE, USA, 2015.
7. NDMC (National Drought Mitigation Center). *The Drought Monitor—Texas*; University of Nebraska-Lincoln: Lincoln, NE, USA, 2013; Available online: http://droughtmonitor.unl.edu/archive.html (accessed on 10 March 2013).
8. Folger, P.; Cody, B.; Carter, N. *Drought in the United States: Causes and Issues for Congress*; Congressional Research Service Report for Congress 7-5700; CreateSpace Independent Publishing Platformb: North Charleston, SC, USA, 22 April 2013.
9. Stotts, D. *Oklahoma Agricultural Losses from Drought More Than $1.6 Billion*; Oklahoma State University Agricultural Communications Services: Stillwater, OK, USA, 2011.
10. Fannin, B. Updated 2011 Texas Agricultural Drought Losses Total $7.62 Billion. Texas A&M AgriLife Today, 2012. Available online: http://today.agrilife.org/2012/ 03/ 21/updated-2011-texas-agricultural-drought-losses-total-7–62-billion/ (accessed on 3 January 2013).
11. Guerrero, B. *The Impact of Agricultural Drought Losses on the Texas Economy*; Briefing Paper; AgriLife Extension, Texas A&M System: College Station, TX, USA, 2011.
12. Ziolkowska, J.R. Socio-Economic Implications of Drought in the Agricultural Sector and the State Economy. *Economies* **2016**, *4*, 19.

13. Texas Water Development Board (TWDB). *2012 State Water Plan*; Texas Water Development Board: Austin, TX, USA, 2012.
14. Oklahoma Water Resources Board (OWRB). *Oklahoma Comprehensive Water Plan*; Executive Report; Oklahoma Water Resources Board: Oklahoma City, OK, USA, 2012.
15. Sesmero, J.P. Cellulosic biofuels from crop residue and groundwater extraction in the US Plains: The case of Nebraska. *J. Environ. Manag.* **2014**, *144*, 218–225.
16. Condon, L.E.; Hering, A.S.; Maxwell, R.M. Quantitative assessment of groundwater controls across major US river basins using a multi-model regression algorithm. *Adv. Water Resour.* **2015**, *82*, 106–123.
17. Foster, T.; Brozović, N.; Butler, A.P. Why well yield matters for managing agricultural drought risk. *Weather Clim. Extrem.* **2015**, *10*, 11–19.
18. Li, Z.; Quan, J.; Li, X.Y.; Wu, X.C.; Wu, H.W.; Li, Y.T.; Li, G.Y. Establishing a model of conjunctive regulation of surface water and groundwater in the arid regions. *Agric. Water Manag.* **2016**, *174*, 30–38.
19. Huo, A.D.; Dang, J.; Song, J.X.; Chen, X.H.; Mao, H.R. Simulation modeling for water governance in basins based on surface water and groundwater. *Agric. Water Manag.* **2016**, *174*, 22–29.
20. Wang, L.; Jackson, C.R.; Pachocka, M.; Kingdon, A. A seamlessly coupled GIS and distributed groundwater flow model. *Environ. Model. Softw.* **2016**, *82*, 1–6.
21. Decuyper, M.; Chávez, R.O.; Copini, P.; Sass-Klaassen, U. A multi-scale approach to assess the effect of groundwater extraction on Prosopis tamarugo in the Atacama Desert. *J. Arid Environ.* **2016**, *131*, 25–34.
22. Minor, T.B.; Russell, C.E.; Mizell, S.A. Development of a GIS-based model for extrapolating mesoscale groundwater recharge estimates using integrated geospatial data sets. *Hydrogeol. J.* **2007**, *15*, 183–195.
23. Singh, P.; Thakur, J.K.; Kumar, S. Delineating groundwater potential zones in a hard-rock terrain using geospatial tool. *Hydrol. Sci. J.* **2013**, *58*, 213–223.
24. Kuria, N.D.; Gachari, M.K.; Macharia, M.W.; Mungai, E. Mapping groundwater potential in Kitui District, Kenya using geospatial technologies. *Water Resour. Environ. Eng.* **2012**, *4*, 15–22.
25. Lia, B.; Rodell, M. Evaluation of a model-based groundwater drought indicator in the conterminous U.S. *J. Hydrol.***2015**, *526*, 78–88.
26. Singh, A. Managing the water resources problems of irrigated agriculture through geospatial techniques: An overview. *Agric. Water Manag.* **2016**, *174*, 2–10.
27. Mashburn, S.L.; Ryter, D.W.; Neel, C.R.; Smith, S.J.; Magers, J.S. *Hydrogeology and Simulation of Ground-Water Flow in the Central Oklahoma (Garber-Wellington) Aquifer, Oklahoma, 1987 to 2009, and Simulation of Available Water in Storage, 2010–2059*; U.S. Geological Survey Scientific Investigations Report 2013; U.S. Geological Survey: Reston, VA, USA, 2013; Volume 5219, 92p.
28. Konikow, L.F. Groundwater depletion in the United States (1900–2008). USGS Scientific Investigations Report 2013–5079; p. 63. Available online: http://pubs.usgs.gov/sir/2013/5079 (accessed on 14 February 2017).
29. Bhanja, S.N.; Rodell, M.; Li, B.; Saha, D.; Mukherjee, A. Spatio-temporal variability of groundwater storage in India. *J. Hydrol.* **2017**, *544*, 428–437.
30. Sakakibara, K.; Tsujimura, M.; Song, X.; Zhang, J. Spatiotemporal variation of the surface water effect on the groundwater recharge in a low-precipitation region: Application of the multi-tracer approach to the Taihang Mountains, North China. *J. Hydrol.* **2017**, *545*, 132–144.
31. Zhang, L.; Brutsaert, W.; Crosbie, R.; Potter, N. Long-term annual groundwater storage trends in Australian catchments. *Adv. Water Resour.* **2014**, *74*, 156–165.
32. Manna, F.; Cherry, J.A.; McWhorter, D.B.; Parker, B.L. Groundwater recharge assessment in an upland sandstone aquifer of southern California. *J. Hydrol.* **2016**, *541*, 787–799.
33. Dakhlalla, A.O.; Parajuli, P.B.; Ouyang, Y.; Schmitz, D.W. Evaluating the impacts of crop rotations on groundwater storage and recharge in an agricultural watershed. *Agric. Water Manag.* **2016**, *163*, 332–343.

34. Xiang, L.; Wang, H.; Steffen, H.; Wu, P.; Jia, L.; Jiang, L.; Shen, Q. Groundwater storage changes in the Tibetan Plateau and adjacent areas revealed from GRACE satellite gravity data. *Earth Planet. Sci. Lett.* **2016**, *449*, 228–239.
35. Ouyang, W.; Liu, B.; Wu, Y. Satellite-based estimation of watershed groundwater storage dynamics in a freeze–thaw area under intensive agricultural development. *J. Hydrol.* **2016**, *537*, 96–105.
36. Uddameri, V.; Karim, A.; Hernandez, E.A.; Srivastava, P.K. Sensitivity of Wells in a Large Groundwater Monitoring Network and Its Evaluation Using GRACE Satellite Derived Information. In *Sensitivity Analysis in Earth Observation Modelling*; Elsevier: Amsterdam, The Netherlands, 2017; pp. 235–256.
37. Seyoum, W.M.; Milewski, A.M. Monitoring and comparison of terrestrial water storage changes in the northern high plains using GRACE and in-situ based integrated hydrologic model estimates. *Adv. Water Resour.* **2016**, *94*, 31–44.
38. Bhanja, S.N.; Mukherjee, A.; Saha, D.; Velicogna, I.; Famiglietti, J.S. Validation of GRACE based groundwater storage anomaly using in-situ groundwater level measurements in India. *J. Hydrol.* **2016**, *543*, 729–738.
39. Jones, W.R.; Spence, M.J.; Bowman, A.W.; Evers, L.; Molinari, D.A. A software tool for the spatiotemporal analysis and reporting of groundwater monitoring data. *Environ. Model. Softw.* **2014**, *55*, 242–249.
40. Brock, F.V.; Crawford, K.C.; Elliott, R.L.; Cuperus, G.W.; Stadler, S.J.; Johnson, H.L.; Elits, M.D. The Oklahoma Mesonet: A Technical Overview. *J. Atmos. Ocean. Technol.* **1995**, *12*, 5–19.
41. Ziolkowska, J.R.; Reyes, R. Geological and Hydrological Visualization Models for Digital Earth Representation. *Comput. Geosci.* **2016**, *94*, 31–39.
42. Palmer, W.C. *Meteorological Drought*; U.S. Weather Bureau Research Paper; U.S. Government Publishing Office: Washington, DC, USA, 1965; Volume 45, p. 58.
43. Alley, W.M. The Palmer drought severity index: Umitation and Assumptions. *J. Appl. Meteorol. Climatol.* **1984**, *23*, 1100–1109.
44. Karl, T.R. Sensitivity of the Palmer drought severity index and Palmer's Z-index to their calibration coefficients including potential evapotranspiration. *J. Appl. Meteorol. Climatol.* **1986**, *25*, 77–86.

CHAPTER 5

Urban Groundwater Quality in India

Nohar Singh Dahariya, Shobhana Ramteke, Bharat Lal Sahu, Khageshwar Singh Patel*

School of Studies in Chemistry/Environmental Science, Pt. Ravishankar Shukla University, Raipur, India

ABSTRACT

The electrical conductivity, alkalinity, salinity, hardness and chemical load of the groundwater in urban area of the country are increased enormously due to excessive urban stresses by making water unsafe for drinking purposes. Therefore, the groundwater quality of Raipur city, capital of Chhattisgarh state, India has been investigated. The physico-chemical characteristic of the ground- water along with the chemical loading variations is described. Various indices were used for rating of groundwater quality for drinking and irrigation purposes. The cluster and factor analysis models were used for source apportion of the contaminants.

Keywords: Groundwater, Quality, Sources, Indices

1. INTRODUCTION

The urban groundwater has emerged as one of the world's most challenging issues [1] . The quality of available groundwater resources is being increasingly degraded by geogenic and anthropogenic activities [2] [3] . Asian countries face serious water problems almost everywhere mainly due to explosive population growth, heavy seasonal rains, massive flooding, decreasing of water levels, mixing of waste water, etc. [4] . In India, groundwater is used intensively for drinking, irrigation and industrial purposes. Several land and water-based human activities are causing pollution of these precious resources. India is now the biggest user of groundwater for agriculture in the World [5] . The most dramatic change in the groundwater scenario in India is that the share of tube wells in irrigated areas is rising tremendously. By now, tube wells have become the largest single source of irrigation water in India. Hence, groundwater has a high incidence of water quality problems such as increase in concentration levels of chemical species i.e. fluoride, iron, salinity, nitrate, arsenic, etc., may be due to mineralization [6] - [15] . In the present work, the urban groundwater quality of the most industrialized city of the country, Raipur is presented. The variations and sources of contaminants and quality assessment of the groundwater are described.

2. MATERIALS AND METHODS

2.1. Study Area

Raipur is the capital of the Chhattisgarh state of the country with vast industrial growth having population of >2.0 million habitants. The groundwater is widely used for domestic, agricultural and industrial purposes. Several sponge iron, steel, alloy and cement plants are running using coal as source of energy. The water quality of city is deteriorated tremendously due to over use of groundwater and input of the industrial and sewage wastes in it [16] .

2.2. Sample Collection

The sampling locations for the groundwater are presented in Figure 1. Fifty six groundwater samples from 28 locations of the city from the residential, urban and industrial area in two seasons (i.e. post and pre monsoon) of years, 2012 were collected by using the established methodology [17] . The water was collected into cleaned and rinsed 1 lit polyethylene flask. The physical parameters i.e. pH, electrical conductivity (EC), dissolved oxygen (DO) and reduction potential (RP) of the groundwater were measured at the spot. The samples were dispatched to the laboratory by subsequent frizzing at $-4^{\circ}C$.

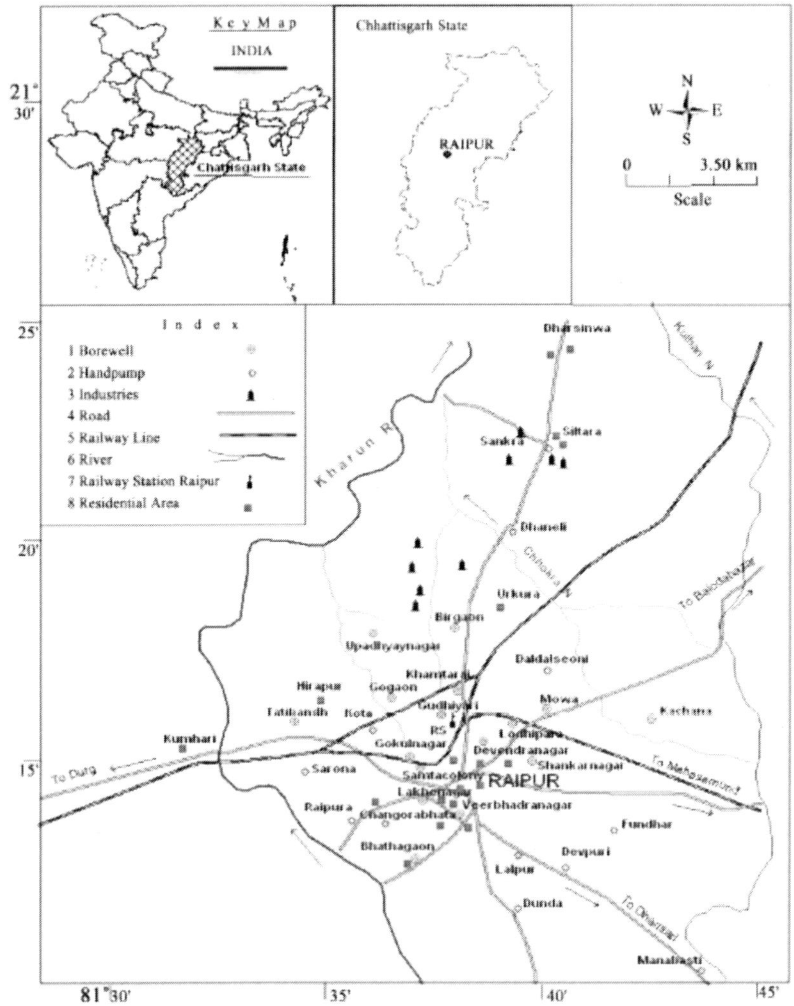

Figure 1. Representation of sampling points in Raipur city, CG, India.

2.3. Analysis

The total dissolved solid (TDS) value of the filtered samples (through glass fiber filter) was determined by the evaporation method [18] . The total hardness (TH) and total alkalinity (TA) values of the groundwater samples were analyzed by titration methods [19] . The CHNSO-IRMS Analyzer, SV Instruments Analytica Pvt. Ltd. was used for analysis of the dissolved organic carbon (DOC) and dissolved inorganic carbon (DIC). The Dionex DX-1100 Ion chromatograph (Dionex Corporation, Sunnyvale, CA, USA) equipped with anion and cation separation columns and conductivity detector was used for analysis of the ions. The Metrohm-781 ion meter was employed for monitoring of F^- content in the presence of the buffer in 1:1 volume ratio. The buffer was prepared by dissolving sodium citrate (300 g), 1, 2-cyclohexanediamine-N-tetraaceticacid (22 g) and NaCl (60 g) in a volume of 1 lit with the de-ionized water by subsequent adjustment of pH value to 5.2 ± 0.2.

The indices *i.e.* sodium hazard (SH), magnesium hazard (MH), sodium adsorption ratio (SAR), permeability index (PI), Kelly's ratio (KR) and water quality index (WQI) were evaluated by using the following equations.

$$SH = \left(\{[Na]+[K]\}/\{[Na]+[K]+[Mg]+[Ca]\}\right)\times 100$$

$$MH = \left([Mg^{2+}]/\{[Mg^{2+}]+[Ca^{2+}]\}\right)\times 100$$

$$SAR = [Na^{+}]/\sqrt{\{1/2([Ca^{2+}]+[Mg^{2+}])\}}$$

$$PI = \left(\{[Na]+\sqrt{[HCO_3^-]}\}/\{[Na]+[Mg]+[Ca]\}\right)\times 100$$

$$KR = [Na^{+}]/\{[Mg^{2+}]+[Ca^{2+}]\}$$

where, all ions are expressed in meq/L

Ten parameters *i.e.* pH, EC, DO, TH, TA, Mg^{2+}, Ca^{2+}, Cl^-, NO^-_3 and SO^2_4 were selected for evaluation of the WQI by using the standard values recommended by BIS and WHO [20] [21] . The weighed arithmetic method was used for calculation of the WQI of the groundwater with the help of following expression.

$$WQI = \sum q_n W_n / \sum W_n$$

where:

$$q_n = 100(V_n - V_{io})/(S_n - V_{io})$$

q_n = Quality rating of the nth water quality parameter

V_n = Estimated value of the nth parameter of a given water

S_n = Standard permissible value of the nth parameter

V_{io} = Ideal value of the nth parameter of pure water (*i.e.*, 0 for all other parameters except pH and Dissolved oxygen (7.0 and 14.6 mg/L, respectively)

W_n = Unit weight for the nth parameter

K = Proportionality constant

The Aquachem water quality software was used for the preparation of Piper diagrams. Multivariate statistical analysis i.e. hierarchical cluster analysis (HCA) and factor analysis (FA) were used for apportion of contaminant sources [22] - [24] . The windows statistical software STATISTICA 7.1 was employed for the multivariate statistical calculations.

3. RESULTS & DISCUSSION

3.1. Geological Characteristics of Tube Wells

The geological characteristics of the tube wells are presented in Table 1. The depth of tube wells was ranged from 30 - 150 m with mean value (p = 0.05) of 90 ± 13 m. The age of tube wells was found in the range of 1.0 - 46 Yr with mean value (p = 0.05) of 10 ± 5 Yr. Generally, younger tube was found to be contaminated with higher content of ions i.e. Na^+, K^+, Mg^{2+}, Ca^{2+}, Cl^-, NO^-_3, and SO^{2-}_4 , may be due to poor scavenging fluxes. The T value of the groundwater was ranged from 20.1 - 22.9°C with mean value (p = 0.05) of 21.6 ± 0.3°C. The slight increase with respect to increasing depth of tube well from 30 - 150 m was observed, may be due to geothermal energy.

3.2. Physico-Chemical Characteristics of Groundwater

The physical characteristics of the groundwater of 28 locations are shown in Table 2. The groundwater was colorless when drawn out from the tube well but after some time (≈12 hr) become turbid due to precipitation of the bicarbonates as carbonates. The pH value of the water was ranged from 6.3 - 8.2 with mean value (p = 0.05) of 6.8 ± 0.2. The lowest pH value, 6.3 of the water was observed in the sewage dumping and industrial areas: Kachana and Siltara, due to presence of excessive levels of anions (i.e. Cl^-,NO^-_3 and SO^{2-}_4). In 92% locations, the pH values of the water were marked in the range of desirable range i.e. 6.5 - 9.2 [20] . The DO and RP values were ranged from 3.7 - 7.2 mg/L and 114 - 144 mV with mean value (p = 0.05) of 4.6 ± 0.5 mg/L and 126 ± 0.03 mV. The DO value of the water observed was above the recommended value of 4.0 mg/L [20] [21] . The RP value of the groundwater was found at least 5-folds lower than the recommended value of 650 mV. Extremely high EC values of the water was marked, ranging from 1419 - 6300 μS/cm with mean value ($p = 0.05$) of 2629 ± 508 μS/cm. The EC value was found to be correlated well ($r = 0.92$) with the sum of total ionic concentration ions (i.e. Cl^-, $NO^-{}_3$, $SO^{2-}{}_4$ and Na^+). The highest EC value of water was marked at the location i.e. Kachana, may be due to input of the sewage waste.

Table 1. Geological characteristics of tube wells.

S. No.	Location	Depth, m	Age, Yr	Environment
1	Manabasti	75	11	R
2	Dunda	30	21	R
3	Devpuri	90	6	R
4	Lalpur	105	10	R
5	Fundhar	75	23	R
6	Bhathagaon	75	6	R
7	Changorabhatha	75	11	R
8	Veerbhadranagar	66	8	R
9	Raipura	75	21	R
10	Upadhyay Nagar	84	20	R
11	Lakhenagar	45	11	R
12	Samta colony	105	3	R
13	Sarona	90	7	R
14	Gokulnagar	60	2	R
15	Devendra	72	4	R
16	Lodhipara	75	4	R
17	Shankar Nagar	75	6	R
18	Tatibandh	90	16	R
19	Kota	45	16	R
20	Gudhiyari	90	1	C
21	Mowa	150	2	R
22	Kachana	45	1	R
23	Gogaon	75	3	R
24	Khamtarai	100	1	C
25	Daldalsiwni	90	46	R
26	Birgaon	135	1	I
27	Dhaneli	75	5	I
28	Siltara	90	5	I

C = Commercial, I = Industrial, R = Residential.

Similarly, extremely high TDS, TH and TA values of the water were observed, ranging from 6338 - 14,568, 241 - 1432 and 209 - 878 mg/L with mean value ($p = 0.05$) of 10,731 ± 603, 519 ± 130 and 319 ± 74 mg/L, respectively. The highest TDS value of the water was seen at the location: Gudhiyari may be due to huge commercial and transportation activities.

However, the highest TA and TH values were recorded at location *i.e.* Kachana, may be due to input of the sewage waste.

Table 2. Mean value (n = 2) of physical characteristics of groundwater in post monsoon period, January, 2012.

S. No.	pH	DO mg/L	T °C	RP, mV	EC, μS/cm	TDS, mg/L	TH, mg/L	TA, mg/L
1	6.7	4.7	21.2	132	2620	10,771	276	411
2	6.5	6.5	20.1	116	3190	8608	989	515
3	6.7	4.8	21.6	132	2380	11,111	431	362
4	6.7	7.2	22.3	123	2600	6338	602	371
5	6.9	4.5	21.4	132	3110	11,204	617	318
6	7.1	4.3	21.6	114	1911	10,667	242	252
7	6.7	4.1	21.6	126	2800	9416	358	345
8	6.7	4.3	21.2	124	2020	11,593	405	209
9	6.9	4.2	22.3	123	1945	9651	241	243
10	6.9	4.3	21.2	134	1557	9950	445	221
11	6.8	4.2	20.4	133	2080	8750	441	256
12	7.0	4.5	22.8	134	2250	12,660	553	211
13	6.8	3.9	21.6	123	2890	9761	402	210
14	6.6	4.6	21.2	132	2640	14,002	529	262
15	7.0	4.6	21.3	124	1419	10,738	474	231
16	6.8	4.4	21.2	127	2290	11,035	341	251
17	6.9	4.2	21.1	121	2020	10,745	446	232
18	7.0	3.7	21.3	123	1970	9747	448	216
19	6.9	4.9	22.4	134	2150	13,039	544	278
20	6.8	5.1	22.2	140	2420	14,568	456	262
21	6.7	4.7	22.9	124	2670	10,770	347	267
22	6.3	6.7	20.4	128	6300	9957	1432	878
23	6.7	3.8	21.2	123	2350	11,708	431	346
24	6.9	4.0	22.6	134	2860	11,463	553	314
25	6.9	4.9	21.5	144	2270	10,268	441	231
26	7.1	4.5	22.7	130	3940	10,667	574	389
27	8.2	4.2	21.4	133	2580	11,043	592	324
28	6.3	3.9	21.9	131	4380	10,248	919	518

The chemical characteristics of the groundwater are summarized in **Table 3**. The concentration of DIC and DOC was ranged from 1300 - 7400 and 4500 - 7900 mg/L with mean value ($p = 0.05$) of 4400 ± 1300 and 6500 ± 900 mg/L, respectively. Similarly, the highest DIC and DOC values were marked in the water at location i.e. Gudhiyari. The DOC value of the water was several folds higher than the recommended value of 2 mg/L, may be observed due to percolation of the industrial and municipal waste into the groundwater aquifer. The higher DIC value of the groundwater of this region was marked, may be due to oxidation of the DOC with the microbes.

The concentration of F^-, Cl^-, NO^-_3, SO^{2-}_4,Na^+, K^+, Mg2+ and Ca^{2+} in the post monsoon period was ranged from 0.4 - 1.6, 88 - 446, 46 - 991, 45 - 173, 31 -

130, 4 - 56, 16 - 84 and 56 - 256 mg/L with mean value ($p = 0.05$) of 1.0 ± 0.2, 224 ± 45, 206 ± 101, 105 ± 20, 60 ± 13, 18 ± 5, 28 ± 7 and 92 ± 22 mg/L, respectively. The highest concentration of all ions (except F^- and Na^+) was seen at location i.e. Kachana, may be due to input of the sewage waste. At two locations i.e. Gogaon and Khamtarai, the concentration of F^- was found to be above the permissible limit of 1.5 mg/L [20] [21] .

Table 3. Mean value (n = 2) of chemical characteristics of groundwater during post monsoon period, January, 2012, mg/L.

S. No.	DIC	DOC	F^-	Cl^-	SO_4^{2-}	NO_3^-	Na^+	K^+	Mg^{2+}	Ca^{2+}
1	3900	6600	1.3	266	173	419	34	7	36	120
2	2500	5600	0.9	362	108	221	35	26	48	148
3	5400	5700	1.4	220	110	351	49	10	34	108
4	1500	4500	0.9	255	78	129	52	16	36	104
5	4400	6800	1.4	340	162	223	72	10	32	90
6	4300	6900	0.8	124	63	67	39	14	23	70
7	3500	5900	0.9	276	118	74	87	26	31	104
8	4500	7700	0.8	142	69	116	49	19	16	60
9	4300	5700	1.0	163	86	64	58	15	21	72
10	4500	5900	1.0	132	62	120	37	9	17	57
11	3200	5700	1.5	134	68	193	62	13	23	72
12	5600	7700	0.5	163	74	144	78	23	17	56
13	4400	5600	1.0	266	128	50	94	24	17	56
14	6700	7900	0.8	269	128	115	67	12	24	80
15	4500	6800	1.1	88	45	131	31	14	22	68
16	4700	6800	1.3	170	93	73	65	13	19	76
17	3700	7600	0.9	127	62	87	38	4	24	64
18	4600	5400	1.0	177	86	216	44	16	18	56
19	5700	7700	0.9	177	88	377	50	11	26	80
20	7400	7900	0.9	234	107	115	53	17	22	76
21	4500	6600	0.9	255	116	46	71	10	22	80
22	1300	6800	0.4	446	170	991	82	56	84	256
23	4200	7700	1.6	251	114	216	37	16	30	100
24	5100	6500	1.6	291	145	181	70	16	22	96
25	4700	5900	0.8	134	56	226	48	32	17	64
26	4600	5700	1.2	304	152	420	109	29	36	112
27	4600	6700	1.3	191	110	58	130	29	29	98
28	4400	5400	1.4	324	160	336	35	14	38	160

Chloride in ground water may originate from various sources including: the dissolution of halite. The recommended tolerance limit for Cl^- reported is 250 mg/L [20] [21] . The Cl^- concentration beyond permissible limit was observed in the water of 50% tube wells. The ionic ratio of $[Cl^-]/[Na^+]$ in the water was ranged from 0.9 - 6.8 with a mean value ($p = 0.05$) of 2.7 ± 0.5 with exceeding of $[Cl^-]$ to $[Na^+]$ in water of the 96% tube wells, may be due to dissolution of halite minerals. The high Na^+ concentration (beyond 20 mg/L) in the water may cause health hazards i.e. high blood pressure, heart disease, kidney problem, etc. In the water of all tube wells, the [Na^+] was found above the limit of 20 mg/L. The ionic ratio of [Na^+]/[K^+] in the water was ranged from 2 - 17 with a mean value ($p = 0.05$) of 7 ± 1 similar to natural water.

High concentration of $SO^{2-}{}_4$ (150 mg/L) in groundwater of 18% tube wells was marked, ranging from 254 - 2330 mg/L with mean value ($p = 0.05$) of 1059 ± 392 mg/L. The water of all tube wells was contaminated with $NO^-{}_3$ above the permissible limit of 45 mg/L [20] [21] . Groundwater of 31% and 62% tube wells of the studied area was found to be contaminated with Mg^{2+} and Ca^{2+} beyond permissible limit of 30 and 75 mg/L, respectively.

The domination of the DOC, DIC and Cl^- in the water was observed. The chemical species i.e. carbons and ions in the water was found to occur in following decreasing order: DOC > DIC > Cl^- > $NO^-{}_3$, > $SO^{2-}{}_4$ > Ca^{2+} > Na^+ > Mg^{2+} > K^+.

3.3. Seasonal Variation

The variation of ion concentration during post (January, 2012) and pre monsoon (May, 2012) periods is presented in Figure 2. The T value of the groundwater ($n = 28$) was increased during the pre monsoon period (i.e. May, 2012) in the range of 26.0 - 28.0°C with mean value ($p = 0.05$) of 27.3 ± 0.2°C. The value of pH, EC, TDS, TH, TA, DOC, DIC, F^-, $SO^{2-}{}_4$, Na^+, Mg^{2+} and Ca^{2+} of the water was found to be increased in the pre-monsoon period, may be due to increase in the water temperature, ≈4°C and shrinking of the water Table up to ≈30 m. However, the concentration of ions i.e. $NO^-{}_3$ and K^+ was found to decrease, may be due to decrease of the anthropogenic activities.

3.4. Sources

In the present study, HCA was used to classify the sample sites into hydrochemical groups in the post monsoon period. The Ward's linkage method was used by using Euclidean distance for similarity measurement in the classification scheme. The result obtained by HCA detected the similar groups, yielding a dendrogram into three statistically significant clusters such as Group-I, -II and -III in the post monsoon period (**Figure 3**). The Group-I presented groundwater sample of location: Kachana which showed the highest values of parameters *i.e.* EC, TA, TH, and major cations (*i.e.* K^+, Mg^{2+} and Ca^{2+}) become as an outlier. The Group-II and Group-III were characterized by moderate and high pollution load of the chemicals in the groundwater, respectively.

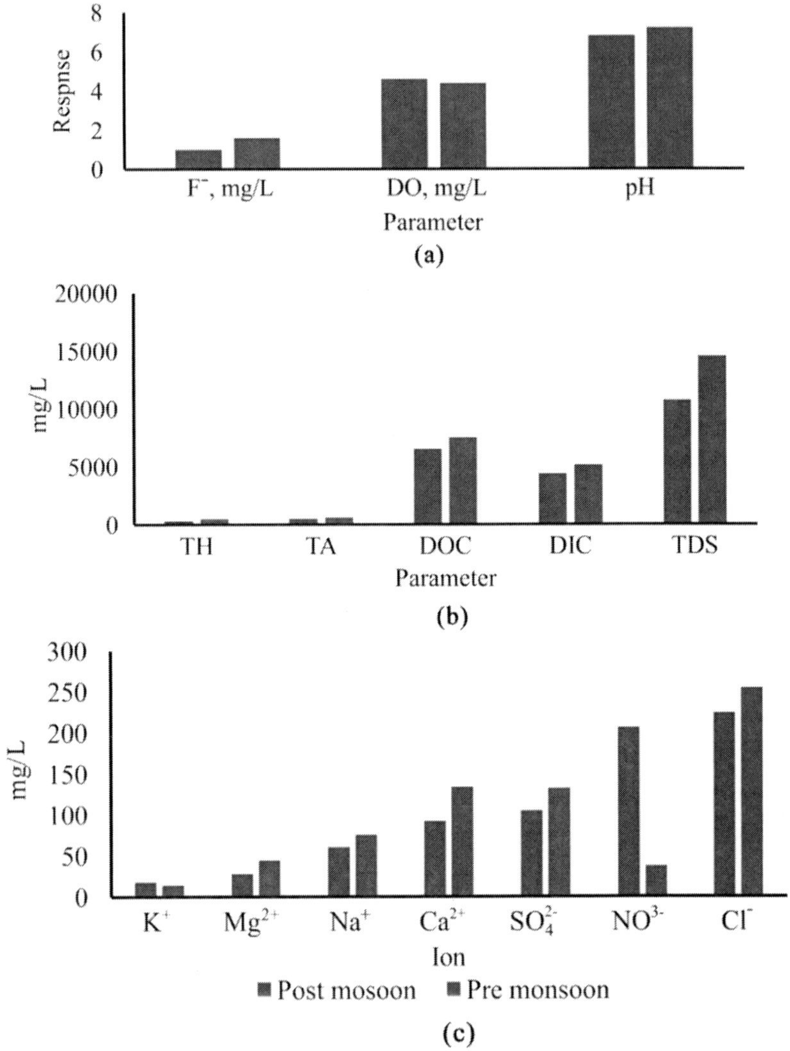

Figure 2. Parameters of groundwater during post and pre monsoon period (i.e. January, 2012 and May, 2012).

The factor analysis (FA) derived a subset of uncorrelated variables called factors that explained the variance observed in the original dataset. In FA, all the variables were standardized at the z-scale (mean and variance set to zero and one, respectively) to minimize the effects of different units and variance of variables [25] . The FA was performed for the set of 28 sample locations with 20 variables in order to establish the association between the chemical and physical variables of the groundwater. Factor extraction was carried out by using principal components based on Kaiser Criterion [26] . Factors with eigenvalues higher than 1 were retained and presented in Table 4. Six factors explaining 84.62% of the total variance were retained after normalized varimax rotation. Factor-1 accounted for 44.18% of the total variance, revealed strong associations (high positive loadings) with TDS, EC, TH, Ca^{2+}, Mg^{2+}, Cl^{-}, NO^{-}_{3}, SO^{2-}_{4} , and TA. This factor was characterized by complex processes derived from anthropogenic sources. Factor-2 explained 11.81% of the total variance, had strong positive loadings on pH and Na+. Factor-2 could represent weathering of Na-bearing minerals. Factor-3 accounted for 9.69% of the total variance, had a negative loading on RP, which confirmed that RP in the groundwater did not contribute significantly to the hydrochemistry. Factor-4 accounted for 8.67% of the total variance, had a negative loading on F−. This factor did not contribute significantly to the variation of the groundwater hydrochemistry. In the same order, Factor-5 which accounted for 5.26% of the total variance, had a negative loading on depth and did not contributed significantly to the groundwater hydrochemistry. At last, factor-6 accounted for 5.01% of the total variance, had a positive and negative loading on DOC and age which did not contribute statistically.

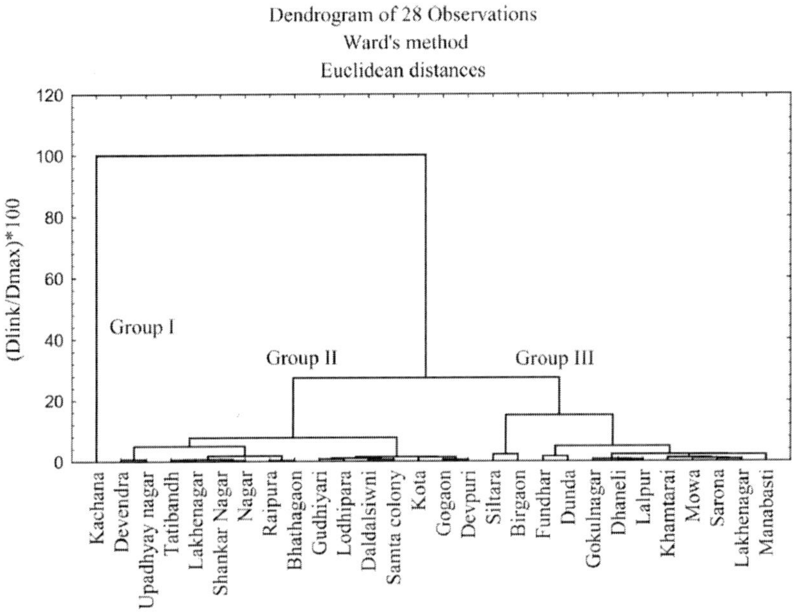

Figure 3. Dendrogram of groundwater samples in the post monsoon period, January 2012.

Table 4. Factor analysis of variables of groundwater during post monsoon period, January, 2012.

Parameter	Factor-1	Factor-2	Factor-3	Factor-4	Factor-5	Factor-6
Depth	−0.15	0.12	−0.01	−0.02	**−0.91**	0.03
Age	−0.22	−0.00	−0.10	0.06	0.27	**−0.74**
T	−0.21	−0.10	−0.62	0.15	0.03	−0.08
pH	−0.44	**0.78**	0.04	−0.10	0.10	0.00
DO	0.49	−0.21	−0.01	0.55	0.03	−0.31
RP	0.00	0.00	**−0.89**	−0.14	0.01	0.14
EC	**0.95**	0.09	0.15	0.06	−0.12	0.04
TDS	**0.96**	0.09	0.15	0.08	−0.10	0.05
Alk	**0.79**	0.06	0.07	0.43	0.23	0.07
TH	**0.94**	−0.09	0.19	0.09	0.14	−0.06
CC	−0.41	0.14	−0.57	−0.21	−0.14	0.47
OC	−0.13	0.04	−0.29	0.17	0.33	**0.81**
F^-	−0.02	0.01	0.00	**−0.93**	0.00	−0.12
Cl^-	**0.89**	−0.05	0.09	−0.13	−0.21	0.02
SO_4^{2-}	**0.79**	0.07	−0.03	−0.44	−0.24	0.17
NO_3^-	**0.85**	−0.06	−0.09	0.12	0.23	−0.04
Na^+	0.25	**0.86**	0.02	−0.01	−0.29	0.09
K^+	0.61	0.47	0.21	0.46	0.08	−0.12
Mg^{2+}	**0.92**	−0.07	0.18	0.15	0.19	−0.07
Ca^{2+}	**0.94**	−0.09	0.18	0.05	0.13	−0.03
Eigen value	8.36	1.72	1.80	1.97	1.44	1.63
Explained Variance %	44.18	11.81	9.69	8.67	5.26	5.01
Cumulative Variance %	44.18	55.99	65.68	74.34	79.61	84.62

Significant loadings > 0.7 (in italic and bold) at $p < 0.05$.

The results of FA in the pre monsoon period are shown in Table 5. Three factors explaining 79.37% of the total variance were extracted. Factor-1 accounted for 56.67% of the total variance, described complex processes such as mineralization and dissolution of evaporate in the water. Factor-2 represented 13.44% of the total variance, had positive and negative loading NO^-_3 on and pH, which did not contribute statistically to the hydrochemistry. Factor-2 was related to anthropogenic activities (i.e. runoff and waste water). Factor-3 was accounted for 9.26% of the total variance, had a positive and negative loading on T (°C) and F^-. Factor-3 characterized the importance of temperature on the dissolution of evaporate.

3.5. Water Quality Assessment

As per Piper diagram, the groundwater in the post monsoon season was mostly classified into two types: Ca-Cl- SO_4-NO_3-HCO_3 and Ca-SO_4-NO_3-HCO_3 of water. The most dominating type of water was Ca-Cl-SO_4-NO_3- HCO_3. In the pre monsoon season, only Ca-SO_4-NO_3-HCO_3 water was observed. The WQI of the water in the post and pre monsoon period was ranged from 218

- 846 and 321 - 889 with mean value (p = 0.05) of 361 ± 45 and 504 ± 56, respectively. The value of TDS, TH, TA, F−, NO^-_3, Cl−, SO^{2-}_4, Mg^{2+} and Ca^{2+} in the pre monsoon period was found above the recommended values of 500, 200, 120, 1.5, 45, 200, 150, 30 and 75 mg/L, respectively [20] [21] . These data showed that the groundwater of Raipur city was found to be unsuitable for the drinking purposes.

The values of SAR, KR, SH, MH and PI values of the groundwater in the post monsoon period were ranged from 1.3% - 6.2%, 0.3% - 2.0%, 26% - 70%, 28% - 38% and 47% - 87% with mean value (p = 0.05) of 3.0 ± 0.5, 0.9 ± 0.2, 48 %± 4%, 33% ± 1% and 71% ± 4%, respectively. Generally, the MH, SAR, KR and PI value of <50, <10, <1 and <25 were considered good for the irrigation purposes [21] [27] - [29] . The SH value in the water of all tube wells was found in the range of 26% - 70%, indicating permissible water quality for the irrigation purpose [29] .

Table 5. Factor analysis of variables of groundwater during pre monsoon period, May, 2012.

Parameter	Factor-1	Factor-2	Factor-3
T (°C)	0.05	0.11	**0.75**
pH	−0.26	**−0.80**	−0.14
EC	**0.93**	0.07	−0.08
Alk	**0.84**	0.29	0.21
TH	**0.84**	0.33	0.12
F^-	0.09	0.06	**−0.81**
Cl^-	**0.95**	0.17	−0.09
SO_4^{2-}	**0.92**	−0.22	−0.13
NO_3^-	−0.02	**0.81**	−0.06
Na^+	**0.90**	−0.03	−0.04
K^+	**0.85**	0.17	0.02
Mg^{2+}	**0.73**	0.55	0.05
Ca^{2+}	**0.82**	0.46	0.04
Eigenvalue	6.85	2.13	1.34
Explained Variance %	56.67	13.44	9.26
Cumulative Variance %	56.67	70.11	79.37

Significant loadings > 0.7 (in italic and bold) at p < 0.05.

4. CONCLUSION

The water of Raipur city is found to be very hard with high EC and TA values. The domination of Ca-Cl-SO_4- NO_3-HCO_3 type water was observed, may be due to huge industrial and urban activities. The quality of water was found to be deteriorated due to excessive mineralization of species i.e. DOC, DIC, Cl^-, NO^-_3, SO^{2-}_4, HCO^-_3,Na^+, K^+, Mg^{2+} and Ca^{2+}, and not found to be suitable for drinking purposes.

ACKNOWLEDGEMENTS

We are thankful to the UGC, New Delhi for awarding the RG scholarship to one of the author, NSD for doing this work.

REFERENCES

1. Tellam, J.H., Rivett, M.O., Israfilov, R.G. and Herringshaw, L.G. (2006) Urban Groundwater Management and Sustainability. Springer, Berlin. http://dx.doi.org/10.1007/1-4020-5175-1
2. Jain, S.K. (2012) Sustainable Water Management in India Considering likely Climate and Other Changes. Current Sciences, 102, 177-188.
3. Schwarzenbach, R.P., Egli, T., Hofstetter, T.B., Gunten, U.V. and Wehrli, B. (2010) Global Water Pollution and Human Health. Annual Review of Environment and Resources, 35, 109-136. http://dx.doi.org/10.1146/annurev-environ-100809-125342
4. Biswas, A.K. and Seetharam, K.E. (2008) Achieving Water Security for Asia. Water Resources Development, 24, 145- 176. http://dx.doi.org/10.1080/07900620701760556
5. Hah, T. (2009) Taming the Anarchy: Groundwater Governance in South Asia. Resources for the Future. Washington, DC, USA and International Water Management Institute, Colombo, 310. http://hdl.handle.net/10568/36566
6. Basavarajappa, H.T. and Manjunatha, M.C. (2015) Groundwater Quality Analysis in Precambrian Rocks of Chitradurga District, Karnataka, India using Geo-Informatics Techdnique. Aquatic Procedia, 4, 1354-1365. http://dx.doi.org/10. 1016/ j.aqpro. 2015.02.176
7. Chakraborti, D., Das, B. and Murrill, M.T. (2011) Examining India's Groundwater Quality Management. Environmental Science and Technology, 45, 27-33. http://dx.doi.org/ 10.1021/es101695d
8. Dagar, J.C. (2005) Salinity Research in India: An Overview. Bulletin of the National Institute of Ecology, 15, 69-80.
9. Dar, I.A., Sankar, K. and Dar, M.A. (2011) Spatial Assessment of Groundwater Quality in Mamundiyar Basin, Tamil Nadu, India. Environmental Monitoring and Assessment, 178, 437-447. http://dx.doi.org/10.1007/s10661-010-1702-2
10. Ramesh, K. and Elango, L. (2012) Groundwater Quality and Its Suitability for Domestic and Agricultural use in TondiarRiver Basin, Tamil Nadu, India. Environmental Monitoring and Assessment, 184, 3887-3899. http://dx.doi.org/10.1007/s10661-011-2231-3
11. Rosin, K.G., Kaur, R., Singh, S.D., Singh, P. and Dubey, D.S. (2013) Groundwater Vulnerability to Contaminated Irrigation Waters—A Case of Peri-Urban Agricultural Lands around an Industrial District of Haryana, India. Procedia Environmental Sciences, 18, 200-210. http://dx.doi.org/10.1016/j.proenv.2013.04.026
12. Rao, N.S. (2008) Iron Content in Groundwater of Visakhapatnam Environs, Andhra Pradesh, India. Environmental Monitoring and Assessment, 136, 437-447.
13. Tank, D.K. and Chandel, C.P.S. (2010) A Hydrochemical Elucidation of the Groundwater Composition under Domestic and Irrigated Land in Jaipur City. Environmental Monitoring and Assessment, 166, 69-77. http://dx.doi.org/10.1007/s10661-009-0985-7
14. Venkateswaran, S. and Deepa, S. (2015) Assessment of Groundwater Quality Using GIS Techniques in Vaniyar Watershed, Ponnaiyar River, Tamil Nadu. Aquatic Procedia, 4, 1283-1290. http://dx.doi.org/10.1016/j.aqpro.2015.02.167
15. Vijay, R., Khobragade, P. and Mohapatra, P.K. (2011) Assessment of Groundwater Quality in Puri City, India: An Impact of Anthropogenic Activities. Environmental Monitoring and Assessment, 177, 409-418. http://dx.doi.org/10.1007/s10661-010-1643-9

16. Patel, K.S., Dahariya, N.S., Chakradhari, S., Sahu, P., Rajhans, K.P., Ramteke, S., Lata, L. and Milosh, H. (2015) Sewage Pollution in Central India. American Journal of Analytical Chemistry, 6, 787-796. http://dx.doi.org/10.4236/ajac.2015.610075
17. Nielsen, D.M. and Nielsen, G.L. (2000) The Essential Handbook of Groundwater Sampling. CRC Press, Boca Raton, USA.
18. APHA (2005) Standard Methods for the Examination of Water and Wastewater. 21st Edition, APHA, AWWA and WEF, Washington DC.
19. Nollet Leo, M.L. and De Gelder Leen, S.P. (2007) Handbook of Water Analysis. 2nd Edition, CRC Press, Boca Raton.
20. BIS (Bureau of Indian Standard) (2009) Drinking Water—Specification. 2nd Edition, New Delhi. http://bis.org.in/sf/fad/FAD25(2047)C.pdf
21. WHO (2011) Guidelines for Drinking-Water Quality. 4th Edition, World Health Organization, Geneva. http://www.who.int/water_sanitation_health/publications/2011/dwq_chapters/en/
22. Liu, C.W., Lin, K.H. and Kuo, Y.M. (2003) Application of Factor Analysis in the Assessment of Groundwater Quality in a Blackfoot Disease Area in Taiwan. Science of the Total Environment, 313, 77-89. http://dx.doi.org/10.1016/S0048-9697(02)00683-6
23. Reghunath, R., Murthy, T.R.S. and Raghavan, B.R. (2002) The Utility of Multivariate Statistical Techniques in Hydrogeochemical Studies: An Example from Karnataka, India. Water Resources, 36, 2437-2442.
24. Shrestha, S. and Kazama, F. (2007) Assessment of Surface Water Quality Using Multivariate Statistical Techniques: A Case Study of the Fuji River Basin, Japan. Environmental Modelling and Software, 22, 464-475. http://dx.doi.org/10.1016/j.envsoft.2006.02.001
25. Singh, K.P., Malik, A., Mohan, D. and Sinha, S. (2004) Multivariate Statistical Techniques for the Evaluation of Spatial and Temporal Variations in Water Quality of Gomti River (India): A Case Study. Water Resources, 38, 3980-3992. http://dx.doi.org/10.1016/j.watres.2004.06.011
26. Kaiser, H.F. (1958) The Varimax Criterion for Analytical Rotation in Factor Analysis. Psychometrika, 23,187-200. http://dx.doi.org/10.1007/BF02289233
27. Kelley, W.P. (1940) Permissible Composition and Concentration of Irrigation Water. Proceedings of the American Society of Civil Engineers, 66, 607-613.
28. Richards, L.A. (1954) Diagnosis and Improvement of Saline and Alkali Soils, Agricultural Handbook 60. USDA and IBH Publishing Co. Ltd., New Delhi, 98-99.
29. Wilcox, L.V. (1955) Classification and Use of Irrigation Waters. USDA, Circular 969, Washington DC.

CHAPTER 6

The Energy-Water Nexus: Spatially-Resolved Analysis of the Potential for Desalinating Brackish Groundwater by Use of Solar Energy

Jill B. Kjellsson [1,*] and Michael E. Webber [2]

[1]*Department of Civil, Architectural and Environmental Engineering, The University of Texas at Austin, 204 E. Dean Keeton St. Stop C2200, Austin, TX 78712, USA*

[2]*Department of Mechanical Engineering, The University of Texas at Austin, 204 E. Dean Keeton St. Stop C2200, Austin, TX 78712, USA*

ABSTRACT

This research looks at coupling desalination with renewable energy sources to create a high-value product (treated water) from two low value resources (brackish groundwater and intermittent solar energy). Desalination of brackish groundwater is already being considered as a potential new water supply in Texas. This research uses Texas as a testbed for spatially-resolved analysis techniques while considering depth to brackish groundwater, water quality, and solar radiation across Texas to determine the locations with the best potential for integrating solar energy with brackish groundwater desalination. The framework presented herein can be useful for policymakers, regional planners, and project developers as they consider where to site desalination facilities coupled with solar photovoltaics. Results suggest that the northwestern region of Texas—with abundant sunshine and groundwater at relatively shallow depths and low salinity in areas with freshwater scarcity—has the highest potential for solar powered desalination. The range in capacity for solar photovoltaic powered reverse osmosis desalination was found to be 1.56×10^{-6} to 2.93×10^{-5} cubic meters of water per second per square meter of solar panel ($m^3/s/m^2$).

Keywords: desalination; groundwater; solar energy; energy-water nexus; solar; photovoltaics; GIS

1. INTRODUCTION

The relationship between the water and energy sectors—for example the energy required to distribute, collect, and treat water and wastewater and the water required to extract and process fossil fuels and to generate electricity—is of increasing interest to policymakers and planners. A growing population has placed a stress on water resources, which is exacerbated by an increased demand for water for energy production. The decreasing water quality and increasing depth and distance to raw water sources increases the strain that is already on the available water supply. As regions look at more energy intensive options such as conveying water greater distances and treating lower quality water with technologies such as desalination, the energy required to treat and transport raw water increases. Since constraints on one of these two resources exacerbate the strains on the other, solutions to the problem of limited resources often address both constraints simultaneously.

As population and water demand increase, there is a need for new freshwater sources. Because brackish water is abundant, desalination of brackish water represents one possible alternative supply. However, desalination is an energy intensive treatment method which remains an inhibitor to its broader adoption. At the same time, concerns about carbon emissions and their effects on the climate favor the use of renewable energy over conventional fossil fuels. However, renewable energy technologies such as wind and solar are intermittent, which can present a challenge to grid operation if large-scale storage (such as batteries, pumped hydroelectric, or compressed air energy storage) is not available [1]. One possible solution is to couple solar photovoltaic (PV) with desalination that can be operated intermittently, such that stored, treated water is a proxy for energy storage. At the same time, using the water facility only when solar power is available reduces fluctuations in power supply to the electric grid, adding value to the integrated facility.

This research evaluates the potential of co-locating desalination plants for inland brackish groundwater treatment and solar PV generation. This project explores the spatial component using Geographic Information Systems (GIS) to assess the co-location of solar and brackish groundwater resources to estimate the potential for an integrated facility to solve several problems simultaneously. Because Texas has an isolated electric system, growing population, water scarcity, and an abundance of solar energy and brackish groundwater, it is used as a testbed for this analysis. However, the approach developed here is applicable outside Texas. Using temporally and spatially resolved solar irradiance and brackish groundwater depth and quality data, a framework was developed to evaluate these resources. This framework is intended to be useful for policy makers, regional planners and project developers as they look towards alternative water supply in water stressed areas.

Literature surrounding the technical and economical aspects of coupling desalination and solar PV exist, as does research on the availability of these resources. The coupling of desalination with renewable energy is also a growing field of research [2,3]. Similar research has also been conducted on the sustainable siting of co-located seawater desalination facilities with solar PV factoring in technical, social and economical impacts such as solar insolation, ocean salinity, water temperature, water stress, population and ability to pay [4], as has research on resource optimization of wind-powered brackish groundwater RO desalination [5]. What is lacking from the literature is methodologies for determining the optimal locations of co-located facilities for brackish groundwater reverse osmosis (RO) and PV based on resource characteristics at a given location. This research aims to fill that knowledge gap. While the authors are aware of the environmental concerns associated with greenhouse gas emissions and disposal of the brine waste stream, the aim of this paper is not to address these issues but rather to deliver a framework for the geographic analysis of the solar and water resources impacting a coupled PV-RO facility.

2. BACKGROUND ON SOLAR-POWERED DESALINATION AND BRACKISH GROUNDWATER

While the population in Texas is expected to grow 82% from 2010 to 2060, water use is estimated to grow only 22% due to decreases in agricultural water use from efficiency improvements and municipal water use from conservation measures [6]. Municipal water demand is expected to increase from 4.9 million acre-feet in 2010 to 8.4 million acre-feet in 2060 while natural fresh water supplies are estimated to decrease 10% over this period [6]. Therefore, at the current rate of production and consumption, future supplies of water will not meet future demand unless alternative sources are tapped. Given the existing strain on freshwater along with a wealth of solar and brackish water resources, this research will focus on Texas as a testbed; however, the research methodology and results will be broadly applicable to areas with similar resources and prevailing conditions.

Texas has an abundance of brackish groundwater, thought to be more than 2.7 billion acre-feet [7], which can possibly be desalinated and used to meet public needs. Brackish groundwater is defined as water with a total dissolved solids (TDS) concentration of 1000 to 10,000 mg/L. For comparison, seawater has a TDS concentration of 35,000 mg/L. The Texas Commission on Environmental Quality (TCEQ) has set a primary standard concentration for TDS of 500 mg/L and a secondary standard of 1000 mg/L for public use [7].

RO is the most common type of desalination technology worldwide, accounting for 60% of desalination capacity in 2010 [8]. In Texas, RO accounts for 80% of desalination systems in operation [9]. RO applies pressure to a solution on one side of a selective membrane to reverse the natural flow of solvent to the side with higher solute concentration. The solute remains, while the pure solvent passes to the other side, thereby producing freshwater. Electrodialysis, a less common desalination method accounting for 3.6% of desalination capacity worldwide in 2010 [8], uses electromotive forces applied to electrodes that are adjacent to both sides of a membrane to purposely move salt ions through the membrane leaving behind freshwater. ED is best applied to treatment of brackish-water with TDS up to 5000 mg/L and is not economical for higher concentrations [10], except in cases where ED is used for partial treatment [11]. Both methods are used coupled with PV systems [10]. RO and ED are the most common forms of desalination to be coupled with PV [10]. Thermal methods of desalination, accounting for 34.8% of desalination worldwide in 2010 [8], are not well-suited for coupling with PV because they are heat-based as opposed to electricity-driven.

PV cells are a rapidly growing technology with costs decreasing over time. The price of solar PV modules in 1987 was roughly $9/W [12]. From 1998 to 2011 the price fell from $4.90/W to $1.28/W [13]. Typical PV levelized cost of energy (LCOE) are in the range of $0.20–$0.40/kWh for low latitudes with high insolation of 2500 kWh/m^2/year, $0.30–$0.50/kWh for 1500 kWh/m^2/year (which is typical of Southern Europe), and $0.50–$0.80 per kWh for high latitudes with 1000 kWh/m^2/year [12].

In Texas, the solar energy potential, or the amount of solar radiation available to a solar PV cell, ranges from 872 to 1310 kWh/m^2/year during the winter and from 2150 to 2884 kWh/m^2/year during the summer [14]. The high end of the range corresponds to the western portion of the state while the low end corresponds to the eastern portion of the state. Annual average global tilt solar radiation across Texas is shown in Figure 1. There is a range from 4.76 kWh/m^2/day to 6.58 kWh/m^2/day from east to west.

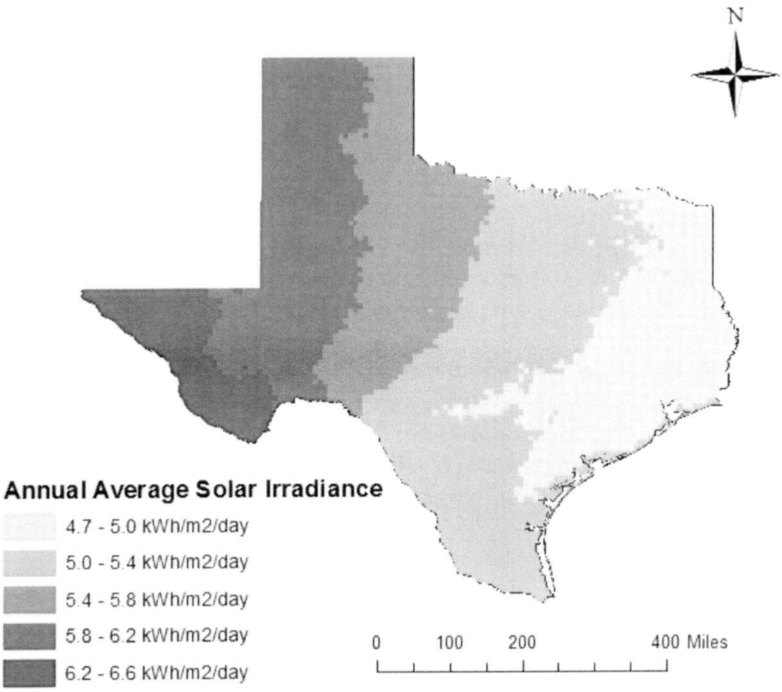

Figure 1. Global tilt solar radiation across Texas from the National Renewable Energy Laboratory (NREL)'s National Solar Radiation Database dataset [15].

The most common combination of renewable energy and desalination employed worldwide is PVRO, and accounts for 31% of renewable energy-powered desalination installation [16]. If solar energy can be located near the desalination plant, using solar panels directly for desalination eliminates the need to incorporate solar energy into the grid, although grid interconnectedness provides support for the system and could allow for continuous operation of the desalination plant.

The energy intensity of desalinating brackish groundwater has been estimated to be 0.5–3 kWh/m^3 of product (or treated) water [17] while other sources estimate this value to be 1–2.5 kWh/m^3 of product water [18]. The energy requirement is proportional to the TDS concentration as well as the depth to the groundwater source. There are currently 46 brackish water desalination plants in Texas [19]. Twelve of these facilities treat brackish surface water and account for a design capacity of 50 million gallons per day (56,000 acre-feet per year) [19]. Thirty-four of these facilities use brackish groundwater as their raw water source, which acounts for a design capacity of 73 million gallons per day (81,760 acre-feet per year) [19]. In the 2012 State Water Plan, 5 out of 16 regional water planning groups recommended brackish groundwater desalination as a water management strategy, accounting for 181,568 acre-feet by the year 2060 [19]. Desalination plants currently in operation in Texas are spread throughout the state, as shown in Figure 2. Only the northern portion of the state, known as the panhandle, does not have desalination facilities.

This research focuses on RO desalination since the majority of desalination facilities in Texas already utilize this technology. PV cells will be the primary focus of solar power generation because PV technology can produce energy from both direct and diffuse radiation as opposed to other concentrating solar power (CSP) technologies, which can only make use of the direct radiation. Diffuse radiation is the radiation that is scattered from the direct beam by the atmosphere. Integrating PV with desalination and using water storage as a proxy for energy storage can advance the implementation of these two technologies by each one solving the problems of the other.

3. DATA AND METHODOLGY

3.1. Data

Data for this research was obtained from the following two sources: The Texas Water Development Board's (TWDB) Groundwater Database [20] and the National Renewable Energy Laboratory's (NREL) National Solar Radiation Database (NSRD) [15]. The TWDB database was used to acquire well location coordinates, well depth, and well total dissolved solids (TDS) concentrations for wells with well depths in the range of 100–12,000 feet and TDS concentrations in the range of 1000–30,000 mg/L. TDS range was selected to correspond to brackish water. Available for download online, NREL data provides monthly and average daily total solar resource averaged over cells of 0.1 degrees in both latitude and longitude, or nearly 10-km in size, using 1998–2005 data and projected using 1983 North American geographic coordinate system. For this research, the global tilt radiation was used, which is the total (direct and diffuse) radiation on a tilted surface.

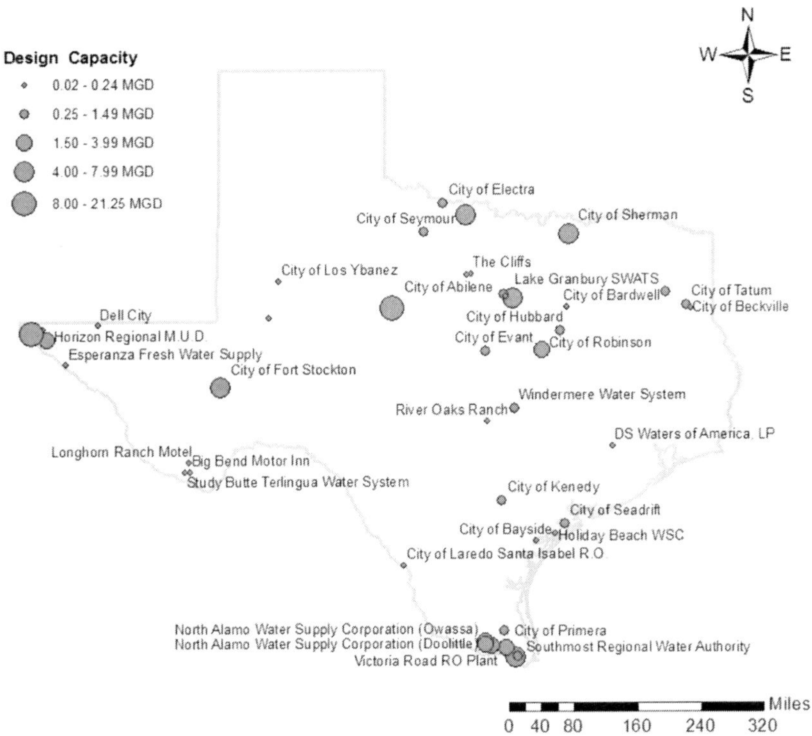

Figure 2. Existing desalination plants and their design capacities in Texas.

3.2. Methodology

This research was carried out by first determining the spatial variability of brackish water resources and of well characteristic data. Next, an integrated analysis was conducted to estimate the solar-powered desalination capacity across Texas as a function of different factors which included solar radiation, well depth and TDS concentration.

3.2.1. Model Assumptions

The following assumptions were made in carrying out this research: the efficiency of solar PV panels was assumed to be 15% for calculating the power generation from solar PV panels as reported by industry as a typical efficiency [13], a pump efficiency of 65% was assumed based on literature [5], and a specific gravity of 9.81×10^3 N/m^3 was assumed for calculating the power requirements of desalination [5].

3.2.2. Spatial Variability of Brackish Groundwater Well Characteristics and Solar Resources

Brackish water wells with depths in the range of 100–12,000 feet across Texas were analyzed using the software program ArcGIS [21]. Locations were provided in decimal degree units and were projected using North American 1983 geographic coordinate system. Using the "Spherical Kriging Interpolation" tool in ArcGIS, the well depth and TDS concentration was estimated for locations across Texas and projected to two different rasters. Using "Extract by Mask" tool and an outline of Texas layer imported from ArcGIS online, the output was fit to Texas. Using the "Project Raster" tool, the rasters were projected to USA Contiguous Albers Equal Area Conic spatial reference in order to preserve the area. This spatial reference will be important in later use of the rasters with the "Raster Calculator" tool. In order to prepare the solar data for raster calculation, the data from NREL was converted to a raster using the "Polygon to Raster" tool.

3.2.3. Potential Capacity of Solar-Powered Desalination

The power requirements of desalination by RO depends on the RO and pumping power requirements. It is assumed that the desalination facility is located at the well so that power required for the transportation of brackish groundwater is negligible. Thus, the power for desalination can be expressed by Equation (1) [5]:

$$W_{desalination} = W_{RO} + W_{pumping} \quad (1)$$

where

$W_{desalination}$= Power needed for desalination, W

W_{RO}= Power needed for RO treatment, W

$W_{pumping}$= Power needed to pump brackish groundwater to the surface, W

The following Equation (2) is an extended version of Equation (1) with the equations for RO and pumping energy substituted into the Equation [5]:

$$W_{desalination} = \left(\frac{v \times \gamma \times WD}{\eta_{pump}}\right) + \left(\frac{1,000\ N\ L}{m^2\ mg} \times v \times (TDS_{in} - TDS_{out})\right) \quad (2)$$

where

γ = specific weight of water, N/m^3

WD = well depth, m

η_{pump}= net efficiency of pump and motor system

TDS_{in}= TDS of feed water, mg/L

TDS_{out}= TDS of treated water, mg/L (desired water quality, 500 mg/L)

v = volumetric flow rate of feed water, m^3/s

The coefficient of the second part of the equation, 1000 N L/m^2 mg , was changed from the cited authors' work to reflect the empirical data for the energy intensity of brackish groundwater desalination in Texas [22]. The authors appear to have used a theoretical minimum value in the equation based on the theoretical power requirements to overcome osmotic pressure. For this research, the authors have chosen to modify the equation to reflect the empirical data from the Kay Bailey Hutchison brackish groundwater desalination plant in El Paso, Texas to more accurately predict the energy requirements. The coefficient was interpreted from the enegy intensity and TDS data.

Equation (3) is used for the calculation of power generation from PV [23]:

$$W_{PV} = \eta_{PV} \times GTR \times A_{PV} \times \frac{day}{24\ h} \quad (3)$$

where

W_{PV}= operating electrical power, W

η_{PV}= Efficiency of electrical power operating subsystem

GTR = Global tilt radiation, Wh/m^2/day

A_{PV}= Surface area of PV cell, m^2

Setting the operating electrical power generated by PV cells (Equation (3)) equal to the power requirements of desalination (Equation (2)) and solving for the flow rate, *v* (assuming there is no battery included in the PV system) yields the following Equation (4):

$$v = \frac{\eta_{PV} \times GTR \times A_{PV} \times \frac{day}{24\ h}}{\left(\frac{\gamma \times WD}{\eta_{pump}}\right) + \left(\frac{1{,}000\ N\ L}{m^2\ mg} \times (TDS_{in} - TDS_{out})\right)} \quad (4)$$

Equation (4) suggests that if well depth is small, incoming total dissolved solids concentration is small, and the solar radiation is high then the amount of water that can be treated is higher than in places where sunshine is less prevalent, wells are deeper, and water is more saline.

Using the "Raster Calculator" tool in GIS, the well depth, TDS concentration, and solar radiation rasters from the previous two steps in this analysis could be used in the equation to solve for the potential treatment (as a flow rate, *v*) for different locations based on their local water and solar conditions. The output is also in raster form. Having the rasters projected to the same spatial reference in the previous steps allowed for the mathematical operations used in raster calculator to be performed on a cell by cell basis, *i.e.*, for every location across Texas.

While the raster shows the variance across Texas, it can be desirable to know a more precise location of the optimal events as well as the values of the variables that resulted in that output. In order to determine these values, the output raster was converted back to points using the "Raster to Point" tool and X, Y coordinates were added using the "Add XY Coordinates" tool.

4. RESULTS AND ANALYSIS

Brackish water wells with depths in the range of 100–12,000 feet and TDS in the range of 1000–30,000 mg/L across Texas are depicted on the maps in Figure 3 and Figure 4. Figure 3 shows the wells with greater well depths with larger blue circles and Figure 4 shows the same wells with the larger red circles representing higher TDS concentrations.

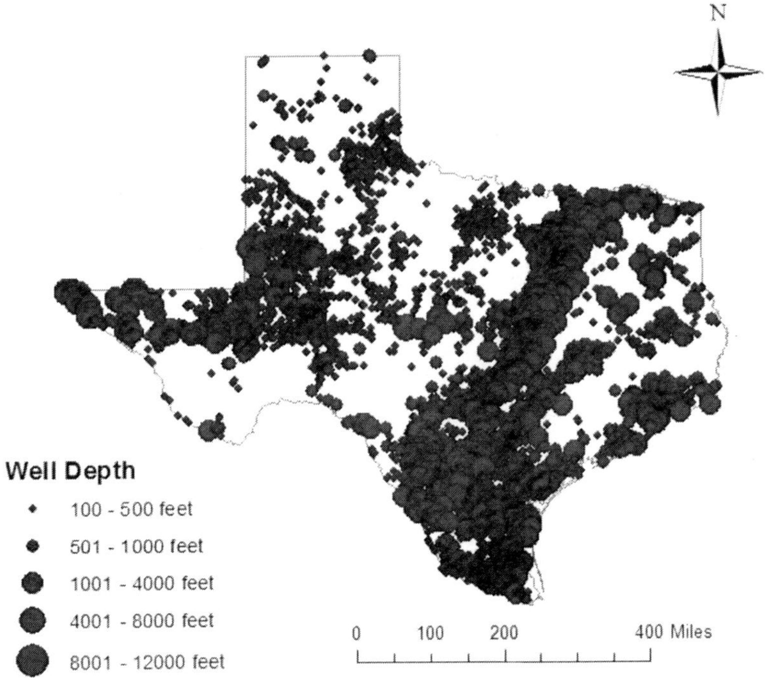

Figure 3. Each dot represents a well from the Texas Water Development Board (TWDB) database. The larger dots represent greater distances to brackish water level.

Wells appear to be concentrated in the southern tip of Texas and along the coast as well as in the northwestern portion of the state. Well depth and TDS concentrations seem to be uniformly spread in these regions. The depth and TDS concentrations of the wells was also depicted after interpolation across the state was performed as a way to predict the depth and TDS concentration at any location across Texas based on the known values for existing wells, as shown in Figure 5 and Figure 6.

Figure 7 shows the capacity (as a volumetric flow rate, v) that one square meter of PV array could produce and the location of the six highest calculated capacities. Table 1 shows the values of the variables that resulted in these highest capacities.

The framework for the optimization performed is important because it helps tease out the non-obvious tradeoffs. While it might appear obvious that low salinity, shallow depths to brackish groundwater and high solar insolation would be ideal, they might not be available at the same location. In instances when they are not, it can be difficult to know which variable plays a less important role and to what degree. Optimization helps to answer these questions.

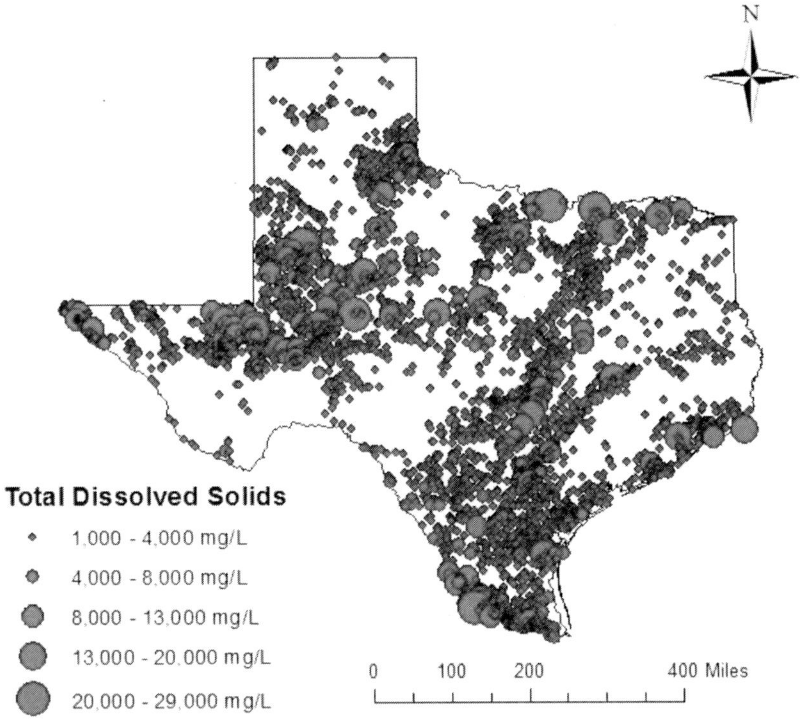

Figure 4. Each dot represents a well from the TWDB database. The larger dots represent higher total dissolved solids concentration for that well.

5. DISCUSSION AND CONCLUSIONS

Using the methodology presented herein, a PV system covering an area of 1 m^2 is capable of producing roughly 2.93×10^{-5} m^3/s in some regions of Texas if located where optimal conditions are present. These optimal conditions include low well depth, low TDS and high solar radiation. By using PV to power the pumping and RO processes that are a part of desalination, the negative impacts of carbon emissions from such an energy-intensive process are reduced compared to if conventional fossil fuel sources are used.

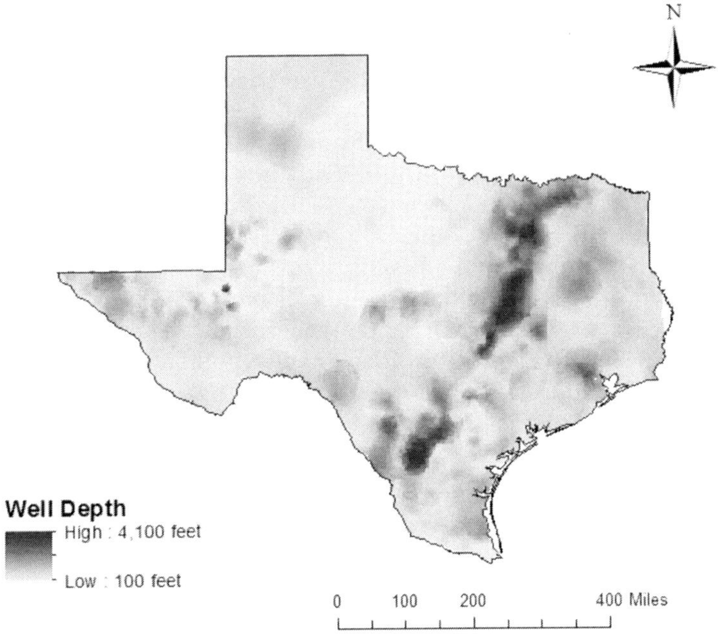

Figure 5. The depth of wells across Texas after interpolation was used to estimate well depth for areas without data.

Interestingly, locations with deeper wells also have higher TDS concentrations. However, this is not always the case, as is evident in the northern region. The area with deeper wells across Texas going from the southwest to the northeast part of the state follows along Interstate 35. While it is not certain, this could be because there is a lot of development along Interstate 35, including some of Texas' largest cities such as Dallas, Fort Worth, Austin and San Antonio. The Trinity and Carrizo-Wilcox aquifers, two of Texas' major aquifers, also run a similar path across Texas. The Balcones Escarpment, or Balcones Fault, runs from the southwest of Texas and up, also following a similar path through Texas. While there might be a correlation between the area of deeper wells across Texas, the natural land and water systems, and the development along it, the authors merely wish to point out these patterns but did not evaluate their relationship.

More than providing information for Texas, this research presents a framework for analyzing the optimal locations of a co-located desalination facility and PV array by optimizing the utilization of the combination of resources available. This method could be applied to other locations where the input conditions are very different to help planners and others who must contemplate options for where to site water treatment facilities. This analysis, being performed on a statewide level, crosses many aquifers. This approach used a kriging interpolation which uses a distance-weighted averaging approach to estimate the value at an unknown point based on neighboring points. The appropriateness of this approach should be tested and other alternative interpolation methods could be analyzed to determine which most accurately predicts well characteristics at locations where measuerments have not been taken in order to improve the results of this analysis. The study could be scaled down to an individual aquifer level, thereby improving our understanding of individual aquifer characteristics and potential for desalination of brackish water from them. Another consideration that needs to be addressed before implementing new solar-powered desalination facilities is whether there is a sufficient supply of water available to protect aquifers from the harmful effects of overuse, such as land subsidence. The disposal of the brine waste stream is also of environmental concern and methods for disposal should also be evaluated.

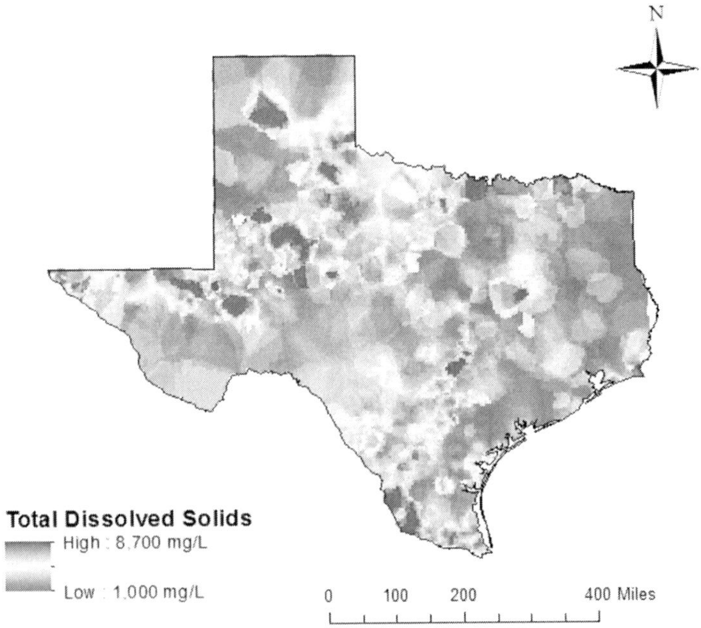

Figure 6. The total dissolved solids concentration of the brackish groundwater across Texas after interpolation was performed to estimate total dissolved solids (TDS) concentrations for areas without data.

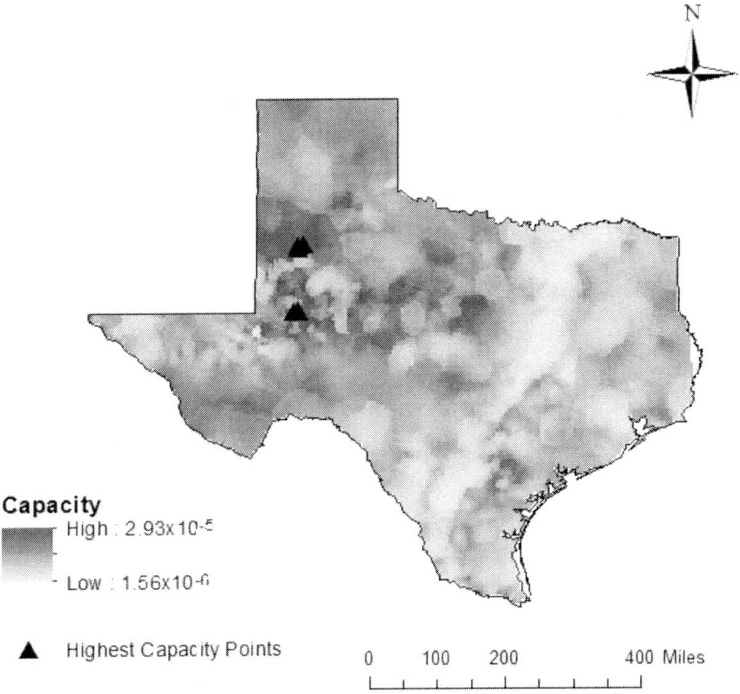

Figure 7. Desalination capacity across Texas calculated using well characteristic and solar radiation data.

Table 1. Variable Values for Six Points with the Highest Capacities.

Longitude (Decimal Degrees)	Latitude (Decimal Degrees)	Capacity (m^3/s)	TDS (mg/L)	Well Depth (m)	GTR ($Wh/m^2/day$)
−102.1313625	33.42166	2.927×10^{-5}	1142	42.4	6008
−102.1313625	33.3767	2.923×10^{-5}	1141	42.5	5998
−102.1313625	32.02784	2.889×10^{-5}	1142	43.4	5992
−102.1313625	32.0728	2.888×10^{-5}	1142	43.4	5992
−101.9964763	33.46663	2.886×10^{-5}	1173	41.0	5968
−102.2212867	32.02784	2.850×10^{-5}	1157	44.0	6024

For this research, a sensitivity analysis would help quantify the trade-offs associated with siting a co-located facility based on the solar resources or water resources available. Considering the large spread of desalination plants currently in Texas, an economic analysis could be performed to determine the cost of installing PV at desalination facilities currently in operation that meet the optimal conditions set forth in this report. Future research might also include analyzing different solar technologies, such as concentrating solar power or solar thermal power, and other renewable energy sources such as wind and geothermal as well as different desalination technologies, *i.e.*, thermal treatment methods. By coupling wind and solar power with RO desalination, the co-located facility could operate for longer hours during the day, since wind is stronger at night [24].

ACKNOWLEDGMENTS

The authors would like to thank the Texas State Energy Conservation Office (SECO) and the Tarrant Regional Water District (TRWD) for their support of this research.

REFERENCES

1. Mbarga, A.A.; Song, L.; Williams, W.R.; Rainwater, K. Integration of Renewable Energy Technologies With Desalination. *Curr. Sustain. Renew. Energy Rep.* **2014**, *1*, 11–18.
2. Clayton, M.E.; Stillwell, A.S.; Webber, M.E. Implementation of Brackish Groundwater Desalination Using Wind-Generated Electricity: A Case Study of the Energy-Water Nexus in Texas. *Sustainability* **2014**, *6*, 758–778.
3. Rainwater, K.; Nash, P.; Song, L.; Schroeder, J. The Seminole Project: Renewable Energy for Municipal Water Desalination. *J. Contemp. Water Res. Educ.* **2013**, *151*, 50–60.
4. Grubert, E.A.; Stillwell, A.S.; Webber, M.E. Where does solar-aided seawater desalination make sense? A method for identifying sustainable sites. *Desalination* **2014**, *339*, 10–17.
5. Venkataraman, K.; Ortegon, J.; Uddameri, V.; Dyke, R. Mapping of Wind-Powered Desalination Potential in South Texas. In Proceedings of the AWRA 2012 Spring Specialty Conference, New Orleans, LA, USA, 26–28 March 2012.
6. Bennett, J.; Dascaliuc, S.; Grossman, N. *Water for Texas 2012 State Water Plan*; Technical Report; Texas Water Development Board: Austin, TX, USA, 2012.
7. *Desalination: Brackish Groundwater*; Technical Report; Texas Water Development Board: Austin, TX, USA, 2004.
8. What Technologies Are Used? Available online: http://www.desalination.com/ market/ technologies (accessed on 11 March 2015).
9. Arroyo, J.; Shirazi, S. *Cost of Water Desalination in Texas*; Texas Water Development Board: Austin, TX, USA, 2009; pp. 1–7.
10. Al-Karaghouli, A.; Renne, D.; Kazmerski, L. Technical and economic assessment of photovoltaic-driven desalination systems. *Renew. Energy* **2010**, *35*, 323–328.
11. McGovern, R.K.; Zubair, S.M.; Lienhard V, J.H. The cost effectiveness of electrodialysis for diverse salinity applications. *Desalination* **2014**, *348*, 57–65.
12. Price, S.; Margolis, R. *2008 Solar Technologies Market Report*; Technical Report January; National Renewable Energy Laboratory: Golden, CO, USA, 2010.
13. SunShot. In *Photovoltaic System Pricing Trends Historical, Recent, and Near-Term Projections*; Technical Report; U.S. Department of Energy: Washington, DC, USA, 2012.

14. Wogan, D.M.; Webber, M.; da Silva, A.K. A framework and methodology for reporting geographically and temporally resolved solar data: A case study of Texas. *J. Renew. Sustain. Energy* **2010**, *2*.
15. Dynamic Maps, GIS Data, and Analysis Tools. Available online: http://www.nrel.gov/gis/solar.html (accessed on 15 May 2015).
16. Papapetrou, M.; Wieghaus, M.; Biercamp, C. *Roadmap for the Development of Desalination Powered by Renewable Energy*; Technical Report; ProDes: Stuttgart, Germany, 2010.
17. Carter, N. *Desalination and Membrane Technologies: Federal Research And Adoption Issues*; Technical Report; Congressional Research Service: Washington, DC, USA, 2013.
18. Papadakis, G. *ADIRA Handbook: A Guide to Autonomous Desalination System Concepts*; Technical Report; European Union: Istanbul, Turkey, 2008.
19. *Desalination: Brackish Groundwater*; Technical Report; Texas Water Development Board: Austin, TX, USA, 2013.
20. Groundwater Database Reports. Available online: http://www.twdb.texas.gov/groundwater/data/gwdbrpt.asp (accessed on 28 November 2012).
21. Environmental Systems Research Institute. In *ArcGIS Desktop: Release 10.2.2*; Environmental Systems Research Institute: Redlands, CA, USA, 2014.
22. MacHarg, J.P. *Energy Optimization of Brackish Groundwater Reverse Osmosis Desalination*; Technical Report; Texas Water Development Board: Austin, TX, USA, 2011.
23. Aybar, H.S.; Akhatov, J.S.; Avezova, N.R.; Halimov, A.S. Solar powered RO desalination: Investigations on pilot project of PV powered RO desalination system. *Appl. Sol. Energy* **2011**, *46*, 275–284.
24. Clayton, M.E.; Kjellsson, J.B.; Webber, M.E. Can renewable energy and desalination tackle two problems at once? Earth, 26 October 2014.

CHAPTER 7

Flow through Multiple Well Points System

Ajay K. Vashisht[1*] and Shri K. Shakya[2]

ABSTRACT

Under natural geographical settings, there are regions all over the world where the native groundwater quality is brackish. However, recharging from the surface water bodies (i.e., rivers and canals) establishes freshwater lenses in the upper portions of these brackish aquifers. Skimming of these freshwater lenses is a viable technique for sustaining livelihood and agricultural practices in these regions. In the present chapter, various skimming methods have been discussed. In addition, one another type of problem has also been developed in certain pockets of these regions. The non-withdrawal of groundwater and the excess use of easily available surface water resources (i.e., canal water) have created severe water logging conditions. Subsequently, it leads to the creation of salt-affected soils. One of such areas located in the southwestern Punjab (India) was taken under study. Keeping the region's problems in mind and considering the merits and demerits of all the skimming methods, a much efficient and affordable technique named multiple well points system (MWPS) has been proposed. The continuous use of MWPS not only reclaimed the area but also improved the groundwater quality. The system was field tested and modified before recommending to the farmers. Later on, the MWPS's feasibility as a groundwater recharging system was also demonstrated in the field and the results have shown that it is a viable technique for reviving freshwater lenses in the region.

Keywords: brackish/saline aquifers, groundwater recharging, multiple well points system, skimming wells, well hydraulics

1. INTRODUCTION

Since ancient times, it is a natural tendency of humankind to settle near the perennial and freshwater resources. But for many millennia, humankind's capability to abstract groundwater was very little compared with the available resources [1]. With time the demand for freshwater increases due to rapid population growth and expanding economies in many parts of the world [2]. The mechanization advancements in the twentieth century, geological knowledge, well drilling, and pump technology made it more convenient to abstract groundwater from deeper aquifers [3, 4]. Globally, groundwater withdrawal ranges between 600 and 700 km^3/y [5]. In the past five-to-six decades, the overexploitation of fresh groundwater beyond sustainability limits has resulted in the widespread and progressive depletion of water level all over the world. As a result, the freshwater resources are becoming scarcer every year. The global warming and climate change is further threatening its security all over the world [6]. Moreover, the heavy withdrawal of freshwater in the regions surrounded by brackish/sea water is responsible for the intrusion of brackish/sea water into the freshwater aquifers [7]. It is not only the anthropogenic reasons that are responsible for the development of brackish aquifers, these may also occur naturally. Under natural geographical settings, there are regions all over the world where the native groundwater quality is brackish [8] and one of the possible reasons is the retention of ions from salt water trapped at the time of ancient marine transgression [9]. The vast stretches in the states of Rajasthan, Punjab, and Haryana of India and Indus basin of Pakistan are having brackish aquifers. Since the groundwater quality is not fit for domestic and agricultural purposes, alternate sources of freshwater were generated to fulfill the demand for potable supply in these regions. Keeping in mind the agricultural productivities of the regions, the governments of both the countries make lots of efforts to lay dense canal network by constructing hydroelectric projects and/or barrages on major rivers and their tributaries. It is a well-known fact that under all practical situations a huge quantity of water is lost through seepage from the canal network (either lined or unlined) before reaching the destination. According to Wachyan and Rushton [10], a well-maintained canal with a 99% perfect lining reduces seepage to about 30–40%, but seepage cannot be controlled completely. By the time water reaches the field, around 45% of the actual volume of water supplied from the canal head seeps out [11]. In Indian conditions, the loss of water through seepage from unlined canals generally varies from 0.3 to 7.0 m^3/s per 10^6 m^2 wetted surface; which depends on the permeability of soil through which the canal passes, location of water table, distance of drainage, bed width, side slope, and water depth inside the canal [12]. Because of the excess seepage from the dense canal network (lined and unlined), deep percolation losses from surface irrigation practices, and in the absence of adequate drainage facilities, the water table has started rising in these regions. Consequently, the water table has risen from 3.65 m in 1930s to less than 1.20 m in 1980s in most of the lower Indus basin [13]. The situation was worse in certain pockets of Indus basin of India where the water table has virtually reached the land surface (**Figure 1**). As a result, the soils become severely alkaline (i.e., alkaline) in nature (**Figure 2**). These conditions have debarred the agricultural practices in the region.

Figure 1. Photograph showing the water table at the ground surface (snapped in early 1980s).

Under these circumstances, the major challenges in front of the water resource managers were

a) to decline the water table and to improve the condition of soil in the areas where water is virtually standing on the surface (is termed as the *first type of problem*), and

b) to arrest the further rise of water table in the areas where water table is near to create water logging conditions (is termed as the *second type of problem*).

Figure 2. Photograph showing the alkaline soils in the region (snapped in early 1980s).

First type of problem: The reclamation process for this type of problem is comparatively difficult and takes longer period of time than the second type of problem. The main reason of high water table in the area is more inflow than the outflow. In general, the aquifers in these areas are very thin (i.e., near about 10 m). High water table conditions are further responsible for the creation of alkaline soils. It is a well-known fact that the presence of high sodium content in alkaline soils makes clay particles to remain in suspension and appear in the form of slurry. It makes the construction of trenches and installation of pipe drainage system difficult at the desired depth and suspects its workability. Therefore, the first requirement for reclaiming such areas was to decline the water table and to improve the soil conditions. In these conditions, only vertical drainage was possible as it was the only way by which the clay particles would be trapped in the upper part of the aquifer. The main advantage of the vertical drainage is that it lowers the water table to any desired depth which is not possible in the case of subsurface and pipe drainage system. Further, the increase in outflow from the aquifer creates space for the storage of seeped-out freshwater from the nearby canal.

Second type of problem: The sustenance of cultivation practices in these regions was vest in the condition that the existing groundwater could be used which will eventually lower the water table. This decision was based on the existence of those freshwater lenses which had established in varied thicknesses over the native brackish water due to the seepage from surface water bodies. Since the aquifer contains water of different densities in two layers, therefore, pumping the aquifer by merely placing the screen of a partially penetrating well in the freshwater zone will not serve the purpose as the development of discharge head in the aquifer may cause up-coning of fresh–brackish water interface to effective pumping zone of the well [14]. This phenomenon can be better explained by the Ghyben–Herzberg principle. Both the researchers have independently recognized the flotation of a steady-state, fresh groundwater lens on top of sea water [15, 16]. If h_f is the freshwater head above sea level and h_s is the depth to the sharp interface (i.e., considering immiscible fluids) below sea level, then for a system in static equilibrium the pressure at the interface due to the overlying column of freshwater must be equivalent to that due to the column of salt water (**Figure 3**). Considering ρ_f is the density of freshwater and ρ_s as that of saltwater, static equilibrium as shown in **Figure 3** can be expressed mathematically as

$$h_s\rho_s = (h_s + h_f)\,\rho_f \quad (1)$$

Rearranging the terms, Eq. (1) changes to

$$h_s = \left(\frac{\rho_f}{\rho_s - \rho_f}\right) h_f \quad (2)$$

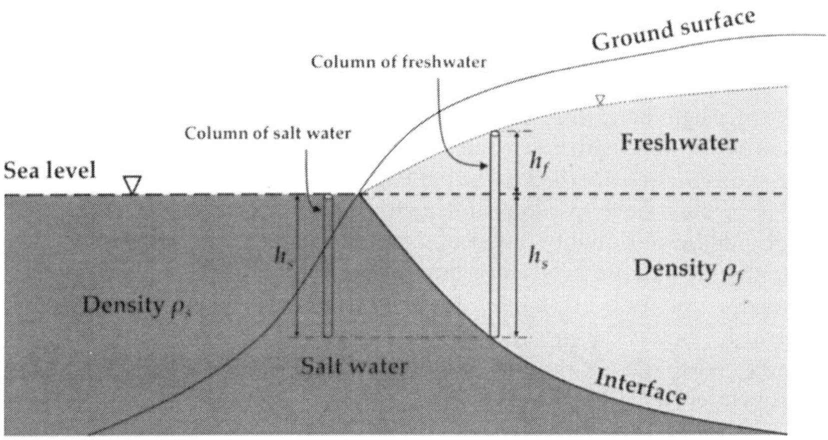

Figure 3. Schematic sketch explaining the Ghyben–Herzberg principle.

According to Eq. (2), the up-coning of the interface depends on the density difference between the fresh and brackish water. For any operational discharge head value, decrease in density difference between freshwater and brackish/saline water leads the interface to rise higher [8], and which is synonymous to Wang's [17] results. Reclamation of such lands for sustaining livelihood is a major challenge for the scientists of India and Pakistan. Researchers from all over the world have also shown keen interest in the reclamation of such areas. Based upon the hydrogeological conditions of the areas, various water management techniques have been proposed. More specifically, the stress has been given to skim the available freshwater from the fresh–brackish/saline aquifers. In the subsequent sections, various types of skimming techniques are discussed.

1.1. Skimming tube well(s)

During the pumping of freshwater from fresh–brackish/saline aquifer, the pressure head in the vicinity of well is lowered which ultimately leads to the rise of brackish/saline water. This phenomenon is known as up-coning and is responsible for the increase in salt content of the pumped water. Obviously, the purpose is to pump the freshwater without much disturbing the brackish/saline zone. Therefore, the term *"skimming tube well"* may be defined as the *tube well to extract freshwater from the fresh–brackish/–saline aquifer by specifying the extraction rate so as to limit the rise of fresh–brackish/saline interface to reach effective pumping zone of the well* (**Figure 4a**). The performance of a skimming tube well depends upon the factors, viz., (i) relative thickness of the freshwater lens with respect to the brackish/saline water zone at the point of extraction, (ii) densities of freshwater and brackish-/saline water, (iii) screen length penetrating the freshwater zone, (iv) distance of skimming well from the recharging source (i.e., canal/river), (v) diameter of the well, (vi) spacing between the adjoining skimming wells, and (vii) the aquifer parameters. Studies show that the single-bore skimming tube well technique is successful only in aquifer with the freshwater zone having thickness more than 30 m [18]. In the regions where the thicknesses of the freshwater zones are less than 30 m, the farmers' have to compromise either with the discharge rate or with the pumped water quality if they abstract groundwater through single-bore well. In such situations, a more practicable technique is to abstract water through a number of smaller capacity tube wells which are installed arbitrarily in circular array and are joined with a single pumping unit (**Figure 4b**). The complete pumping setup may be termed much appropriately as *"skimming tube wells system."* Field experience shows that the number of tube wells in such type of system may vary from 2 to more than 16 [18]. However, based upon the performance studies on skimming tube wells system with reference to quantity and quality of pumped freshwater it was found that the double-strainer skimming well is most efficient [19]. Its basic reason is explained in Section 1.6.

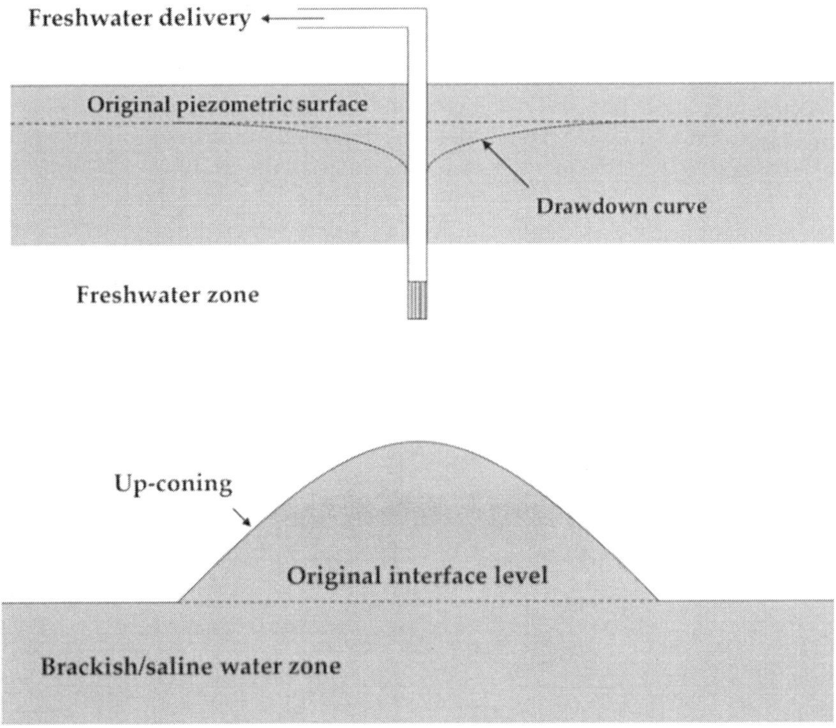

Figure 4A. Schematic sketch showing a skimming tube well.

1.2. Scavenger well system

To restrict the rise of fresh–brackish/saline interface to reach the effective pumping zone of the well, the adoption of scavenger well system is one of the possible solutions in which pumping from the production well is accompanied by brackish/salt water pumping from a scavenger well in the vicinity [20, 21]. In a scavenger well system, two wells (i.e., production well and scavenger well) are installed either in a single-bore hole (**Figure 5a**) or side-by-side bore holes (**Figure 5b**). The production well screens the freshwater zone whereas scavenger well taps the brackish/saline water zone. The development of discharge heads due to simultaneous pumping of these wells restricts the mixing of freshwater with brackish/saline water.

The discharge rates of the two wells are adjusted in such a way that the up-coning caused by pumping from the production well could be countered by the down-coning of the interface caused by pumping from the scavenger well [20]. The resultant level of interface at any radial distance from the well can be evaluated from the difference of hydraulic heads of both production and scavenger wells considering both are working independently (**Figure 5a** and **b**). Ideally, this difference should be zero for keeping the interface at original level. Considering its pumped water quality, around 400 scavenger wells were installed in the lower Indus basin of Pakistan [13]. However, the concerns such as the mobilization of deep salts, disposal of pumped saline water, and the long-term negative environmental impacts due to seepage of saline water during its disposal were the major limitations on the sustainability of scavenger wells [13].

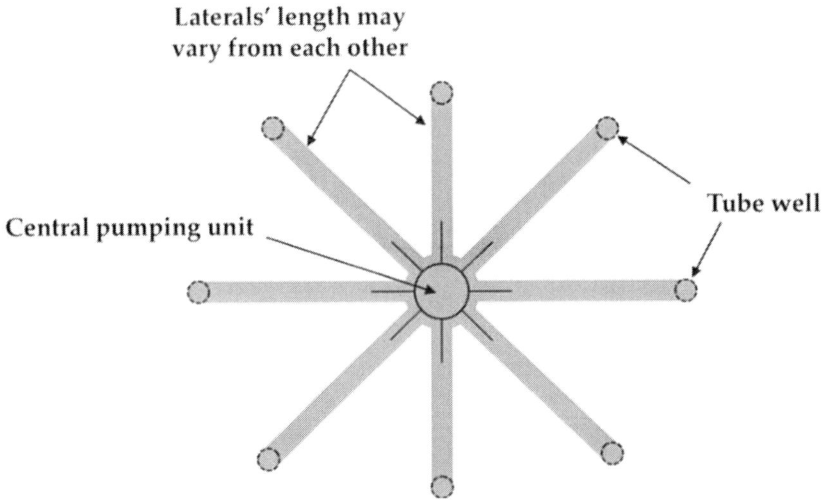

Figure 4B. Schematic sketch showing skimming tube wells system in which smaller capacity tube wells are installed arbitrarily in circular array and are joined with a single pumping unit.

No doubt, the answer to these major limitations was given days back in 1965 by Jacob who proposed and patented a doublet well system (**Figure 6**) which was different from the present day scavenger well system in the sense that the scavenging discharge is pumped back to deeper formation [22], but its acceptability was restricted to thick aquifers only. From the above discussion it is clear that the adaption of scavenger well system for inlands depends upon the safe disposal of pumped brackish water. On the other hand, scavenger well system may prove to be best fresh-saline aquifer management practice in the coastal regions as the disposal of pumped saline water is comparatively easy by constructing reasonably long seepage-proof conveyance channels.

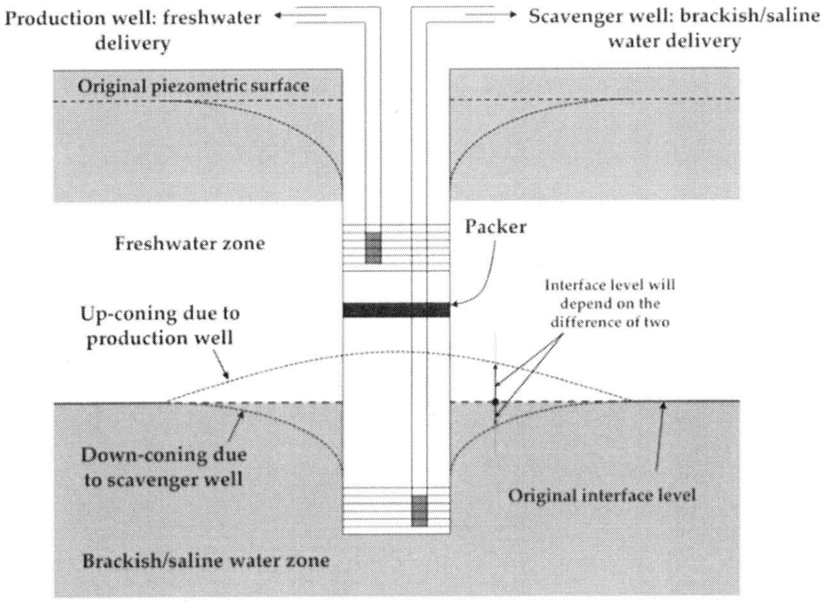

Figure 5A. Schematic sketch showing scavenger well system installed in a single-bore hole.

1.3. Recirculation well

The design of a recirculation well is somewhat similar to scavenger well (**Figure 7**). Unlike scavenger well, its lower screen is also kept in the freshwater zone. A part of the freshwater pumped from the upper screen is recirculated under gravity pressure into the lower screened chamber of the well which is separated from the upper through a packer. The injection of recirculated freshwater into the aquifer counters the up-coning of brackish/saline water mound beneath the well. Depending upon the position of lower screen with respect to interface, intermixing of injected freshwater may occur with the saline water. The establishment of flow pattern from the lower to upper screen takes this mixed quality water to freshwater zone, thus deteriorates the pumped water quality. Though the technique has been used in the petroleum industry [23], but the recirculation well's performance results in the water industry are not yet available [19].

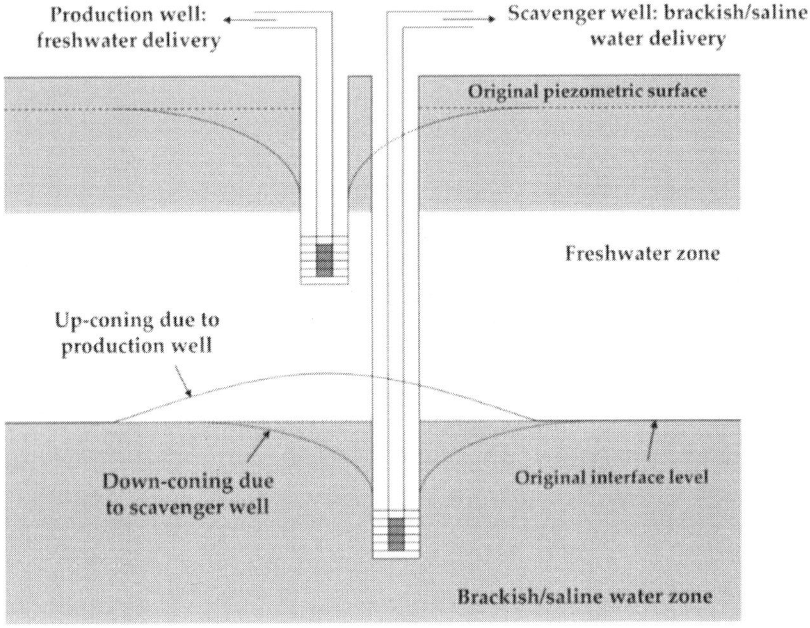

Figure 5B. Schematic sketch showing scavenger well system installed in side-by-side bore holes.

1.4. Horizontal wells

To control the rise of interface to reach the effective pumping zone of the abstraction well, the only feasible option is to limit its discharge head [8]. For the purpose, horizontal wells can be used to abstract groundwater. Previously, the horizontal well is introduced for the remediation of contaminated groundwater [24] as in comparison to vertical well, horizontal well can influence a large planar area during remediation [25]. However, now these wells are gaining popularity for use in water supply and drainage fields. Rushton and Brassington [26] have grouped the horizontal wells into two configurations. The first one is the radial collector well in which a number of horizontal perforated pipes are connected to a shaft or caisson. The second configuration consist of a single perforated pipe which is placed in shallow unconfined aquifer with a suction pump located either toward the center or at one end of the horizontal well [27].

The horizontal wells can be differentiated from the fully penetrating vertical wells on the basis of the flow components around the wells [28]. The vertical flow components are more dominant near a horizontal well whereas flow patterns around a fully penetrating vertical well are horizontal (or two dimensional). No doubt, in case of partially penetrating or partially screened wells, the flow in the near vicinity of pumping/recharging well is strictly three dimensional in nature [29]. From the above discussion, it can be envisaged that the radial collector wells may be the suitable option in the regions under study. But, its installation is difficult and requires extreme precision. Field experiences show that the construction of shaft, particularly in alkaline soils is a very hard task [30].

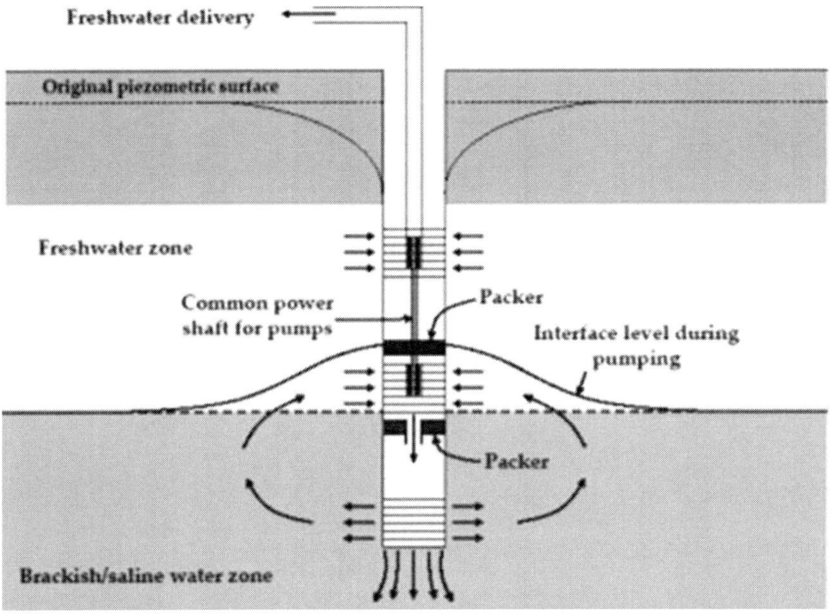

Figure 6. Schematic sketch showing the doublet well.

1.5. Dug wells

The areas where freshwater zones are very thin (less than 5 m) and where soils are not alkaline in nature, dug well can be constructed for skimming freshwater. Sometimes, these are also referred as shallow hand-dug wells [31]. Though the yield of these wells is very less, even then the available water can be efficiently utilized by opting well-recognized pressurized irrigation systems. The yield of a dug well can be increased by increasing its diameter and penetration. Large diameter of well facilitates water storage and its construction. However, the well penetration is limited by the fresh–brackish/saline interface.

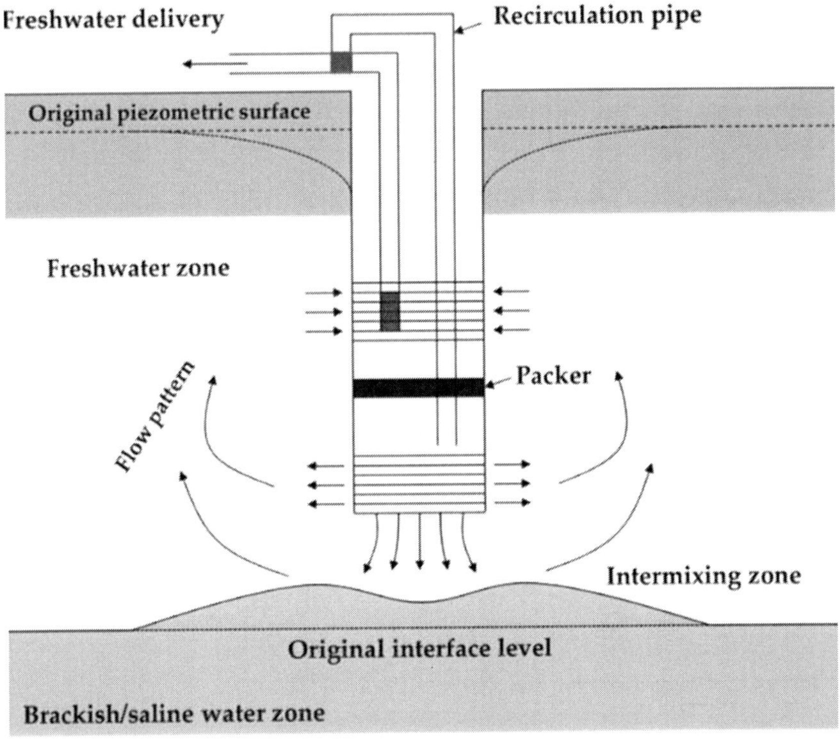

Figure 7. Schematic sketch showing recirculation well and its probable behavior while pumping.

1.6. Multiple well points system

In Sections 1.1–1.5, various skimming techniques are discussed. The performance of any skimming technique depends on the persisting hydrogeological conditions of area and every technique may not be suitable under all set of conditions. The major drawbacks in all the above-mentioned techniques are stated below:

a) Skimming tube well (i.e., single-bore well) is not feasible in aquifer having very thin freshwater zone.

b) In skimming tube wells system, various studies have concluded that the groundwater procured with fewer number of bore holes is of better quality in comparison to more number of bore holes [19, 32]. With the increase in the number of skimming wells, various types of geometrical patterns are possible in which these wells can be installed in the field. All the patterns create different flow conditions in the aquifer and hence affect the freshwater skimming. However, the adoption of a particular configuration without considering the resultant flow pattern toward the wells becomes responsible for the deterioration of pumped water quality. It is explained schematically in **Figure 8**. It is a well-known fact that during pumping, the adjoining portion of the well screen is recuperated by the aquifer formation under influence. It is obvious that during the pumping of the system the recuperation of the aquifer formation with freshwater in the enclosed portions (as shown in **Figure 8b–d**) is minimal as the major portion of the freshwater moving toward the system is trapped by the wells before reaching the enclosed portions. Moreover, the interference of wells causes the drawdown to increase to maximum extent in the enclosed portion which in turn raises the interface to effective pumping zone of the wells. Combined effect of both these factors which comes into action with the start of pumping is responsible for the up-coning of brackish/saline water. On the other hand, in case of two and in-line tube wells system (**Figure 8a** and **e**), recuperation rate of freshwater is much higher than the other enclosed patterns and thus the procured groundwater quality is noticeably better than the other patterns.

c) The major limitations on the sustainability of the scavenger wells are the mobilization of deep salts, disposal of pumped saline water, and seepage of saline water from its disposal system. The acceptability of the doublet well system is also restricted to thick aquifers only.

d) Though the recirculation wells are more efficient than scavenger wells, the intermixing of brackish/saline water with the injected freshwater and the rising of this mixed quality water to freshwater zone are few undesirable features.

e) Extreme precision is required in the installation of horizontal wells. Moreover, its construction (specifically in alkaline soils) is a very hard task.

f) The low-yield of the dug wells restricts its adaption in surface irrigation practices. The available water's utilization is limited to well-recognized pressurized irrigation systems only.

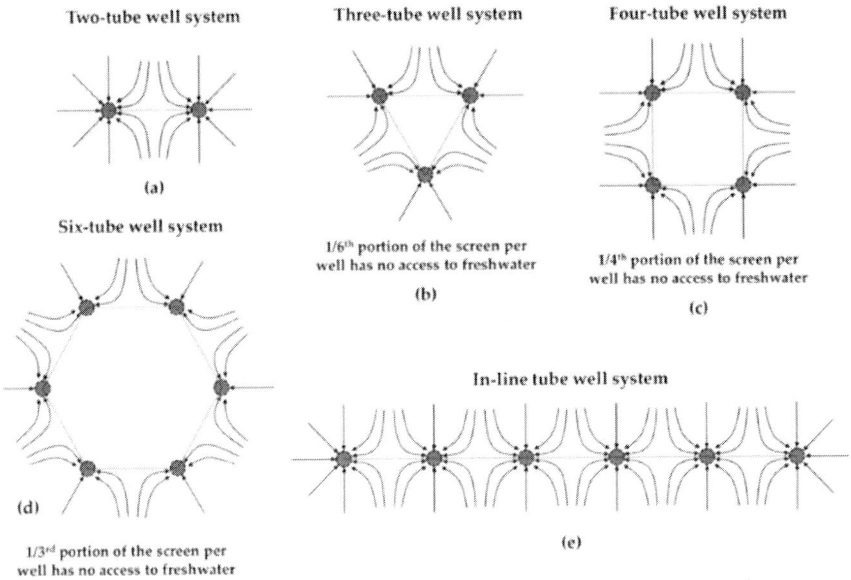

Figure 8. Schematic sketches showing general geometrical patterns in which skimming wells can be installed.

Keeping the region's problems in mind and considering the merits and demerits of all the above-mentioned skimming techniques, Shakya et al. [33] proposed the skimming of thin freshwater layers by a much efficient and affordable technique named multiple well points system (MWPS). The proposed system was capable to reclaim both types of problems (as mentioned in Section 1) persisted in the region. No doubt, skimming of freshwater was the basic objective but in comparison to other drainage technologies it can drain water logged soils by lowering the water table in wide spread region without disturbing much the deeper brackish water zone. The proposed MWPS was different from the skimming tube wells system [34] as mentioned in Section 1.2. In skimming tube wells system, different wells are installed almost in a circular array and are attached to the central pumping unit with separate lateral pipes of varied lengths (**Figure 4b**); whereas, MWPS constitutes several smaller capacities well points partially penetrating the aquifer and are arranged in a line connected to each other below the ground surface through single lateral and pumped centrally (**Figure 9**).

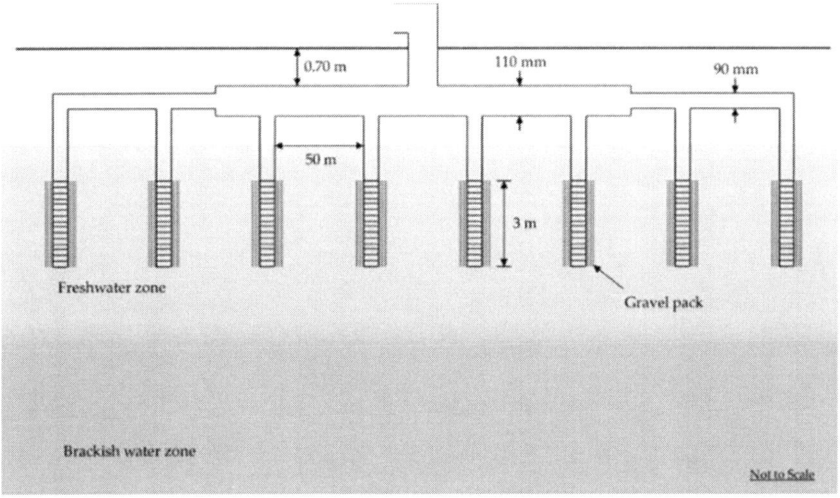

Figure 9. Schematic sketch showing multiple well points system with eight bores.

2. STUDY AREA

To examine the performance of conceived MWPS, an area which was out of cultivation was considered (as shown in **Figures 1** and **2**). The area is situated at a distance of 15 km from Ferozepur city on Ferozepur-Faridkot state highway No. 15 (**Figure 10**). The land of the project area belongs to three villages namely Golewala, Chak Kalan Tola, and Jhariwala (District Faridkot, Punjab, India). Project area falls in the arid zone having the minimum and maximum temperatures of 10 and 46°C during the months of January and June, respectively. The average values of the minimum and maximum relative humidity are 37 and 65%, respectively. The mean annual rainfall received in the region is about 400 mm. The topography of the area is leveled in general.

3. CHARACTERIZATION OF UNDERGROUND WATER

Before the installation of MWPS, depth-wise analysis of the groundwater quality was performed. In **Figure 11** the change in electrical conductivity (EC) and residual sodium carbonate (RSC) with depth is shown. Perusal of **Figure 11** shows that the groundwater quality deteriorates with depth.

Water can be considered fit for irrigation only if EC and RSC are less than 2 dS/m and 2.5 me/l, respectively. Though the salinity is within range but because of higher RSC, groundwater of the study area was not fit for irrigation. Hence, the project area was under the first type of problem.

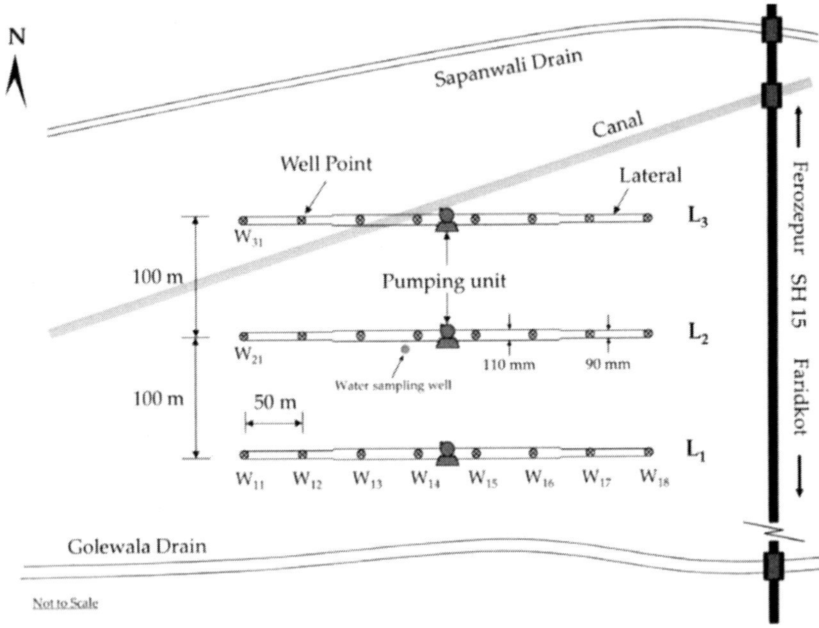

Figure 10. Schematic sketch showing the study area.

4. PUMPING TEST OF THE AQUIFER

A pumping test was conducted in a farmer's well which was existing in the project area. From the well log data, it was found that the depth of the soil at the location was 2.5 m and the aquifer was 10 m thick. Diameter of the well was 7.5 cm and it was installed in the said aquifer at a total depth of 10.5 m out of which 4.5 m was blind pipe of galvanized iron (GI). The well had a coir strainer of length 6 m. At a radial distance of 28.8 m from this well, another irrigation well of certain specifications was available which was used as a piezometer. Other than this available tube well, two more piezometers each of 9.0 m deep were installed at radial distances of 60 and 90 m from the test well, respectively. During the pumping test, the well was pumped at the rate of 10.82 l/s and drawdowns were recorded in all the three piezometers with respect to time. The change in drawdowns with respect to time at the three piezometric locations is presented in **Figure 12**. The recorded data were analyzed by the Walton method [35]. From the pumping test data, the average values of transmissivity *T*, hydraulic resistance *c*, and storage coefficient *S* were evaluated equal to 570 m^2/d, 17.04 d, and 0.0142, respectively [30]. The hydraulic conductivity of the overlying soil layers ranged from 1.4 to 2.0 cm/d.

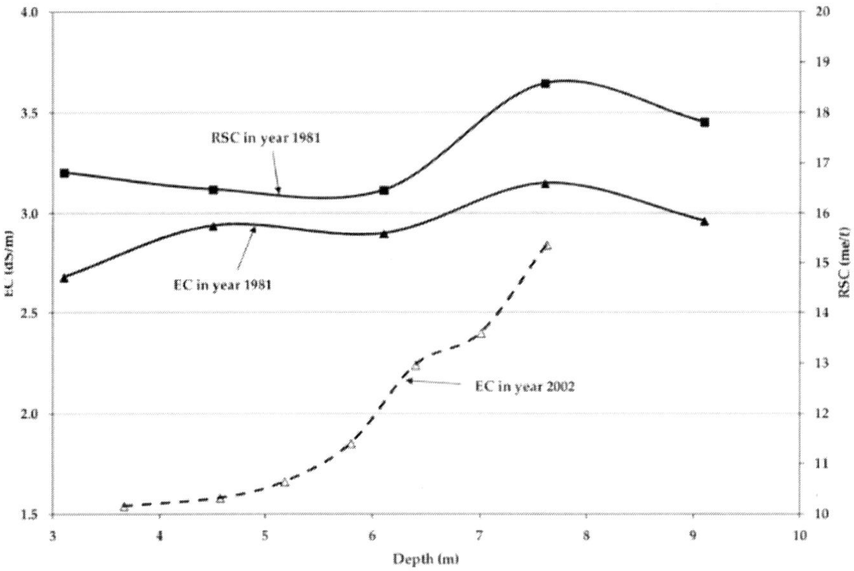

Figure 11. Change in EC and RSC of the groundwater with depth.

5. INSTALLATION OF MULTIPLE WELL POINTS SYSTEM

It was proposed to install a battery of 24 wells constituting three laterals L_1, L_2, and L_3 spaced 100 m apart. The well spacing along the laterals was considered to be 50 m with 8 wells on each lateral (**Figure 9**). The wells were installed by the local drillers. These wells were connected to each other at about 70 cm below the ground surface using 90 and 110 mm diameter PVC pipes by digging trenches. The screen of each well point had an effective perforated area of about 16% which is surrounded by pea size, clean, river bed, well rounded gravel pack of 30 mm thickness. In the middle of each 350 m long lateral, a T-section was provided which was raised to the ground surface such that it remained in level with the ground after placement of reflux valve. The centrifugal pumps were connected with the aforesaid outlet points of three laterals. In order to prime the lateral, a reciprocating pump was placed over the delivery of each centrifugal pump. Initially, the air from the lateral was removed through the reciprocating pump and when there was an adequate flow of water, centrifugal pump was started. Whenever, there was disruption of flow, reciprocating pump was operated and was stopped when there was continuity in flow. Before covering the laterals with soil, the complete system was tested for any leakage. From the desired discharge rate and available head values, the power requirement to operate the system was evaluated equal to ~4 hp (considering the working efficiencies of both centrifugal pump and diesel engine equal to 80% each). Therefore, three diesel engine operated pumps of capacity 5 hp each (nearest available higher size in the market) were installed on each lateral.

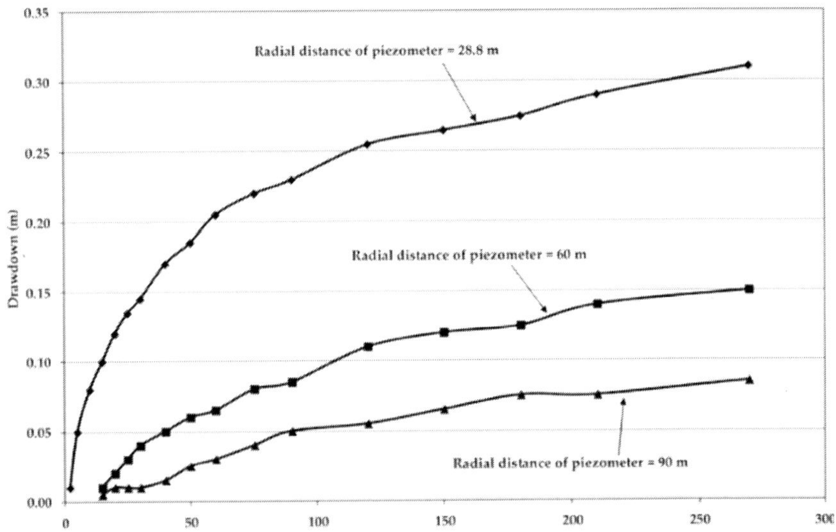

Figure 12. Change in drawdown with time at three piezometric locations while performing pumping test.

6. BEHAVIOR OF MULTIPLE WELL POINTS SYSTEM AND ITS APPLICABILITY

For monitoring the drawdown distribution in an aquifer while pumping through MWPS, piezometers were installed close to each well point and at the midpoint of the well spacing along the laterals. Each piezometer was of diameter 40 mm and was installed at a depth of 6 m. The piezometers installed near to well points were given names in accordance with their nearness to the respective well points. The symbol W_{LN} indicates the Nth well point on the Lth lateral. The piezometers adjoining to the respective well points are named as P_{LN}. For example, P_{23} indicates the piezometer near to the third well point on the second lateral (i.e., W_{LN}). The piezometers installed at the midpoint of two adjacent well points along the lateral are named as $P_{LN-L(N+1)}$. For example, P_{23-24} represents the piezometer installed midway between the well points W_{23} and W_{24}. The lateral L_1 was discharging at a rate of 3110.4 m^3/d whereas the laterals L_2 and L_3 were delivering at a rate of 2419.2 m^3/d each. The available drawdown data were plotted and is presented in **Figure 13**. It is evident from **Figure 13** that the lowering of the water table is minimal at the four locations (i.e., in the center of the laterals at the outer boundary of the system). On the other hand, maximum lowering has been noticed around the central well points of middle lateral. Considering a section along the lateral lines L_1 or L_3 for the water table profiles, it is evident that the difference between the maximum water level at the midway between the wells and that of inside the wells was varied from 0.5 m at outer well to 0.72 m for central wells. These variations for L_2 lateral were found to vary from 0.47 m at outer well to 0.87 m for central wells.

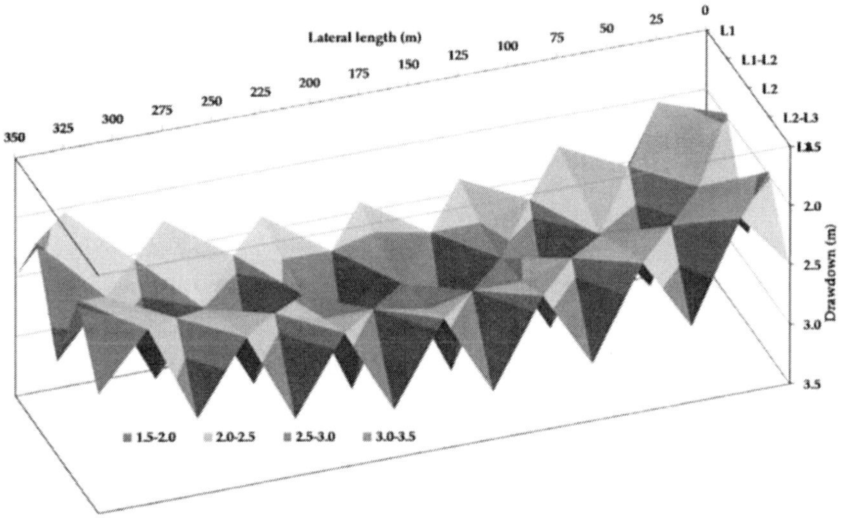

Figure 13. Behavior of multiple well points system while pumping.

Since the first type of problem was persisting in the project area, there was a need to control the water table first. Therefore, the pumping test was done with a schedule of 8 h of pumping and 16 h of recovery in a day for a week to observe the water table behavior in aquifer and soil. A residual drawdown of 17 cm in aquifer and 24 cm in soil was observed. It was also observed that the rise of water table in the soil during recovery is delayed with the passage of time. It could be explained as, once the water from soil is drained, rise would take place only after the pores of the soil are filled with water. It has imparted the delayed effect in rise of water tables in the soil. These encouraging results motivated us and long duration pumping of the aquifer was performed. Eventually, with the lowering of water table agricultural practices were started with the cultivation of paddy. Marginal quality water was used for irrigating the paddy which was achieved by mixing pumped water supply (through MWPS) with canal water. Studies show that the irrigation with marginal quality water does not affect the yield of paddy to much extent [30]. The motive was to continue pumping from the aquifer. In due course, groundwater quality in the upper portion of the aquifer started improving. In year 2002, groundwater samples were again analyzed for EC depth-wise and are presented in **Figure 11**. Perusal of the comparison clearly indicates that the there is a drastic improvement in the groundwater quality in the upper portion of the aquifer. The excellent performance by the MWPS has proven that it is a viable technology for the region and needs promotion.

7. RECOMMENDATION BASED ON EXPERIMENTAL RESULTS

In order to propagate the use of MWPS in the region, a smaller system with four-well points spaced at 6 m in a line connected horizontally at about 1 m below the land surface, pumped centrally was found to give technically satisfactory performance (i.e., no mixing of floating good quality water with the underlying poor quality groundwater) and economically viable results. Each well point has a screen length of 3 m which has 16% effective perforation and surrounded by pea size, well rounded clean river bed gravel envelope. Blind pipe may be adjusted according to the depth of aquifer. The details are shown in **Figure 14**. After watching the enthusiastic results of the project area, farmers of the region have adopted the recommended MWPS with four-well points on a large scale. As per the survey of Faridkot district in year 2000–2001, around 1400 MWPSs have already been installed [36]. From the analysis of water quality samples taken from some selected MWPSs, it was found that these are withdrawing good quality water that is fit for irrigation. Farmers themselves have admitted that they are using MWPS from last 10 years and have not noticed any deterioration in pumped water quality. Moreover, the adoption of MWPS has also lowered the water table in the region.

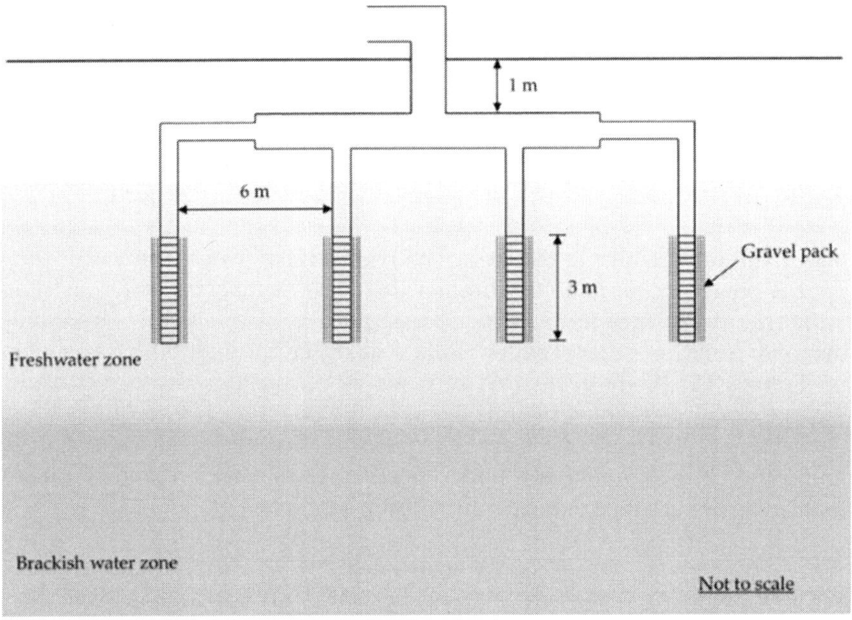

Figure 14. Multiple well points system with four bore holes.

8. GROUNDWATER RECHARGING THROUGH MWPS UNDER CONSTANT HEAD CONDITIONS

It is already mentioned in Section 7 that farmers of the region have adopted the MWPS in big way and their dependency on canal water has declined considerably. It is also true that these freshwater lenses are very limited and are now under stress. For long-term sustainability of these limited freshwater lenses, enhancement in freshwater recharging rate is required. Therefore, we have conceived an idea to recharge the freshwater zones with surplus canal water when available during the monsoon season through the existing MWPSs in the region [37]. Aquifer storage recovery (ASR) is worldwide recognized technique for managing brackish/saline aquifers [38] in which freshwater when available is stored in the aquifer and later on recovered for use. However, our proposal was little different from this basic technique. In the ASR technique, recharging of the brackish/saline aquifer is generally performed with the fully penetrating well of high capacity in which the injected freshwater displaces the saline water. As a result, considerable portion of the stored freshwater gets salinized before its recovery. The native groundwater movement and the buoyancy effect are the two factors which are responsible for the irrecoverable portion of stored freshwater. With the use of multiple partially penetrating wells in a single bore hole, the freshwater losses can be reduced considerably [39]. In this modified ASR technique, freshwater is injected in the deeper portion of the aquifer whereas its recovery is made from the shallow well. On the other hand, as the MWPS operates under limited head conditions and it penetrates only the upper portion of the freshwater zone of the aquifer through smaller capacity wells, it causes minimum disturbance to the denser native brackish water which is flowing in the lower portion of the aquifer [8]. The recharged water gets an opportunity to store in the wide-spread portion of the aquifer. Hydraulic head distribution in the aquifer while recharging through MWPS is shown in **Figure 15**. While evaluating this, we have considered the recharging rates through individual laterals equal to the pumping rates as mentioned in Section 6. Perusal of **Figure 15** indicates that the freshwater got an opportunity to store in the upper portion of the aquifer and it causes minimum disturbance to deeper brackish water zone.

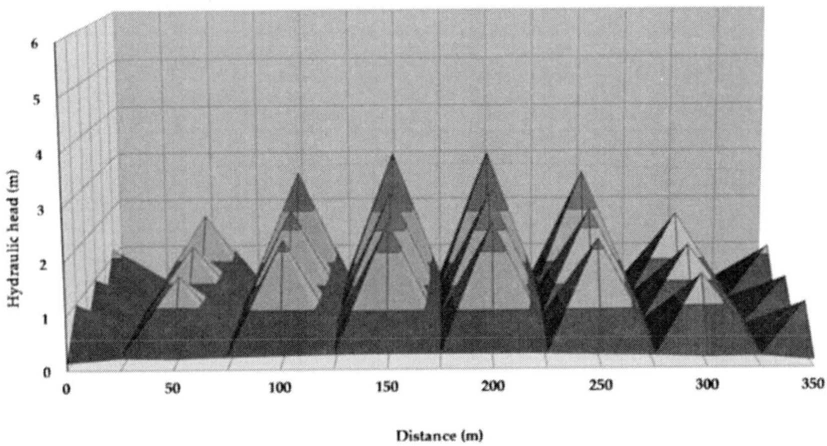

Figure 15. Hydraulic head distribution in the aquifer while recharging through MWPS.

Other than the theoretical investigations, field study was also conducted to monitor the change in quality of aquifer while recharging with freshwater. In the field study, canal water was allowed to recharge the aquifer through

MWPS under constant head conditions. Radial flow filter was used for removing turbidity from the canal water (**Figure 16**).

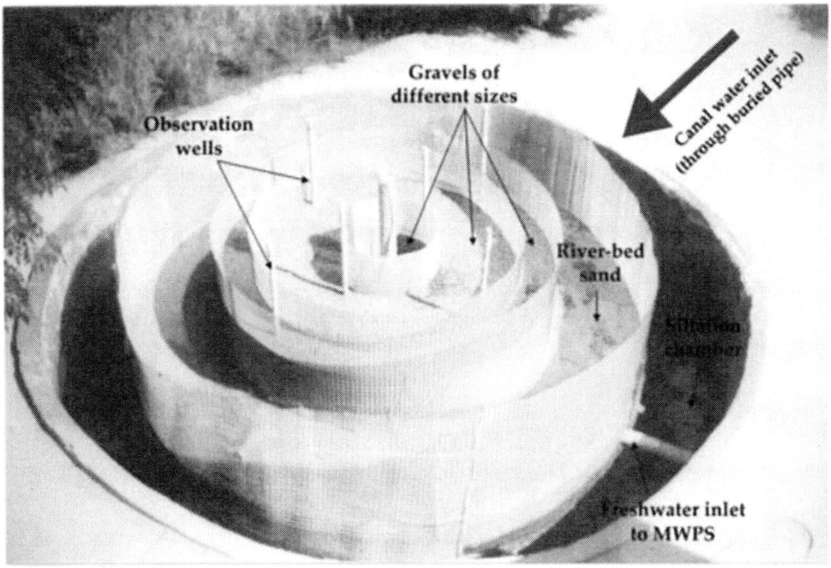

Figure 16. Radial flow filter for removing turbidity of canal water before recharging through MWPS.

For monitoring the change in quality of groundwater due to recharging, a water sampling well of length 9.144 m having diameter 10.16 cm was installed at a radial distance of 3.27 m from the well point W_{24}(**Figure 10**). The perforated length of the observation well was 6 m. For critically analyzing the effect of recharging on the quality of brackish water, it was necessary to collect water samples depth-wise with minimum disturbance. For this purpose, a specialized water sampler was fabricated. In spite of this, for achieving accuracy, considerable time was given between consecutive samplings. Before starting the recharging operation, groundwater samples from various depths were collected and analyzed for EC and pH. The recharging operation was continued for net 65 days and on its completion, water samples were collected from the same depths and were again analyzed for EC and pH. Comparison of the change in groundwater quality due to freshwater recharging is shown in **Figures 17** and **18.** Perusal of both the figures indicates that the recharging of freshwater in fresh–brackish aquifer through MWPS is a viable technique for reviving freshwater lenses in the region. However, for planning any recharging project through MWPS, theoretical investigations are required *a* *priori*. Realizing this need, Vashisht and Shakya [8] have proposed a semianalytical solution for evaluating hydraulic head distribution and recharging rate in a single leaky aquifer while recharging freshwater under constant head conditions through MWPS. The proposed solution was based on the introduction of a new relation termed as "*position-oriented opportunistic proportion*." The most important advantage of recharging through MWPS is that it covers the wide-spread region of the aquifer without disturbing much the native brackish water. The developed hydraulics for MWPS may be considered as the backbone for planning and execution of policies for managing brackish groundwater aquifers in the region.

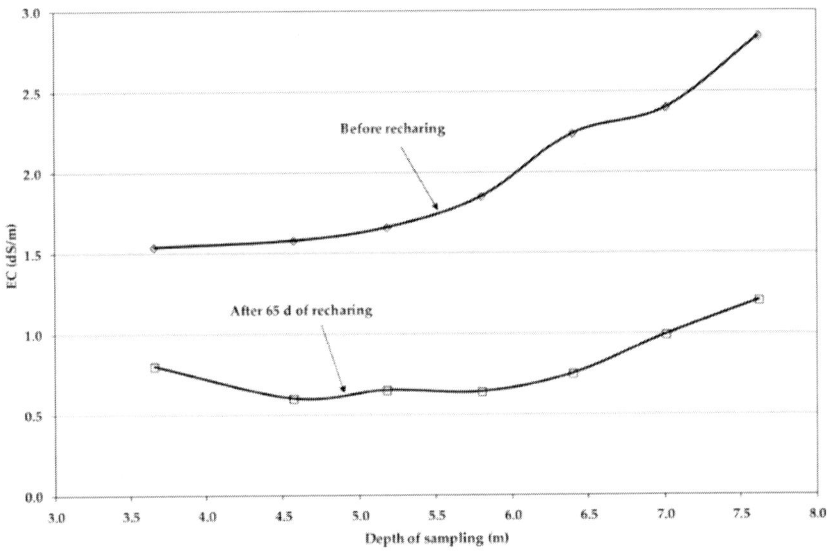

Figure 17. Change in EC of groundwater depth-wise after recharging canal water through MWPS.

9. CONCLUSION

Long-term field investigations were made on the MWPS for checking its performance as a drainage system, a skimming system, and a recharging system. On the basis of the field results and theoretical investigations, it is concluded that MWPS is a viable technique for managing brackish aquifers of a region.

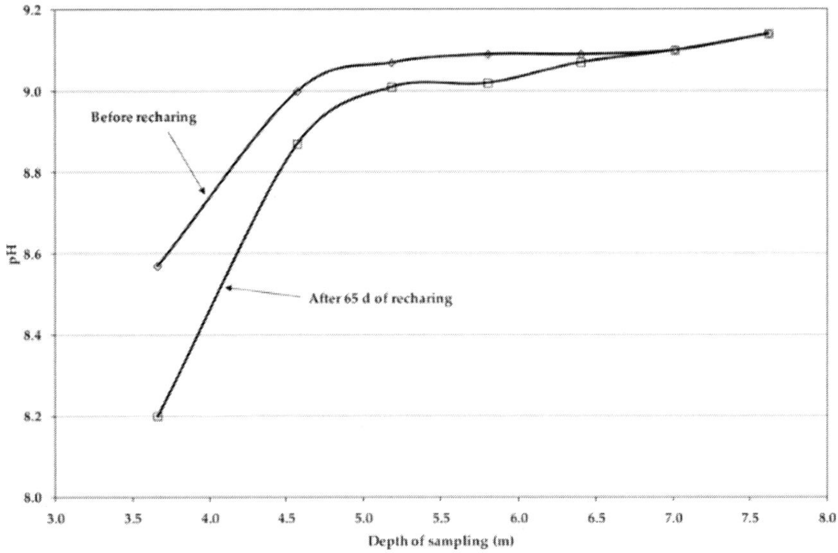

Figure 18. Change in pH of groundwater depth-wise after recharging canal water through MWPS.

ACKNOWLEDGEMENTS

The authors are grateful to the All India Coordinated Research Project on Agricultural Drainage under Actual Farming Conditions on Watershed Basis (Sponsored by Indian Council of Agricultural Research, New Delhi), Department of Soil and Water Engineering, Punjab Agricultural University, Ludhiana, Punjab, India for providing the relevant data related to MWPS for completing this chapter.

REFERENCES

1. Foster SSD, Chilton PJ. Groundwater: the processes and global significance of aquifer degradation. Philosophical Transactions of the Royal Society of London B: Biological Sciences. 2003; 358(1440): 1957–1972.
2. Shaffer DL, Yip NY, Gilron J, Elimelech M. Seawater desalination for agriculture by integrated forward and reverse osmosis: improved product water quality for potentially less energy. Journal of Membrane Science. 2012; 415: 1–8.
3. Foster SSD, Chilton PJ, Moench M, Cardy WF, Schiffler M. Groundwater in rural development: facing the challenges of supply and resource sustainability. World Bank Technical Paper 463; 2000.
4. Vashisht AK. Will the flow end? Artesian wells of Uttarakhand's Tarai. Economic and Political Weekly. 2014; 49(6): Web exclusives.

5. Zektser IS, Lorne E. Groundwater resources of the world and their use. In: IHP Series on Groundwater, No. 6, UNESCO; 2004.
6. Vörösmaty CJV, Green P, Salisbury J, Lammers RB. Global water resources: vulnerability from climate change and population growth. Science. 2000; 289(5477): 284–288.
7. Michaeli A. Brackish water resources in Israel. In: Proceedings of the Ninth Israel Symposium on Desalination, Israel; 1972. pp. 25–33.
8. Vashisht AK, Shakya SK. Development of semi-analytical solution for recharging through multiple well points system under constant head conditions. ISH Journal of Hydraulic Engineering. 2016; 22(1): 50–58.
9. Hahn J. Aspects of groundwater salinization in the Wittmund (East Friesland) coastal area. In: Hydrogeology of Salt Water Intrusion—A Selection of SWIM Papers, Hannover; 1991. pp. 251–269.
10. Wachyan E, Rushton KR. Water losses from irrigation canals. Journal of Hydrology. 1987; 92(3–4): 275–288.
11. Sharma HD, Chawla AS. Manual on canal lining. Technical Report No. 14. Central Board of Irrigation and Power, India; 1975.
12. Indian Bureau of Standards. Measurement of seepage losses from canals—code of practice. IS 9452 (part 1); 1993. pp. 1–6.
13. Ali G, Asghar MN, Latif M, Hussain Z. Optimizing operational strategies of scavenger wells in lower Indus basin of Pakistan. Agricultural Water Management. 2004; 66: 239–249.
14. Reilly TE, Goodman AS. Analysis of saltwater upconing beneath a pumping well. Journal of Hydrology. 1987; 89(3–4): 169–204.
15. Badon-Ghyben W. Nota in verband met de voorgenomen putboring nabij Amsterdam (Notes on the probable results of well drilling near Amsterdam). Tijdschrift van het Koninklijk Instituut van Ingenieurs, The Hague. 1888; 9: 8–22.
16. Herzberg A. Die Wasserversorgung einiger Nordseebäder (The water supply on parts of the North Sea coast in Germany). Journal für Gasbeleuchtung und verwandte Beleuchtungsarten, sowie für Wasserversorg. 1901; 44: 815–819.
17. Wang EC. Approximate theory of skimming well formation in the Indus Plain of West Pakistan. Journal of Geophysical Research. 1965; 70(20): 5055–5063.
18. Saeed MM, Asghar MN, Bruen M. Options for skimming fresh groundwater in the Indus basin of Pakistan: a review. Journal of Groundwater Hydrology (Japanese Association of Groundwater Hydrology). 2003; 11(3): 1–13.
19. Sufi AB, Latif M, Skogerboe GV. Simulating skimming well techniques for sustainable exploitation of groundwater. Irrigation and Drainage Systems. 1998; 12: 203–226.
20. Saravanan K, Kashyap D, Sharma A. Model assisted design of scavenger well system. Journal of Hydrology. 2014; 510: 313–324.
21. Jones CRC, van Wonderen JJ. Exploitation of fresh groundwater lenses: implementation of scavenger wells in Pakistan. In: Proceedings of the International Conference on Groundwater – Drought, Pollution and Management. Brighton, UK; 1994.
22. Wickersham G. Review of C.E. Jacob's doublet well. Groundwater. 1977; 15(5): 344–347.
23. Smith CR, Pirson SJ. Water coning control in oil wells by fluid injection. Society of Petroleum Engineers Journal. 1963; 3(04): 314–326.
24. Cleveland TG. Recovery performance for vertical and horizontal wells using semianalytical simulation. Groundwater. 1994; 32(1): 103–107.
25. Sawyer CS, Lieualen-Dulam KK. Productivity comparison of horizontal and vertical groundwater remediation well scenarios. Groundwater. 1998; 26(1): 98–103.
26. Rushton KR, Brassington FC. Hydraulic behaviour and regional impact of a horizontal well in a shallow aquifer: example from Sefton coast, northwest England (UK). Hydrogeology Journal. 2013; 21: 1117–1128.
27. Brassington FC, Preene M. The design, construction and testing of a horizontal well point in a dune sands aquifer as a water source. Quarterly Journal of Engineering Geology and Hydrogeology. 2003; 36: 355–366.
28. Langseth DE, Smyth AH, May J. A method for evaluating horizontal well pumping tests. Groundwater. 2004; 42(5): 689–699.

29. Vashisht AK, Shakya SK. Hydraulics of a drainage well fully penetrating a leaky aquifer through a multisection screen. Journal of Hydraulic Engineering (ASCE). 2013; 139(12): 1258–1264.
30. Shakya SK. Agricultural drainage under actual farming conditions on watershed basis. Department of Soil and Water Engineering. Punjab Agricultural University, Ludhiana, India; 2002. p 108.
31. Mace RE. Estimation of hydraulic conductivity in large-diameter, hand-dug wells using slug-test methods. Journal of Hydrology. 1999; 219: 34–45.
32. Hafeez A, Piracha ZA, Ahmad N. Multi-strainer tubewells for skimming top layer of freshwater underlain by saline aquifer. Mona Reclamation Experimental Project, Water and Power Development Authority (WAPDA), Bhalwal, Pakistan; 1986. Publication No. 152. p. 50
33. Shakya SK, Singh SR, Panda SN, Gupta PK. Drainage by multiple well point system for land and aquifer management. In: Proceedings of the National Symposium on Drainage and Reclamation of Salt Affected Waterlogged Land, CSSRI Karnal; India; 1986.
34. Saeed MM, Ashraf M. Feasible design and operational guidelines for skimming wells in the Indus basin, Pakistan. Agricultural Water Management. 2005; 74: 165–188.
35. Kruseman GP, De Ridder MA. Analysis and evaluation of pumping test data. Institute for land reclamation and improvment, Wageningen, The Netherlands; 1970. Publication No. 11.
36. Vashisht AK, Shakya SK. Managing saline/sodic soils and brackish water aquifers using multiple well point system. Success Story—II. Agricultural Engineering Today. 2005; 29(3–4): 67–72.
37. Vashisht AK. Modeling brackish water management by recharging through multiple well points. Ph.D. Thesis. Punjab Agricultural University, Ludhiana, Punjab, India; 2004.
38. Bakker M. Radial Dupuit interface flow to assess the aquifer storage and recovery potential of saltwater aquifers. Hydrogeology Journal. 2010; 18(1); 107–115.
39. Zuurbier KG, Zaadnoordijk WJ, Stuyfzand PJ. How multiple partially penetrating wells improve the freshwater recovery of coastal aquifer storage and recovery (ASR) systems: a field and modeling study. Journal of Hydrology. 2014; 509: 430–441.

CHAPTER 8

Understanding the Role of Groundwater in a Remote Transboundary Lake (Hulun Lake, China)

Hongbin Gao 1, M. Cathryn Ryan 2, Changyou Li 1,* and Biao Sun 1

[1] *College ofWater Resources and Civil Engineering, Inner Mongolia Agricultural University, 306 Zhaowuda Rd., Hohhot 010018, China*

[2] *Department of Geoscience, University of Calgary, Calgary, AB T2L 1N4, Canada*

ABSTRACT

Hulun Lake, located in a remote, semi-arid area in the northeast part of Inner Mongolia, China, shares a transboundary basin with Mongolia and supports a unique wetland ecosystem that includes many endangered species. Decadal scale decreases in the lake stage and increased salinity make an understanding of the lake's water and salt sources critical for appropriate design of strategies to protect and manage the lake. Multiple tracers (chloride, and $\delta^{18}O$ and $\delta^{2}H$ in water) in samples collected from lake water, rivers, and nearby water wells were used in conjunction with an annual water balance based on historic data to better understand the lake's major water and salt sources. The average annual water balance was conducted for two time periods: 1981–2000 and 2001–2013. The contribution of river discharge to the annual lake input decreased by half (from 64% to 31%) between the two time periods, while the volumetric contribution of groundwater discharge increased four-fold (from about 11% to about 50% of the total lake input). Significant evaporation was apparent in the stable isotope composition of the present-day lake water, however, evaporation alone could not account for the high lake water chloride concentrations. Limited domestic well water sampling, a regional salinity survey, and saline soils suggest that high chloride groundwater concentrations exist in the region south of the lake. The chloride mass balance suggested that groundwater currently contributes more than 90% of the annual chloride loading to the lake, which is likely four times greater than the earlier period (1981–2000) with lower groundwater input. The use of water and chloride mass balances combined with water isotope analyses could be applied to other watersheds where hydrologic information is scarce.

Keywords: groundwater; evaporative loss; stable isotope; chloride; salt; mass balance; Hulun Lake

1. INTRODUCTION

Lakes and reservoirs are critical water resources for local ecosystem conservation and socioeconomic development, especially in arid and semi-arid regions. However, climatic changes and recent overallocation of water resources have affected the hydrological cycle of these water systems, leading to decreased water availability, decreasing water levels (in groundwater and surface water), water quality deterioration, and land desertification [1,2]. Sustainable management of water resources requires an understanding of the inputs and outputs to a water system and the dynamics of how groundwater and surface water interact with each other.

Hulun Lake, located in a remote, semi-arid area in the northeast part of Inner Mongolia, China (Figure 1a), is the fifth largest lake in China. It supports a unique wetland ecosystem that includes many endangered species, and is a Ramsar Wetland of International Importance included within UNESCO's World Network of Biosphere Reserves [3]. An approximately four-meter water level decline (observed since 2001) has resulted in a rapidly shrinking lake size [4]. Sparse discharge data indicate inflow from the two rivers (Kelulun River and Wuerxun River) that flow into the lake has decreased from 1.75 billion m^3 in 1999 to 0.25 billion m^3 in 2011 [5]. Low river flows have contributed to decreased water levels, which have caused Hulun Lake to become endorheic (i.e., lacking any surface water outflow). No outflow has occurred through Hulun Lake's outlet (the Xinkai River) since 2000 [3], and lake water salinity has increased from 1466 mg/L in 1999 to 2395 mg/L in 2011 [6].

Moreover, the lake has suffered from visible eutrophication, impacting its ecological function [7].

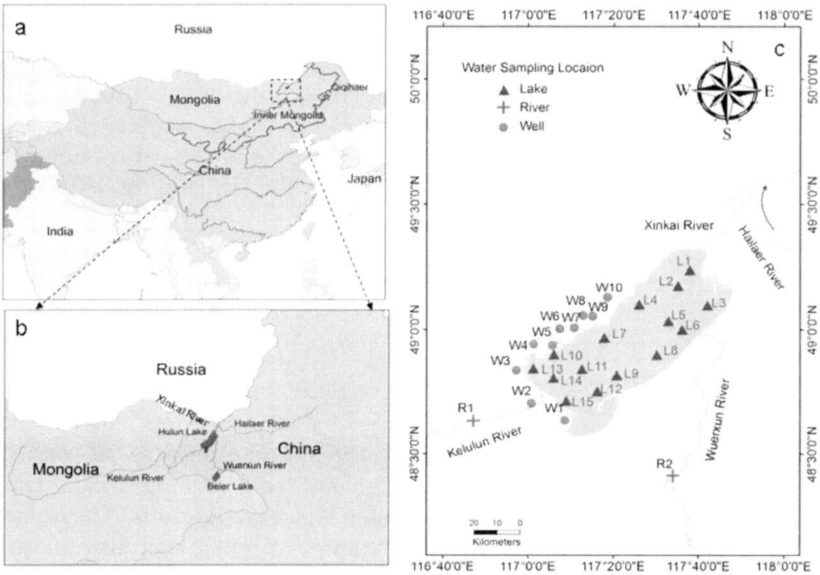

Figure 1. (**a**) Study area in country-scale map, with a green star indicating the location of the International Atomic Energy Agency's Global Network of Isotopes in Precipitation (Qiqihar station); (**b**) Hulun Lake basin showing two major rivers (Kelulun River and Wuerxun River) that flow into Hulun Lake, and the Xinkai River, which intermittently provides an outlet for Hulun Lake to the Hailaer River; (**c**) River, domestic water well, and lake sampling site locations. The arrow near Hailaer River shows the flow direction of Hailaer River. The Xinkai River is an intermittent connection between Hulun Lake and the Hailaer River.

Concern over the declining lake water level and deteriorated water quality have prompted research [3,5,8,9,10,11] which has linked decreased water level to climate change. However, few lake water budget studies were conducted due to sparseness of available hydrologic and geologic information. Although a significant groundwater contribution to Hulun Lake's annual water balance has been implicated [8,9,10], no clear lines of evidence have been used to evaluate the relative magnitude of groundwater contribution to lake input, nor the potential groundwater contribution to increasing lake salinity. In particular, although the stable isotopes of water can provide good insight into lake water balances and associated processes [12,13,14], they have not been applied to the Hulun Lake watershed.

Isotopic investigations are widely acknowledged as powerful approaches to study the hydrologic cycle in ungauged or poorly gauged watersheds [15]. A combination of hydrochemical tracers and stable isotopes approaches can provide insight into hydrological processes, including apportionment between water sources (including groundwater contribution), and their contribution to lake water balances [16,17,18,19,20,21,22]. In particular, the increase in the values of stable isotopes in water (δ^2H and $\delta^{18}O$) with the increasing extent of evaporative losses [23] can be incorporated into isotope-mass balance determinations to quantify water balances [24]. Similarly, conservative ions, such as chloride, can provide an independent line of evidence to estimate evaporative loss when it is reasonable to assume increased chloride concentrations result from evaporation only [25].

The objective of this study was to advance the understanding of the role of groundwater in the Hulun Lake water balance and how this role has changed over time, and to evaluate the relative sources of salt, or chloride, to the present-day lake. The sparse hydrologic data were applied to construct water budgets over two time periods. The sparse historic data were combined with stable isotopic composition and chloride concentrations in samples collected from the lake, nearby wells, and the two inflowing rivers in 2014.

2. MATERIALS AND METHODS

2.1. Study Area

Hulun Lake, located in a remote, semi-arid area in the northeast part of Inner Mongolia, China (Figure 1a), is the fifth largest lake in China, with a maximum area of 2339 km^2 and a maximum water depth of 8 m in 1963 [26]. Although the lake is located in China, about 64% of the lake's total basin is located in Mongolia [27]. Lake level data between 1981 and 2013 showed variable, but overall steady levels between 1981 and 2000, and steadily decreasing levels between 2000 and 2011, with a total lake level decrease of about four meters [4].

Although largely ungauged, discharges from Kelulun and Wuerxun Rivers (Figure 1b) are the main lake water sources, and have been historically thought to be the main control on the lake water level and size [9]. Most of the Kelulun River's basin is located within the Khentii Mountains in Mongolia. The Wuerxun River's source is Beier Lake, which is an important natural fishery (Figure 1). The intermittent Xinkai River, located at the northeastern extent of Hulun Lake (Figure 1c), provides a sporadic outlet from Hulun Lake, transmitting the lake discharge to the Hailaer River, which flows north from its intersection with the Xinkai River, marking the border between China and Russia. The Xinkai River outlet occurs when Hulun Lake's elevation exceeds 544.8 m a.s.l. (above sea level) [21]. At lower lake water levels, there is no lake outlet, and Hulun Lake is a closed, or endorheic basin, with evaporation providing the main source of water loss. There has been no outflow from Hulun Lake since 2000 [3].

The study area is located in a semi-arid steppe area of the mid-temperate zone, with a climate influenced by the prevailing westerly winds and the East Asian monsoon. In the past 50 years, mean annual precipitation ranged between 247 and 319 mm, with 80%–86% of the annual precipitation falling between June and September. Mean annual air temperature is 0.3 °C, with the highest average monthly temperature in July 20.3 °C, and the lowest average monthly temperature in January −21.2 °C. Mean annual evaporation ranged between 1400 and 1900 mm, which is 5–6 times greater than the mean annual precipitation [28].

Hulun Lake is mainly surrounded by broad lacustrine and alluvial plains scattered with stable and semi-stable sand dunes along the southern and eastern shores. Mesozoic volcanic rocks form low mountains (maximum elevation of 850 m a.s.l.) that are located along a fault system on the northwest axis of the lake [28]. Extensive wetlands in the peripheral lake areas and shallow (hand-dug) domestic water wells suggest that a shallow groundwater table is present in the lake periphery.

2.2. Field Sampling and Analytical Methods

Water samples were collected from the lake, the two inflowing rivers, and the domestic water wells west and south of the lake between June and August of 2014 (Figure 1c). The samples were field filtered through 0.45-μm filters and stored in tightly capped plastic bottles for transport. Chloride analyses were conducted by ion chromatography using an ICS-90 chromatograph (Dionex, Sunnyvale, CA, USA) with an analytical uncertainly of ±1%. The hydrogen and oxygen stable isotope analyses were conducted using a CO_2–H_2 equilibration unit coupled with an Isotope Ratio Mass Spectrometer (IRMS; Finnigan MAT Delta S, ThermoFisher, Waltham, MA, USA). The standard deviation in this analysis was 0.1 for $\delta^{18}O$ and 1.0 for $\delta^{2}H$. The isotope values are expressed as deviations in per mill (‰) relative to standard values of Vienna Standard Mean Ocean Water (VSMOW), shown as below:

$$\delta[‰] = \left[R_{sample}/R_{standard} - 1\right] \times 1000 \quad (1)$$

Although the Kelulun River was sampled at one location (Figure 1) on four dates, the Wuerxun River (which is not easily accessed) was sampled only once.

2.3. Average Annual Lake Water Balance

An average annual water balance was conducted on Hulun Lake using available hydrologic data to evaluate the role of groundwater in the lake hydrological system. Historic river discharge data for the Kelulun and Wuerxun Rivers, monthly pan evaporation, and precipitation data were available from the Hulunbeier Hydrological Station in Inner Mongolia, China (Supplementary data). Surface water runoff in the Hulun Lake basin was assumed to be negligible given the high degree of vegetation coverage, permeable soils in this area, and absence of visual evidence. The lake inflow thus includes precipitation, two rivers (Kelulun and Wuerxun), and groundwater. Lake water outflow consists of mainly evaporation and periodic discharge via the Xinkai River (i.e., when the Hulun Lake stage elevation exceeds 544.8 m a.s.l.). The average annual water balance equation of Hulun Lake can be expressed as:

$$\Delta S = (P - E) \cdot A + Q_{in} - Q_{out} + \Delta V \quad (2)$$

where ΔS is the variation in annual lake water storage (m^3), estimated as the change in lake level multiplied by the lake area; A is the lake area (m^2) interpreted by Landsat image; P is the annual volume of precipitation (m); E is the annual volume of lake evaporation (m); Q_{in}, is total annual river discharge (m^3) from the Kherlen and Wuerxun Rivers as interpolated by sparse data; Q_{out} is the annual discharge of the Xinkai River (which was assumed to have an annual discharge of 1.44×10^8 m^3 when the lake level exceeded 544.8 m [26]); and ΔV (m^3) is the annual groundwater component of the water balance, which is the only component that cannot be directly measured or indirectly estimated using available data.

Hulun Lake evaporation was estimated by using monthly pan-evaporation data from the Hulunbeier Hydrological Station, and an empirical correlation coefficient of 0.61 estimated from measured pan evaporation and open lake evaporation during a long-term field experiment at Hulun Lake from 1974 to 1984 [8].

2.4. Estimation of Evaporative Loss Based on the Stable Isotope Composition

The use of stable isotopes in water to quantify evaporative losses is essential in remote and semi-arid areas, where evaporative losses are significant and data are scarce. This approach takes advantage of the fractionation effect on hydrogen and oxygen that occurs at the interface between liquid and vapour during evaporation. The calculation of evaporative loss based on stable isotopes in residual water was initially founded on the Craig–Gordon model [29], which described the change in stable isotope composition of residual water and vapour during the evaporation process. The Craig–Gordon model has been reformulated to express these changes as a ratio of evaporation (*E*) to inflow (*I*) [30,31,32,33]. In this reformulated case, the ratio of evaporation to inflow in a lake is:

$$E/I = \left[\frac{(\delta_L - \delta_I)}{(\delta^* - \delta_L) \times m}\right] \tag{3}$$

where δ_L is the isotopic composition of residual lake water (i.e., Hulun Lake, ‰), δ_I is the isotopic composition of inflowing water (‰), δ^* is the limiting isotope composition (or the isotopic composition after which further evaporation does not result in increased isotopic enrichment of the residual water, defined below, in ‰), and *m* is the slope calculation factor (also defined below) [34,35].

The limiting isotopic composition enrichment factor, δ^*, which is a function of air humidity (*h*), the isotope composition of moisture in ambient air (δ_A), and the enrichment factor (ε), is defined as [34,35]:

δ∗=h×δA+εh−ε1000 (4)

Since the isotopic composition of moisture in ambient air (δ_A in Equation (4)) is difficult to measure in the field, the value is usually estimated from the isotopic composition of local precipitation (δ_P) based the isotopic equilibrium between δ_P and δ_A [31,32,33,34,35,36].

$$\delta^* = \frac{h \times \delta_A + \varepsilon}{h - \frac{\varepsilon}{1000}} \tag{5}$$

where α^+ is the equilibrium liquid–vapour isotope fractionation factor, and ε^+ is the temperature-dependent equilibrium isotope fractionation factor.

The slope calculation factor (m) in Equation (3) is defined [37,38] as:

$$\delta_A = (\delta_P - \varepsilon^+)/\alpha^+ \tag{6}$$

where ε_k is the kinetic isotope fractionation factor [30].

In Equations (4) and (6), ε is a "total isotope fractionation factor, equal to the sum of the equilibrium isotope fractionation factor (ε^+) and the kinetic isotope fractionation factor (ε_k)" [39]:

$$m = \frac{h - \frac{\varepsilon}{1000}}{1 - h + \frac{\varepsilon_k}{1000}} \tag{7}$$

The temperature-dependent equilibrium isotope fractionation factor, ε, is defined as $\varepsilon = 1000(1 - \alpha^+)$, where α^+ is given below as a function of the interface temperature (°K) [40] for each isotope as follows:

$$10^3 \ln(\alpha^+) = 1158.8 \times \left(\frac{T^3}{10^9}\right) - 1620.1 \times \left(\frac{T^2}{10^6}\right) + 794.84 \times \left(\frac{T}{10^3}\right) - 161.04 + 2.9992 \times \left(\frac{10^9}{T^3}\right) \tag{8}$$

for δ^2H, and

$$10^3 \ln(\alpha^+) = -7.685 + 6.7123 \times \left(\frac{10^3}{T}\right) - 1.6664 \times \left(\frac{10^6}{T^2}\right) + 0.35041 \times \left(\frac{10^9}{T^3}\right) \tag{9}$$

for $\delta^{18}O$.

The kinetic enrichment factors for each isotope, ε_k, are dependent on the boundary layer conditions and the humidity deficit:

$$\varepsilon_k = 12.5(1-h) \times 10^{-3} \tag{10}$$

for δ^2H, and

$$\varepsilon_k = 14.2(1-h) \times 10^{-3} \tag{11}$$

for $\delta^{18}O$.

A freely available and validated software, *Hydrocalculator*, was used to calculate the evaporative loss from lake water based on the equations above [41]. The stable isotopic composition of precipitation (δ_P) was estimated as average values measured in precipitation from the nearest

International Association of Atomic Energy's Global Network of Isotopes in Precipitation station (Qiqihar, China (Figure 1); $\delta_P{}^2H$ = −113.94‰, $\delta_P{}^{18}O$ = −15.00‰ [42]). The isotopic compositions of inflowing water for Hulun Lake was estimated as the intersection of the evaporation line with the LMWL (Local Meteoric Water Line) ($\delta_I{}^2H$ = −88.94‰, $\delta_I{}^{18}O$ = −11.70‰), which was taken to represent the weighted mean composition of water input to the lake [24]. Values for mean air temperature (T = 0.3 °C) and relative humidity (h = 0.6) measured at the Hulunbeier Hydrological Station were used. The stable isotopic composition of moisture in ambient air (δ_A) was estimated by ε^+ and α^+, where ε^+ was 111.31‰ for δ^2H and 11.78‰ for $\delta^{18}O$, α^+ was 1.1113 for δ^2H and 1.0118 for $\delta^{18}O$.

3. RESULTS AND DISCUSSION

3.1. Changes in Hydrological Budget and Net Groundwater

Lake water level variations between 1981 and 2013 were separated into two periods: 1981–2000 (when lake levels were relatively stable [4]) and 2001–2013 (when lake levels showed an overall decline of about 4 m) for the water balance calculations (Table 1). The water was balance reflected in the lake level changes, and was significantly different for the two periods. In particular, the annual difference in water storage (ΔS) changed from a surplus in the earlier period (when outflow to the Xinkai River occurred sporadically) to a deficit in the later period, when outflow to the Xinkai River was infrequent, and lake levels gradually declined. The contribution of river water to the annual lake inputs (Q_{in}) decreased from 63.6% in the earlier period to 30.7% in the later period. The relative contribution of precipitation to the annual water budget decreased slightly from 24.9% in the earlier period to 19.4% in the later period. In contrast, evaporation increased by 8%, from 20.4×10^8 m^3 in the earlier period, to 22.0×10^8 m^3 in the latter period.

Although the calculated contribution of groundwater (ΔV) increased from 11.4% in the earlier period to 49.9% in the later period (Table 1), the increase was apparently insufficient to mitigate lake water storage losses (ΔS) in the face of significantly decreased river flows, decreased precipitation, and increased evaporation. The increased contribution of groundwater in the latter period, when the lake level was lower, reflected an increased hydraulic gradient between regional groundwater table and lower lake level.

Table 1. Average annual water budget components in the entire period of 1981–2013, and also separated into two periods (1981–2000 (when lake stage was stable), 2001–2013 (when lake stage was declining)) for Hulun Lake (in 10^8 m^3). The change in lake storage (ΔS), volume of river discharge into the lake (Q_{in}), volume of river outflow discharge (Q_{out}), contribution from precipitation (P), and loss from evaporation (E) are considered. The residual amount is the estimated contribution from groundwater discharge into the lake ΔV (GW). Negative values for storage mean a net decrease in storage over the year. The fraction of total lake input is included for the three main lake water sources (Q_{in}, P, and ΔV (GW)).

Year		ΔS	Q_{in}	Q_{out}	P	E	ΔV (GW)
1981–2013	Water budget (10^8 m^3)	−0.6	10.8	0.3	4.8	21.1	5.2
	% of total lake input	– *	52.2	–	23.0	–	24.8
1981–2000	Water budget (10^8 m^3)	1.8	14.5	0.6	5.7	20.4	24.8
	% of total lake input	–	63.6	–	24.9	–	11.4
2001–2013	Water budget (10^8 m^3)	−4.3	5.4	–	3.5	22.0	8.9
	% of total lake input	–	30.7	–	19.4	–	49.9

* – indicates that the values do not apply.

3.2. Isotopic Composition and Chloride Concentrations in Present Day Hulun Lake and Its Water Sources

The chloride concentrations and isotopic ($\delta^{18}O$ and δ^2H) compositions of lake, river, and well water are shown in Table 2, and a dual isotope plot of $\delta^{18}O$ and δ^2H is shown in Figure 2. The LMWL is defined by the precipitation data from the nearest International Association of Atomic Energy's Global Network of Isotopes in Precipitation station in Qiqihar, China [42]. The three types of water samples are plotted distinctly on the dual isotope plot. The domestic water well samples are plotted along the LMWL, reflecting meteoric water that has recharged the groundwater without undergoing any significant precipitation. The wide range in domestic well water isotopic composition (i.e., $\delta^{18}O$ values ranging from −15.91‰ to −9.41‰, and δ^2H values ranging from −120.87‰ to −72.02‰) suggested groundwater has been recharged at varying temperatures or elevations, with the samples with lower values (e.g., W9 and W10) likely representing groundwater that was recharged at significantly higher elevation [43], perhaps in the mountains northwest of Hulun Lake. Six domestic well water samples form a cluster on the dual isotope diagram (i.e., with $\delta^{18}O$ values ranging from ~−12.33‰ to −11.40‰, and δ^2H values ranging from −93.05‰ to −86.86‰) near the river water samples, and seem to represent more local recharge. The lake water samples fall on a line that originates from the cluster of domestic well water and river water samples, with a slope (5.32) that is less than that of the meteoric water line (7.59) and is consistent with lake evaporation in a semi-arid climate [44].

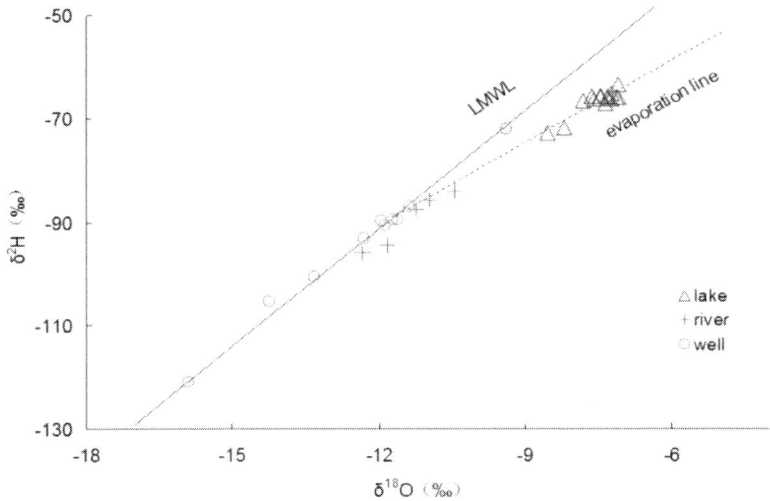

Figure 2. Dual isotope diagram of $\delta^{18}O$ and $\delta^{2}H$ in lake, river, and domestic water well samples. The local meteoric water line (LMWL) is shown for the International Atomic Energy Agency's Global Network of Isotopes in Precipitation, Qiqihar station [42].

A cross plot of chloride concentrations vs. $\delta^{18}O$ values of lake, river, and domestic well water samples is shown in Figure 3. Higher chloride concentrations and $\delta^{18}O$ values were found in the lake relative to the rivers, which is consistent with more extensive evaporation of the lake water relative to the rivers. The domestic well water samples, which did not show any evidence of evaporation (Figure 2), are plotted in two zones in Figure 3. Zone I contains domestic well water samples with a wide variety of isotopic compositions (thought to represent varying recharge elevation) but consistently low chloride concentrations (i.e., less than 70 mg/L). These samples (W4 to W10 in Table 2) are from domestic water wells on the west side of Hulun Lake. In Zone II, relatively high chloride values (>330 mg/L) occurred in three wells south of the lake (W1 to W3 in Table 2), however, the stable isotope values are not different from the samples in Zone I (Figure 2). The high chloride concentrations in the Zone II water samples reflect another source of chloride such as rock–water interaction in regional groundwater flow. A similar effect was studied in Australia, where researchers distinguished between evaporation-induced increases in chloride concentrations and rock–water interaction using a suite of isotopic and geochemical tools [45,46,47]. The southern region of Hulun Lake has salt-rich solonchak soil, suggesting the discharge of, and potential presence of, salt-rich groundwater [26]. The possibility of high chloride concentrations in the south of Hulun Lake is discussed after the chloride mass balance below.

Table 2. Measured $\delta^{18}O$, $\delta^{2}H$ and Cl for lake, river, and domestic well water samples. Sample locations are shown in Figure 1. Lake water samples were collected on 4 July 2014, and domestic well water samples were collected on 30 July 2014.

Sample Type (and Date)	$\delta^{18}O$ (‰)	$\delta^{2}H$ (‰)	Cl (mg/L)
Lake water			
L1	−72.83	−8.55	140.6
L2	−71.87	−8.21	141.9
L3	−66.23	−7.27	189.3
L4	−66.59	−7.83	177.3
L5	−65.88	−7.16	172.6
L6	−65.71	−7.66	181.1
L7	−66.36	−7.61	179.1
L8	−65.97	−7.34	173.7
L9	−65.71	−7.33	187.2
L10	−65.7	−7.47	182.9
L11	−65.91	−7.12	180.3
L12	−67.24	−7.38	173.5
L13	−65.75	−7.49	176.8
L14	−63.44	−7.12	116.9
L15	−65.21	−7.25	181.6
Kelulun River			
R1 (25 June 2014)	−11.84	−94.48	5.0
R1 (26 June 2014)	−12.35	−95.99	5.7
R1 (4 July 2014)	−10.98	−85.72	25.8
R1 (3 August 2014)	−10.46	−84.00	21.0
Wuerxun River			
R2 (4 July 2014)	−11.24	−87.56	12.6
Domestic well water (south of Hulun Lake)			
W1	−11.74	−89.43	330.7
W2	−11.90	−90.52	400.7
W3	−11.98	−89.61	357.8
Domestic well water (west of Hulun Lake)			
W4	−11.40	−86.86	35.6
W5	−9.41	−72.02	28.7
W6	−12.33	−93.05	49.8
W7	−13.34	−100.54	17.0
W8	−11.64	−89.38	65.6
W9	−14.25	−105.24	17.2
W10	−15.91	−120.87	38.4

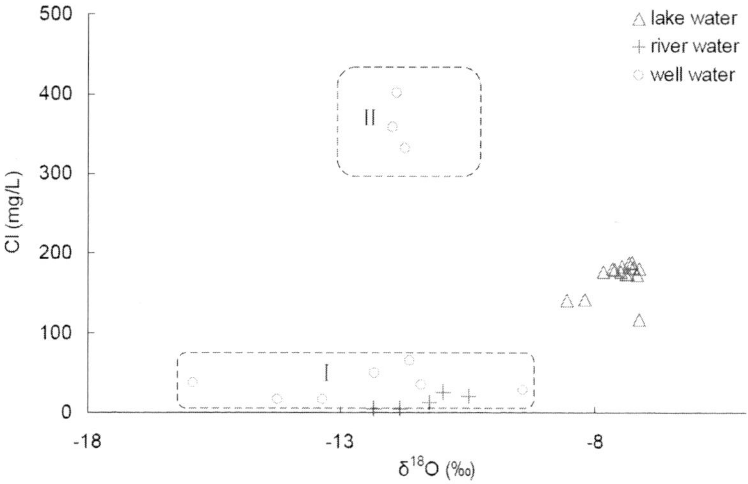

Figure 3. A cross plot of $\delta^{18}O$ values and chloride concentrations in lakes, rivers and wells. Zone I and Zone II are divided based on the significant difference in chloride concentrations in western and southern wells, respectively.

3.3. Evaporative Loss Calculations as an Indication of Saline Groundwater Discharge to Hulun Lake

3.3.1. Evaporative Loss Calculations and Their Uncertainty

The ratio of evaporation to inflow (*E/I*) was calculated using each of δ^2H and $\delta^{18}O$ (Table 3). The mean *E/I* ratios ranged from 0.20 to 0.51 in samples taken from different locations on the lake, which are consistent with relatively high evaporation rates relative to inflow in lakes [41]. Although the *E/I* values vary between δ^2H and $\delta^{18}O$, they were linearly correlated with one another (with an R^2 value of 0.72), and differences between *E/I* values calculated with δ^2H and $\delta^{18}O$ are common studies on evaporative loss [34,35,36,37,38,39,40]. In general, the values calculated using $\delta^{18}O$ (in this case, ranging between 0.20 and 0.33; Table 3) are considered more reliable than those calculated using δ^2H [40,41], and are used below.

The *E/I* values calculated using *Hydrocalculator* [41] were not particularly sensitive to climatic values. The calculated values of *E/I* only differed by ~0.009 for δ^2H, and ~0.001 for $\delta^{18}O$, for each degree of temperature change. A similar, small range of uncertainty occurred when humidity was varied, where difference of 1% resulted in a difference in the E/I value of ~0.005 for $\delta^{18}O$ and ~0.017 for δ^2H. The assumption of isotopic equilibrium between precipitation (δ_P) and moisture in ambient air (δ_A) is a source of uncertainty in E/I. However, the dependence lies primarily in the limiting isotope enrichment factor (δ^*) and moisture in ambient air (δ_A) [41,42]. In our case, change in δ_A estimation of 10‰ in δ^2H input values caused a difference in the E/I value of 0.147, and a change in δ_A input values of 1‰ in $\delta^{18}O$ caused a difference in the E/I value of 0.020.

Table 3. The evaporation to inflow (E/I) ratios calculated by δ^2H and $\delta^{18}O$ values in Hulun Lake with mean values and potential error estimated as absolute value of differences between E/I for δ^2H and $\delta^{18}O$. The calculated Cl is the average chloride concentration of all inputs, as deduced by the E/I ratios of $\delta^{18}O$.

Site Name	E/I			Calculated Total Inflow [Cl] (mg/L)
	Calculated Using δ^2H	Calculated Using $\delta^{18}O$	Mean	
L1	0.33	0.20	0.27	112.5
L2	0.36	0.23	0.30	109.3
L3	0.56	0.32	0.44	128.7
L4	0.55	0.26	0.41	131.2
L5	0.58	0.33	0.46	115.6
L6	0.59	0.28	0.44	130.4
L7	0.56	0.28	0.42	129.0
L8	0.57	0.31	0.44	119.8
L9	0.59	0.31	0.45	129.2
L10	0.59	0.3	0.45	128.0
L11	0.58	0.33	0.46	120.8
L12	0.52	0.31	0.42	119.7
L13	0.58	0.29	0.44	125.5
L14	0.69	0.33	0.51	78.4
L15	0.61	0.32	0.47	123.5
Average	0.55	0.29	0.42	120.1

3.3.2. Evidence of Groundwater Discharge with High Salinity to Hulun Lake

Chloride can be used as a conservative tracer to indicate evaporation of water since it is not affected by geochemical processes except at very high concentrations, i.e., when chloride mineral solubilities are exceeded. Assuming that dissolved chloride and stable isotopes are transported in a similar manner in the lake, and their concentration variations depend only on evaporation and mixing of the lake water sources, then the relationship between calculated evaporative loss from lake water (E/I) and chloride concentration can be estimated as follows [48]:

$$E/I = [1 - Cl_I / Cl_L \quad (12)$$

where Cl_I and Cl_L are chloride concentrations in the inflow and lake water respectively. Using the *E/I* ratios calculated from the water isotope data, the initial weighted mean chloride concentration in inflowing water from all water sources can be estimated as:

$$Cl_I = (1 - E/I) \times Cl_L \tag{13}$$

Since the use of $\delta^{18}O$ is more reliable than δ^2H for evaporation calculations (as discussed above), the *E/I* ratios calculated using $\delta^{18}O$ (Table 3) were used to estimate the average chloride concentrations of Hulun Lake's total inflow. The resulting calculated chloride concentrations for the combined input from rivers, precipitation, and groundwater ranged from 78.4 to 131.2 mg/L, with an average value of 120.1 mg/L (Table 3).

The estimated average chloride concentration for total inflow to the lake (i.e., 120.1 mg/L) is much higher than the measured river chloride concentrations (i.e., average of 14.4 mg/L (std. dev. 10.6, n = 4) in the Kelulun River, and 12.6 mg/L for a single sample the Wuerxun River; Table 1), and also higher than chloride concentrations in precipitation (e.g., 5.8 mg/L measured in northern China) [48]. Recent lake sediment sampling in Hulun Lake revealed low sediment chloride concentrations do not indicate lake bottom sediments as a chloride source [49]. The remaining component of lake water input that could be responsible for relatively high lake water chloride concentrations is groundwater discharge, which is estimated to be 24.8% of total annual input (Table 1, 1981–2013). As discussed above, the three domestic water wells sampled south of the lake (W1, W2, and W3; Figure 1) had elevated chloride concentrations ranging from 330 to 400 mg/L (Table 2), with an average value of 363 mg/L, suggesting groundwater discharge to Hulun Lake could be a significant source of chloride. This is discussed below in the context of a chloride mass balance.

3.4. The Chloride Mass Balance on Hulun Lake Inflow

Inflow to Hulun Lake is thought to be comprised of three main water sources: precipitation, the two rivers and groundwater. As the chloride concentration in precipitation and the two rivers were measured, and the estimated volumetric contribution of each of the three main sources of water to the lake have been estimated (2000–2013; Table 1); a chloride mass balance can be used to estimate an average chloride concentration in groundwater discharge to the lake as follows:

$$Cl_{GW} = \frac{Q_I \times Cl_I - Q_R \times Cl_R - Q_P \times Cl_P}{Q_{GW}} \tag{14}$$

where Cl_I, Cl_{GW}, Cl_R, Cl_P are chloride concentrations of the total combined inflow, groundwater, river water, and precipitation, respectively. Similarly, Q_I, Q_{GW}, Q_R, Q_P are the relative contributions (%) of total inflow, groundwater, river water ($Q_R = Q_{in}$), and precipitation, respectively. The annual water budget estimated the relative contributions of river discharge (52.2%), precipitation (23.0%), and groundwater (24.8%) for the period of 1981 to 2013 (Table 1), and an average chloride concentration of 120.1 mg/L for total inflow (Table 3) was previously calculated. The average chloride concentration in river water average was 12.5 mg/L (Table 2) and in precipitation the average value was estimated as 5.80 mg/L [49].

An average chloride concentration of 452.5 mg/L was calculated for groundwater based on the chloride mass balance (Equation (14)), which is reasonably similar to the average observed value of 363 mg/L in groundwater in the southwest of the lake. Thus, groundwater apparently contributes most of the annual loading of chloride to Hulun Lake (93.5%) relative to precipitation (1.1%) and river flow (5.4%). Assuming the average chloride groundwater concentration has not changed since 1981, the four-fold increase in groundwater discharge to Hulun Lake in the period from 2001 to 2013 has been associated with a four-fold increase groundwater loading of chloride (Table 1). This corresponds to an increase from an estimated 1.18×10^5 tons (1981–2000) to 4.03×10^5 tons (2001–2013).

Finally, when the range of chloride concentrations and $\delta^{18}O$ values for the combined lake input (i.e. the minimum, maximum, and average; Table 3) are subjected to evaporation (which affects chloride concentration and water isotopes differently), the range of predicted values is similar to that measured in lake samples (Figure 4), consistent with the water and chloride balances presented here.

3.5. Additional Evidence for High Chloride in Groundwater Supplies South of Hulun Lake

The evidence of high chloride concentrations occurring in groundwater south of Hulun Lake is supported by a previous study that mapped salinity in groundwater and lake water by field measurements of salinity in domestic water well samples and in Hulun Lake (Figure 5). This study measured relatively high salinity in domestic water wells in the region south of the lake, which is consistent with high chloride concentrations in groundwater in this region, and a significant contribution of chloride in groundwater discharge to Hulun Lake. Although the source of salinity is not known, its widespread nature suggests rock–water interaction.

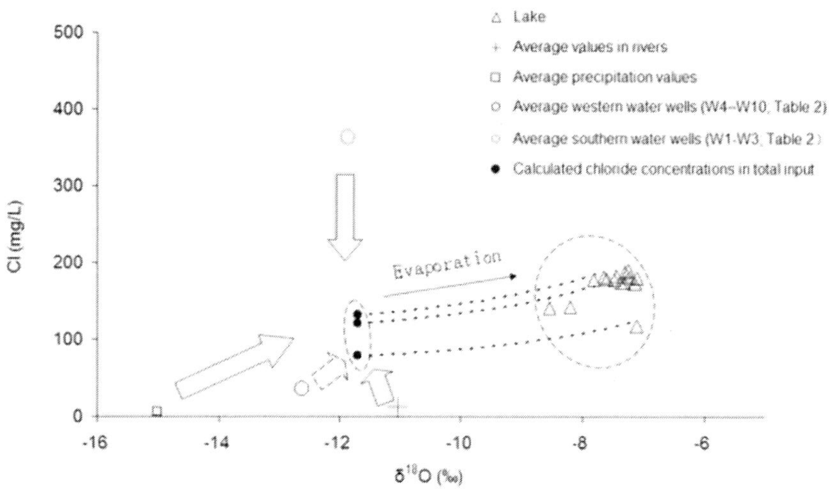

Figure 4. $\delta^{18}O$ vs. chloride concentrations including the minimum, average, and maximum values for total input into Hulun Lake (filled black symbols; Table 3), measured Hulun Lake values, and average values for each of the main three inputs (precipitation, river flow, and groundwater) that are thought to contribute to the combined input (with solid arrows indicating likely inputs, and the dashed arrow represents potential sources). The dashed lines represent $\delta^{18}O$ and chloride concentration evolution with increasing evaporation modeled by *E/I* ratios as described in Section 2.4 (with T = 0.3 °C, and humidity = 60%).

4. CONCLUSIONS

1) The estimated average annual water budget for Hulun Lake was significantly different between the two time periods that were evaluated (1981–2000 and 2001–2013). The later time period was characterized by decreased average annual contributions of river discharge (from 63.6% to 30.7%) and precipitation (from 24.9% to 19.4%) to the lake, an increased evaporative loss from the lake (from 20.4 to 22.0 × 10^8 m^3), and decreased lake storage. As a result of the changes in lake inputs, the estimated proportion of the annual contribution of groundwater to the lake increased almost four-fold (from 11.4% to 49.9%) between the two periods.

2) Although the relatively high $\delta^{18}O$ and $\delta^{2}H$ values for the present-day lake were consistent with significant evaporation of water from combined total input water from rivers, precipitation, and groundwater, the chloride concentrations were higher than could be accounted for by evaporation alone.

3) A chloride mass balance suggested high chloride concentrations in groundwater discharge is responsible for more than 90% of the annual chloride loaded into the lake. Elevated chloride concentrations in domestic water well samples from the region south of Hulun Lake and a regional groundwater salinity survey suggest high groundwater chloride concentrations occur in the region south of Hulun Lake.

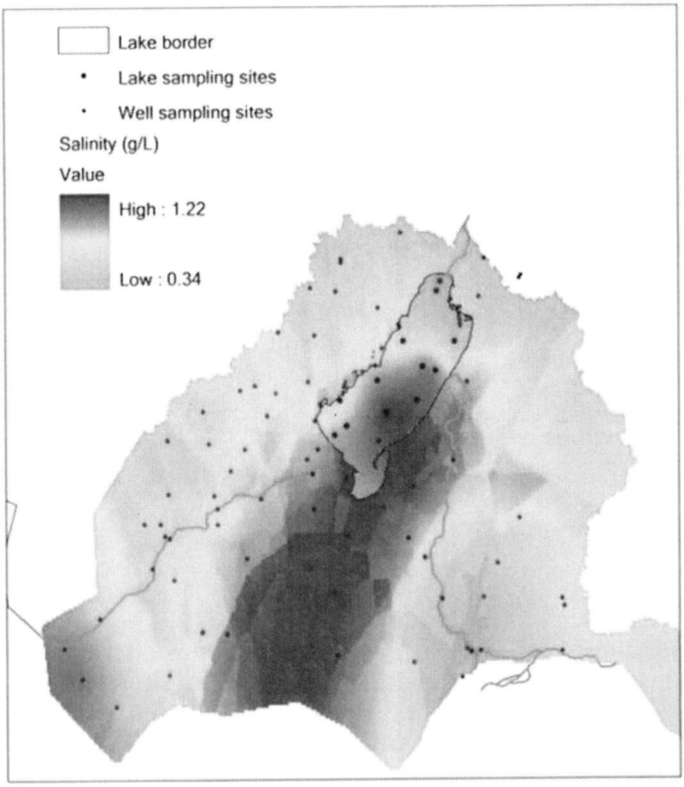

Figure 5. The spatial distribution of salinity (g/L) in domestic water well samples around Hulun Lake. Unpublished data (collected in 2008), College of Water Resources and Civil Engineering, Inner Mongolia Agricultural University.

4) The combined use of stable isotopes in water, in combination with water and chloride mass balances, could be applied to other watersheds where hydrologic information is scarce.

ACKNOWLEDGMENTS

We would like to thank Cairui Fan, Jinyan Guo, Jingjie Wang and Yao Fu from the Water Environment Protection and Restoration Group of Inner Mongolia Agriculture University for sampling assistance. This work was supported by the Chinese National Natural Science Foundation (Nos. 51169011, 51339002, 51169017, 51069007, 51269016 and 51269017), Natural Science Foundation of Inner Mongolia (Nos. 2012MS0612, 20091408, 20080105), and China Scholarship Council (No. 201408150013). David Bethune (University of Calgary, Canada) is acknowledged for providing valuable input, and Dongnan Zhu for many collegial discussions. Finally, three anonymous reviewers and editor (Rachel Lu) are acknowledged for providing professional revising advice and helpful comments.

REFERENCES

1. Helfer, F.; Lemckert, C.; Zhang, H. Impacts of climate change on temperature and evaporation from a large reservoir in Australia. *J. Hydrol.* **2012**, *475*, 365–378.
2. Song, C.Q.; Huang, B.; Ke, L.H.; Richards, K.S. Remote sensing of alpine lake water environment changes on the Tibetan Plateau and surroundings: A review. *ISPRS. J. Photogramm.* **2014**, *92*, 26–37.
3. Cai, Z.; Jin, T.; Li, C.; Ofterdinger, U.; Zhang, S.; Ding, Aizhong.; Li, J. Is China's fifth-largest inland lake to dry-up? Incorporated hydrological and satellite-based methods for forecasting Hulun lake water levels. *Adv. Water Resour.***2016**, *94*, 185–199.
4. Chen, X.; Chuai, X.; Yang, L.; Zhao, H. Climatic warming and overgrazing induced the high concentration of organic matter in Lake Hulun, a large shallow eutrophic steppe lake in northern China. *Sci. Total Environ.* **2012**, *431*, 332–338.
5. Chuai, X.; Chen, X.; Yang, L.; Zeng, J.; Miao, A.; Zhao, H. Effects of climatic changes and anthropogenic activities on lake eutrophication in different ecoregions. *Int. J. Environ. Sci. Technol.* **2012**, *9*, 503–514.
6. Wang, J.; Feng, W.; Zhang, L.; Zong, Z.; Meng, H.; Peng, B.; Li, Z. Monitoring and evaluation on water quality and biology resource quantity in Hulun lake. *J. Hydroecology* **2011**, *32*, 64–68. (In Chinese).
7. Li, Z.; Zhang, Z.; Jiang, F.; Wang, S.; Wang, F. *Annals of Hulun Lake (II)*; Inner Mongolia Culture: Inner Mongolia, China, 2008. (In Chinese)
8. Li, C.; Ma, W.; Ye, B.; Liao, W. Estimation of water evaporation and water balance in ungauged Hulun lake. *J. China Hydrol.* **2006**, *26*, 41–44. (In Chinese). [Google Scholar]
9. Yan, D.; Yan, H.; Deng, W.; Wang, M. Influences of eco-hydrological process to water environmental system in Hulun Lake basin. *Bull. Soil Water Conserv.* **2001**, *21*, 1–5. (In Chinese).

10. Yu, G.; Shen, H. Lake water changes in response to climate change in northern China: Simulation and uncertainty analysis. *Quatern. Int.* **2010**, *212*, 44–56.
11. Zhang, G.; Xie, H.; Yao, T.; Kang, S. Water balance estimates of ten greatest lakes in China using ICESat and Landsat data. *Sci. Bull.* **2013**, *58*, 3815–3829.
12. Zuber, A. On the environmental isotope method for determining the water balance of some lakes. *J. Hydrol.* **1983**, *6*, 409–427.
13. Krabbenhoft, D.P.; Bowser, C.J.; Anderson, M.P.; Valley, J.W. Estimating groundwater exchange with lakes 1. The stable isotope mass balance method. *Water Resour. Res.* **1990**, *26*, 2445–2453.
14. Gibson, J.J.; Prepas, E.E.; McEachern, P. Quantitative comparison of lake throughflow, residency, and catchment runoff using stable isotopes: Modelling and results from a survey of boreal lakes. *J. Hydrol.* **2002**, *262*, 128–144.
15. Bennett, K.E.; Gibson, J.J.; McEachern, P. Water yield estimates for critical loadings assessment: Comparisons of gauging methods vs. an isotopic approach. *Can. J. Fish. Aquat. Sci.* **2008**, *65*, 83–99.
16. Alyamani, M.S. Isotopic composition of rainfall and groundwater recharge in the western province of Saudi Arabia. *J. Arid Environ.* **2001**, *49*, 751–760.
17. Gibson, J.; Edwards, T.; Birks, S.; Amour, N.; Buhay, W.; McEachern, P.; Wolfe, B.; Peters, D. Progress in isotope tracer hydrology in Canada. *Hydrol. Process.* **2005**, *19*, 303–327.
18. Carol, E.; Kruse, E.; Mas-Pla, J. Hydrochemical and isotopical evidence of ground water salinization processes on the coastal plain of Samborombo´n Bay, Argentina. *J. Hydrol.* **2009**, *365*, 335–345.
19. Wassenaar, L.I.; Athanasopoulos, P.; Hendry, M.J. Isotope hydrology of precipitation, surface and ground waters in the Okanagan Valley, British Columbia. *Can. J. Hydrol.* **2011**, *411*, 37–48.
20. Kanduc, T.; Mori, N.; Kocman, D.; Stibilj, V.; Grassa, F. Hydrogeochemistry of Alpine springs from North Slovenia: insights from stable isotopes. *Chem. Geol.* **2012**, *300*, 40–54.
21. Cui, B.; Li, X. Characteristics of stable isotope and hydrochemistry of the groundwater around Qinghai Lake, NE Qinghai-Tibet Plateau, China. *Environ. Earth Sci.* **2014**, *71*, 1159–1167.
22. Currell, M.J.; Dahlhaus, P.; Ii, H. Stable isotopes as indicators of water and salinity sources in a southeast Australian coastal wetland: Identifying relict marine water, and implications for future change. *Hydrogeol. J.* **2015**, *23*, 235–248.
23. Clark, I.D.; Frjtz, P. *Environmental Isotopes in Hydrogeology*; Lewis Publishers: New York, NY, USA, 1997.
24. Gibson, J.J.; Edwards, T.W.D.; Bursey, G.G. Estimating evaporation using stable isotopes. *Nord. Hydrol.* **1993**, *24*, 79–94.
25. Kirchner, J.; Tetzlaff, D.; Soulsby, C. Comparing chloride and water isotopes as hydrological tracers in two Scottish catchments. *Hydrol. Process.* **2010**, *24*, 1631–1645.
26. Xu, Z.; Jiang, F.; Zhao, H.; Zhang, Z; Sun, L. *Annals of Hulun Lake*; Jilin Literature and History Publishing House Press: Jilin, China, 1989; pp. 691–692. (In Chinese)
27. Sun, B.; Li, C.; Zhang, S.; Zhao, S. Analysis of Hulun digital basin based on spatial information technology. *Environ. Pollut. Control.* **2010**, *32*, 5–19. (In Chinese).
28. Wen, R.; Xiao, J.; Chang, Z.; Zhai, D.; Xu, Q.; Li, Y.; Itoh, S.; Lomtatidze, Z. Holocene climate changes in the mid-high-latitude-monsoon margin reflected by the pollen record from Hulun Lake, northeastern Inner Mongolia. *Quatern, Res.* **2010**, *73*, 293–303.
29. Craig, H.; Gordon, L.I. Deuterium and oxygen 18 variations in the ocean and the marine atmosphere. In *Stable Isotopes in Oceanographic Studies and Paleotemperatures*; Ongiorgi, E., Ed.; Laboratorio di Geologia Nucleare: Pisa, Italy, 1965; pp. 9–130.
30. Gonfiantini, R. Environmental isotopes in lake studies. In *Handbook of Environmental Isotope Geochemistry*; Volume 2: The Terrestrial Environment; Fritz, P., Ed.; Elsevier: Amsterdam, The Netherlands, 1986; pp. 113–163.
31. Gat, J.R. Stable isotopes of fresh and saline lakes. In *Physics and Chemistry of Lakes*; Lerman, D.I., Ed.; Springer: Berlin, Germany, 1995; pp. 139–165.
32. Gibson, J.J. Short-term evaporation and water budget comparisons in shallow Arctic lakes using non-steady isotope mass balance. *J. Hydrol.* **2002**, *264*, 242–261.

33. Horita, J.; Rozanski, K.; Cohen, S. Isotope effects in the evaporation of water: A status report of the Craig–Gordon model. *Isot. Environ. Health Stud.* **2008**, *44*, 23–49.
34. Gat, J.R.; Levy, Y. Isotope hydrology of inland Sabkhas in the Bardawil area. *Sinai. Limnol. Oceanogr.* **1978**, *23*, 841–850.
35. Gat, J.R. Properties of the isotopic species of water: the "isotope effect". In *Stable Isotope Hydrology, Deuterium and Oxygen-18 in the Water Cycle*; Gat, J.R., Gonfiantini, R., Eds.; IAEA Tech. Rep. Ser. No. 210; IAEA: Vienna, Austria, 1981; pp. 7–19.
36. Gibson, J.J.; Reid, R. Water balance a chain of tundra lakes: a 20-years isotopic perspective. *J. Hydrol.* **2014**, *519*, 2148–2164.
37. Welhan, J.A.; Fritz, P. Evaporation pan isotopic behaviour as an index of isotopic evaporation conditions. *Geochim. Cosmochim. Acta.* **1977**, *41*, 682–686.
38. Allison, G.B.; Leaney, F.W. Estimation of isotopic exchange parameters using constant-feed pans. *J. Hydrol.* **1982**, *55*, 151–161.
39. Gibson, J.J.; Reid, R. Stable isotope fingerprint of open-water evaporative losses and effective drainage area fluctuations in a subarctic shield watershed. *J. Hydrol.* **2010**, *381*, 142–150.
40. Horita, J.; Wesolowski, D.J. Liquid-vapor fractionation of oxygen and hydrogen isotopes of water from the freezing to the critical temperature. *Geochim. Cosmochim. Acta* **1994**, *58*, 3425–3437.
41. Skrzypek, G.; Mydłowski, A.; Dogramaci, S.; Hedley, P.; Gibson, J.J.; Grierson, P.F. Estimation of evaporative loss based on the stable isotope composition of water using Hydrocalculator. *J. Hydrol.* **2015**, *523*, 781–789.
42. IAEA Global Network of isotopes in precipitation. The GNIP database, 2017. Available online: http://www.iaea.org/water (accessed on 16 May 2017).
43. Mulligan, B.M.; Ryan, M.C.; Cámbara, T.P. Delineating volcanic aquifer recharge areas using geochemical and isotopic tools. *Hydrogeol. J.* **2011**, *19*, 1335–1347.
44. Craig, H. Isotope variations in meteoric waters. *Science* **1961**, *133*, 1702–1703.
45. Herczeg, A.L.; Dogramaci, S.S.; Leaney, F.W.J. Origin of dissolved salts in a large, semi-arid groundwater system: Murray Basin, Australia. *Mar. Freshwatyer. Res.* **2001**, *52*, 41–52.
46. Cartwright, I.; Weaver, T.R.; Fifield, L.K. Cl/Br ratios and environmental isotopes as indicators of recharge variability and groundwater flow: An example from the southeast Murray Basin, Australia. *Chem. Geol.* **2006**, *231*, 38–56.
47. Dogramaci, S.; Firmani, G.; Hedley, P.; Skrzypek, G.; Grierson, P.F. Evaluating recharge to an ephemeral dryland stream using a hydraulic model and water, chloride and isotope mass balance. *J. Hydrol.* **2015**, *521*, 520–532.
48. Lin, D.; Jin, M.; Liang, X.; Zhan, H. Estimating groundwater recharge beneath irrigated farmland using environmental tracers fluoride, chloride and sulfate. *Hydrogeol. J.* **2013**, *21*, 1469–1480.
49. Gao, H.; Li, C.; Sun, B.; Shi, X.; Fan, C. Solutes transport between lake water and sediment pore water during the lake water level fluctuation. *J. Lake Sci.* **2017**, (unpublished; accepted).

CHAPTER 9

Groundwater Origin and Its Hydrogeochemistry through GIS Maps in Linares Region, Mexico

René Alberto Dávila Pórcel*, Héctor de León-Gómez

Instituto de Ingeniería Civil (IIC), Facultad de Ingeniería Civil (FIC), Universidad Autónoma de Nuevo León (UANL), Monterrey, México

ABSTRACT

The North-East region of Mexican Republic has a semi-arid weather condition; this area covers partially Tamaulipas, Nuevo León and Coahuila states. Pablillo River Basin (PRB) is located close to Nuevo León south border and its catchment area involves Linares City (LC), Hualahuises City (HC) and Cerro-Prieto dam (CP). This reservoir captures drinking water to Monterrey Metropolitan Zone (MMZ) the major urban center on the north-east Mexican region. More than 50% of the potable water for LC comes from groundwater stockpiles. A combination of GIS maps and major ion chemistry analysis has been assessed, to know the origin, geochemical evolution, and hydraulic interconnection of groundwater. During 2008-2009, 46 groundwater samples were taken and analyze around LC and HC. GIS technology was use to analyze the spatial distribution of the constituents of groundwater through thematic maps. The major ion analysis and its results suggest the presences of two aquifers; shallow porous aquifer on the top and fractured aquifer on the bottom. General water mineralization was discovered, including dissolution of halite, dolomites and carbonates through the aquifers. Water quality varies widely through the aquifers, the general pathways goes from southwest to northeast direction. The overexploitation of groundwater increases the water mixing of groundwater between shallow and deeper aquifers. Natural origin of groundwater and groundwater pollution by anthropogenic activities should be considerate in the groundwater quality analysis for drinking water purpose.

Keywords: Groundwater; hydrogeochemistry; GIS; Mexico

1. INTRODUCTION

In the northeast of Mexico, surface and ground water resources play an important role for the agricultural and urban development. In the southern portion of Nuevo León State are located LC and HC. Groundwater is the most essential water source used to supply the growing demand for irrigational, domestic and industrial purposes in importance order [1], also an intensive use of irrigation makes groundwater a crucial resource nowadays for human activities [2] and careful management is imperative for ecological and environmental reasons [3]. In relation to the water quality, the chemical composition of groundwater is controlled by various factors that consist of precipitation, geological structure and/or rock mineralogy of aquifers and geological processes inside the aquifer [4]. For these reasons, increase the knowledge of aquifer systems and groundwater quality in arid and semiarid regions could conduct to improve understanding of hydrochemical systems in such areas, leading a sustainable development of water resources and effective management of groundwater resource [2].

In the region of LC and HC the situation of the natural processes that govern the hydrochemical composition of the groundwater is not clearly and the factors that have effect on its quality are undetermined. The aim of this paper is to understand the hydrochemical evolution of groundwater in the aquifers using analysis of the major ions of groundwater and thematic maps developed in a GIS environment. The use of both methodologies of assessment in Latin-American and in development countries is uncommon, although its necessity being an important instrument for the decision makers in public and private sectors [5]. In the present study, a specify investigation was carried out with the objective to discover hydrochemical processes and their relation to groundwater quality and aquifer characteristics in the analysis of the geographic distribution of the water constituents.

2. STUDY AREA

The PRB is inside of Nuevo León State, the northeast portion of the basin is the study area analyze in this paper. This area belongs to the Gulf of Mexico Coastal Plain (GMCP), inside the study area (north-eastern PRB) are sited LC and HC. The main access is the National Highway 85 since MMZ ~130 km into the south direction. Approximately 12 km from LC is positioned the drinking water reservoir CP an important source of potable water for Metropolitan-Zone of Monterrey, the most important city in the northeastern territory of Mexico (population ~3.8 Million inhabitants) [6]. CP water reservoir captures the water of the catchment area of Pablillo River, which crosses LC and Hualahuises River crosses HC which joins to Pablillo River in the income of LC. During the last decade, LC suffers an important urban population growth, as well as the agricultural and industrial development. Groundwater resources in this region supplies more than the 50% of the total water for consumption of LC and HC population more than 63,000 inhabitants [6]. However, all the waste water and industrial effluents goes downstream in direction to the CP, hence, the possible pollution of surface and ground water is high. The aquifers of this area have a shallow water-table around LC and HC, this aspect increase the risk to pollution of groundwater stockpiles. Upstream of LC is located the citrus region and HC which has municipal activities, all these areas could be important sources of pollution and can affect the aquifers, due to the presence of private farms, citric plantations, sewage water, wastewater and a variety of little farms. In the downstream portion between LC and CP dam (northeast direction) is the wastewater treatment plant of Linares, also, continues the citric plantations, farms and small human centers (Rio Verde, La Escondida y Guadalupe). The main aims to this study are: 1) identify key processes that govern the hydrochemical condition of groundwater in the aquifers in the north-east part of Pablillo River Basin 2) discover if there is any presence of pollutants inside the study area.

3. GEOLOGY AND HYDROGEOLOGICAL SETTINGS

In the middle of two important geologic provinces: the Sierra Madre Oriental (SMO) and the GMCP is located the PRB, the SMO contains the Southwestern portion of the basin and the GMCP includes the northeastern part of the basin. For this paper, the research area is center on the portion enclosed by GMCP highlighted in **Figure 1**. After field skills like: identification of Geological Formations and structural geology data collection and literature review of previous works, three important formations were studied and are the most recent to oldest: 1) Fluvial

Terraces and Lacustrine Sediments namely Alluvium-Conglomerate Formation (ACF) with a given age period from Tertiary to Quaternary, this formation has a calculate thickness of ~25 m since surface, the origin of its sediments are the mountain canyons of the SMO; 2) Méndez Formation (MF) with a given age period from Campanian to Maastrichtian, this geological formation is the most powerful and prevalent in the area, it has a maximum thickness around 1500 m, consisting in dark gray shale interbedded by olive-green and calcareous shale; finally, the oldest one 3) San Felipe Formation (SFF) with a given age period from Turonian to Conacian, it has a maximum thickness ~150 m, it is formed by heterogeneous sequence of shale, calcareous shale, silicified limestone, marl and sandstone. San Felipe is present in a small area located between the north side of CP and the north border of the catchment area of Pablillo River [7,8]. According by Strahler (1957) PRB is a fourth order catchment classification, [9]. The study area covers an area about 819.28 km^2 (North-East portion of PRB in the GMCP) and the total catchment area of PRB is ~1734.55 km^2.

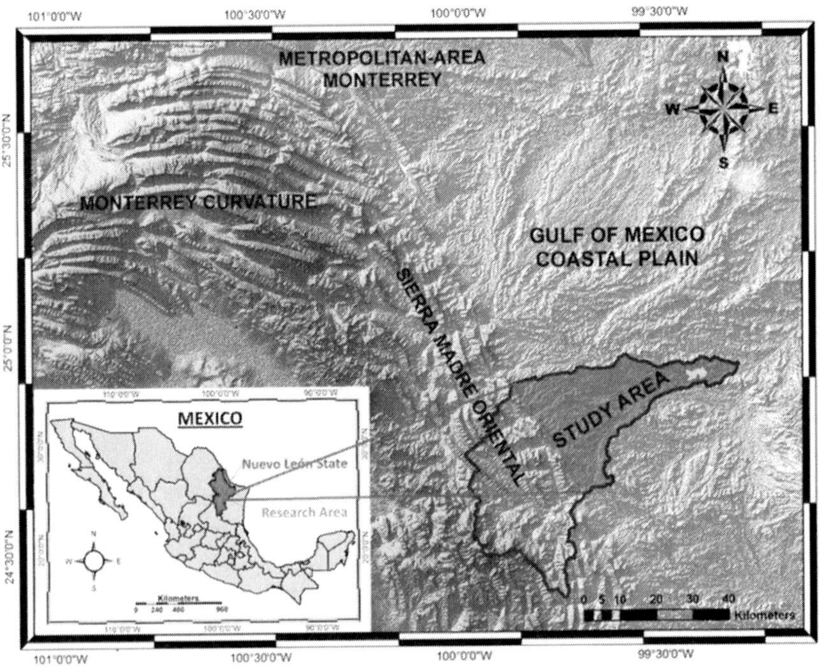

Figure 1. Localization of the study area.

Some researchers had been studied the hydrogeology in this region, in general the main flow direction goes from the high hidraulic heads to the low hidraulic head (from SMO to CP), the main hydrostratigraphic units are MF and ACF [8,10-18]. None of the available literature demonstrates the hydrochemical origin of the groundwater or the aquifers interconnectivity of the basin.

Factors like geology, geographical position and water-table measurements of groundwater levels were realized in 23 hand-dug wells and pipe Wells, in two field visits of sampling on April 2008 and March 2009. A shallow unconfined aquifer was identified which has mainly two hydrologically connected layers: on the top a highly permeable and porous aquifer consisting mostly of gravels and various grades of sand, silt and clay, represented by ACF and in the Bottom a fractured and weathered aquifer mainly represented by the presences of MF and in a small area by SSF. **Figure 2** shows the spatial distribution of the Geologic Formations presents in the study area.

The shallow porous aquifer has a depth that varies from a few meters to ~25 m and the water-table levels in the alluvium do not significantly change throughout the aquifer, therefore the rainfall season does not affect in great manner. The principal variation of groundwater levels is due to topographic morphology of bedrock and abstractions over the area. As a general situation, the water-table average is ~10 m depth measured from the surface. Just as in other shallow aquifers, groundwater is stored in the fractured and weathered bedrock in the lower aquifer, the upper alluvial sediments permits the aquifer recharge throughout its body [19]. The main direction of groundwater flow is from southwest to northeast; additionally the structural geological data confirms this general pathway. In general, in arid and semi-arid regions, the recharge to shallow alluvial aquifers is a result of infiltration from the surface [17]. Therefore, the principal process of recharge in the study area is the surface recharge along GMCP over the aquifers. The amount of recharge is function of rainfall frequency which is characterized by clear spatial and temporal variations over the year. The shallow water-table and coarse alluvial deposit increases the chances of direct recharge into the alluvial aquifer from surface. A value of 1035 mm is the annual average of rainfall (has been recorded month to month for 25 years) over the study area, from this amount of precipitation a range between 18 and 35% converts to direct recharge, finally, the discharge of the aquifer takes place in CP dam and through pumping in LC, HC and its neighboring areas.

4. SAMPLING PROCEDURE AND HYDROGEOCHEMISTRY METHODOLOGY

The data used in this research comes from 46 samples of 23 groundwater-sampling points of hand-dug wells and pipe wells during April 2008 and March 2009 on two field visits (**Figure 2**). Multi-parameter measuring equipment "WTW-340" was used to realize in situ measurements of pH, temperature, and electrical conductivity (EC). Glass fiber paper through 0.45 μm was used to filter the water samples. The samples were acidified to pH = 2 with ultrapure nitric acid and they were cold stored in precleaned bottles (1-1 polyethylene) during the transport to the laboratory.

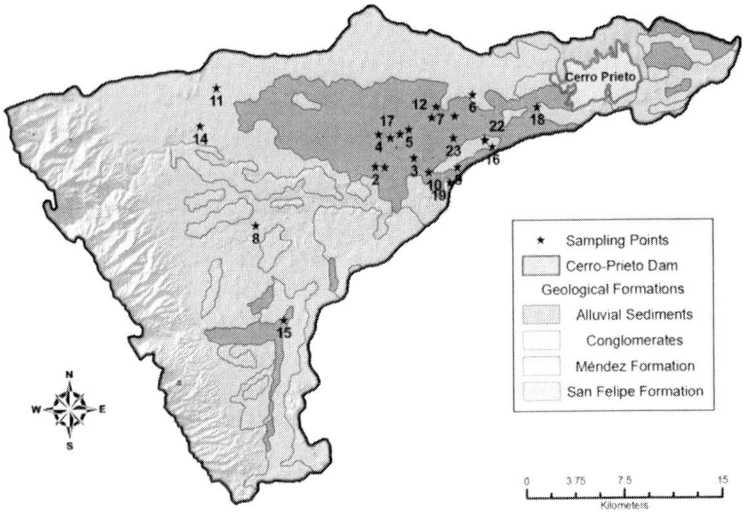

Figure 2. Local geology and sampling points' location.

Alkalinity was measured using titration method in the chemical laboratory of Facultad de Ciencias de la Tierra on the same day of sampling. The measurement of cations concentrations was performed using of Dionex ICS-90 IC (Ion Chromatography), this equipment was used to measures: sodium (Na^+), potassium (K^+), calcium (Ca^{2+}), magnesium (Mg^{2+}). On the other hand, major anions likes: bicarbonate (HCO_3^-), chloride (Cl^-), nitrate (NO_3^-) and sulfate (SO_4^{2-}) were measured using Dionex ICS-120 IC in the Chemistry Laboratory of the Institut für Angewandte Geowissenschaften of the Technische Universität Darmstadt (Germany), all this values constitutes the hydrochemical database to use in the Major Ion Hydrochemistry and GIS-database to generate the thematic maps.

Table 1. Hydrochemical data for groundwater (April 2008).

Sample	pH	Na^+ [mg/l]	K^+ [mg/l]	Ca^{2+} [mg/l]	Mg^{2+} [mg/l]	Cl^- [mg/l]	SO_4^{2-} [mg/l]	HCO_3^- [mg/l]	Sum Anions [meq/l]	Sum Cations [meq/l]
8-NCC	7.06	85.00	2.00	194.37	25.43	162.61	181.51	390.36	15.69	15.56
8-NCCA	7.33	59.00	5.05	220.24	36.68	139.09	393.06	311.23	17.25	16.78
8-NCCP	6.93	30.50	3.00	154.30	13.20	29.5	240.70	313.90	11.20	10.19
8-NCN	6.63	185.00	1.55	258.61	35.53	331.23	259.30	403.53	22.44	23.91
8-NCSJ	6.76	148.00	1.45	200.21	37.28	116.58	327.62	458.02	19.25	19.56
8-NELE	7.64	280.00	3.95	479.68	129.90	758.82	366.72	267.03	46.74	46.99
8-NERV	7.05	43.00	1.00	120.96	18.98	80.78	139.94	281.40	10.20	9.49
8-NLC	7.41	20.00	0.90	63.80	18.30	8.80	29.60	330.50	6.29	5.59
8-NPM	7.09	34.00	1.45	126.80	13.51	36.23	193.86	271.82	9.77	8.95
8-NQP	7.29	43.00	2.45	138.48	4.50	24.48	190.15	238.29	8.67	9.21
8-NRC	7.72	15.25	0.85	74.25	13.64	12.73	27.16	281.40	5.75	5.52
8-NRCH	6.90	42.50	1.30	162.70	26.40	47.00	310.30	364.00	13.93	12.17
8-NREB	7.14	43.00	1.30	128.47	12.50	67.05	154.34	241.88	9.36	9.34
8-NRN	7.32	9.25	0.60	68.82	13.33	7.83	23.05	241.88	5.30	4.95
8-NSR	7.08	15.50	1.70	74.20	23.80	10.80	44.90	356.80	7.27	6.38
8-NST	6.76	79.00	7.60	159.34	30.33	81.06	190.98	445.44	13.94	14.08
8-NZI	6.92	69.00	1.65	167.68	28.15	147.35	155.60	417.70	15.02	13.75
8-PFCT	7.10	47.50	5.15	170.18	16.18	48.53	328.86	273.01	12.97	12.05
8-PGSA	6.93	134.00	2.68	105.11	30.51	101.83	86.43	522.68	14.14	13.71
8-PICCA	6.74	37.00	0.95	188.53	28.01	40.53	298.81	380.78	13.83	13.37
8-PJQ	6.98	26.00	1.45	146.82	10.98	26.02	197.56	270.62	9.48	9.40
8-POR	7.04	23.00	1.60	128.47	12.98	24.77	181.11	275.41	9.13	8.52
8-PRL	7.03	35.00	1.65	174.35	17.37	34.02	278.23	337.68	12.51	11.69

Table 2. Hydrochemical data for groundwater (March 2009).

Sample	pH	Na^+ [mg/l]	K^+ [mg/l]	Ca^{2+} [mg/l]	Mg^{2+} [mg/l]	Cl^- [mg/l]	SO_4^{2-} [mg/l]	HCO_3^- [mg/l]	TDS [mg/l]	Sum Anions [meq/l]	Sum Cations [meq/l]
9-NCC	7.48	103.32	23.18	225.70	24.49	216.81	165.30	291.70	1085.60	15.23	19.26
9-NCCA	8.12	41.86	2.99	157.77	21.04	135.22	343.16	63.80	815.00	12.09	11.58
9-NCCP	7.02	27.46	3.23	181.49	9.65	24.47	240.16	258.30	677.60	10.17	11.18
9-NCN	6.67	176.03	2.75	375.26	36.77	521.08	307.68	315.00	2187.40	27.49	29.71
9-NCSJ	6.82	126.87	1.17	270.90	30.28	128.63	349.71	380.70	1423.00	19.94	21.70
9-NELE	6.76	271.64	17.14	542.76	143.85	708.23	201.42	326.20	6817.20	50.41	51.66
9-NERV	6.91	67.30	1.29	149.74	14.26	64.13	94.88	334.00	621.00	9.73	11.69
9-NLC	7.46	12.91	1.00	84.33	7.37	3.67	17.63	252.10	270.60	4.63	5.42
9-NPM	7.27	43.55	2.21	208.42	13.31	80.36	267.38	252.80	870.20	12.53	13.50
9-NQP	7.30	19.37	2.32	131.21	7.64	17.96	163.91	200.40	512.00	7.31	8.12
9-NRC	7.94	19.58	0.57	90.15	6.69	10.99	40.12	219.30	349.00	5.06	5.94
9-NRCH	7.74	41.67	0.93	192.45	19.63	54.68	351.34	197.30	854.20	12.53	13.12
9-NREB	7.50	32.43	1.27	148.75	13.924	44.40	180.25	218.20	650.60	9.00	10.06
9-NRN	7.48	7.88	0.62	90.44	6.02	3.06	26.15	243.20	286.80	4.65	5.38
9-NSR	6.93	14.49	1.18	125.34	6.632	5.68	26.72	323.90	340.20	6.21	7.49
9-NST	6.68	49.78	7.26	201.10	11.70	57.73	169.72	359.80	782.40	11.72	13.41
9-NZI	6.78	75.90	2.89	243.60	23.60	157.60	171.57	347.30	1051.80	14.77	17.58
9-PFCT	7.03	43.37	4.34	209.59	17.39	54.32	345.21	236.20	900.60	13.10	13.95
9-PGSA	6.61	93.00	1.72	159.97	23.44	89.05	65.73	421.30	720.40	11.63	14.14
9-PICCA	6.71	32.37	0.60	221.24	15.11	26.77	249.39	324.50	955.00	11.43	13.77
9-PJQ	6.98	22.31	1.58	159.53	8.51	22.20	183.16	238.00	505.80	8.64	9.71
9-POR	6.90	22.36	1.13	145.72	8.49	18.70	167.06	226.60	545.40	7.89	9.00
9-PRL	6.85	24.56	1.18	189.82	9.67	23.57	215.16	288.60	670.40	10.07	11.40

Table 3. Statistical summary of concentrations of chemical constituents.

p	Minimum	Maximum	Average	SD	Number
pH	6.61	8.12	7.11	0.36	46
Na^+ [mg/l]	7.88	280.00	62.47	62.72	46
K^+ [mg/l]	0.57	23.18	2.91	4.09	46
Ca^{2+} [mg/l]	63.80	542.76	178.51	93.28	46
Mg^{2+} [mg/l]	4.50	143.85	23.41	26.14	46
Cl^- [mg/l]	3.06	758.82	104.50	164.44	46
SO_4^{2-} **[mg/l]**	17.63	393.06	194.40	108.21	46
HCO_3^- **[mg/l]**	63.80	522.68	304.23	81.95	46
TDS [mg/l]	270.60	6817.20	947.29	1030.87	46
Sum of Anions [meq/l]	4.63	50.41	13.18	8.96	46
Sum of Cations [meq/l]	4.95	51.66	13.69	9.21	46

Tables 1and 2 shows the results of the laboratory analyses carried out in 2008 and 2009 respectively and **Figure 2** shows the geographic location of sampling points in the study area.

The selection of wells was made in attention to: 1) access to the well or hand-dug-well; 2) geographical location; 3) presence of human activity; and 4) current well exploitation. Four points of sampling were select upstream to know the background chemical quality of groundwater before the human impact by agricultural and industrial activities, these points are located closely to the major recharge area over the study area.

A Geographic Information System (GIS) was created on the base of hydrochemical, geological and hydrogeological data. The use of these data is directed to understand the spatial distribution of the water chemistry and its constituents. On the other hand, a statistical analysis was made to understand, the relationship between hydrochemical water quality and aquifer characteristics to know the chemical evolution of groundwater. All the results of this analysis are show on thematic maps, scatter plots and tables.

5. RESULTS AND DISCUSSION

The summary statistics for each water quality parameter are show in **Table 3**. The pH value in groundwater ranges from 6.65 to 7.83 with an average value of 7.11, demonstrating the alkaline nature. The concentration of Na^+ ranges from 8.6 to 275.02 mg/l; concentration of K^+ ranges from 0.61 to 12.59 mg/l; concentration of Ca^{2+} ranges between 74.06 and 51.22 mg/l; concentration of Mg^{2+} ranges from 6.07 to 136.87 mg/l with average of 62.47, 2.91, 178.51, 23.41 mg/l respectively. The relative concentrations of cations occurs in the order of $Ca^{2+} > Na^+ + K^+ > Mg^{2+}$. The sum of cations is 13.70 meq/l, otherwise the concentration of anions are: Cl^- ranges between

5.44 and 733.52 mg/l, SO_4^{2-} ranges from 23.61 to 368.11 mg/l and HCO_3^- ranges from 187.52 to 471.99 mg/l, with averages of 104.50, 194.4 and 304.23 mg/l respectively. The relative concentrations of anions are produced in the following order of HCO^{2-} > SO_4^{2-} > Cl^-. The sum of anions is 13.18 meq/l this situation represents a cation predominance in the general chemical balance of the groundwater. Finally, the TDS variation goes from 290.40 to 5122.60 mg/l denoting less concentration near to the recharge area (close to SMO in the southwest sector) and higher concentration near to de CP in the northeast portion of the study area after the presence of human activity in the citric regions, industrial activities and cities (LC and HC).

This situation explains the hydrochemical mineralization through the pathway of regional groundwater flow. Inverse Distance Weight (IDW) procedure was used to interpolate the chemical database created in GIS environment. A second polynomial order equation allowed obtaining interpolations of major cations and anions, the average values of each constituent were used to create the GIS-database.

Figure 3 represents the spatial variation of the piezometric water-table levels expressed in meter above sea levels, the values showed are the averages between two measure rounds, the maximum value ~450 masl is located in the west side of the study area and the lower value is around 295 masl around CP in the east part.

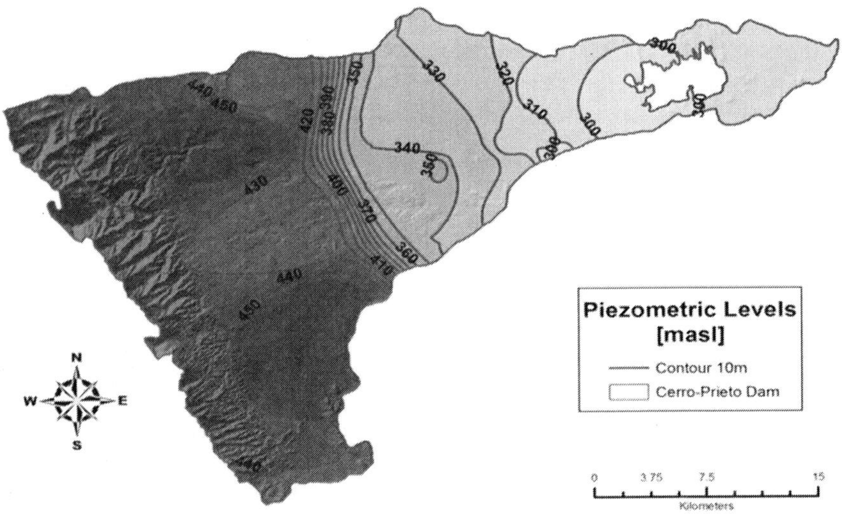

Figure 3. Piezometric maps of the study area.

Figure 4 shows the distribution of Total Dissolved Solids (TDS) the lower values are located close to de SMO where human activity is no present, these values are increased from west to east progressively and the major value is present close LC in the "Ejido La Escondida" reaching more than 6000 mg/l.

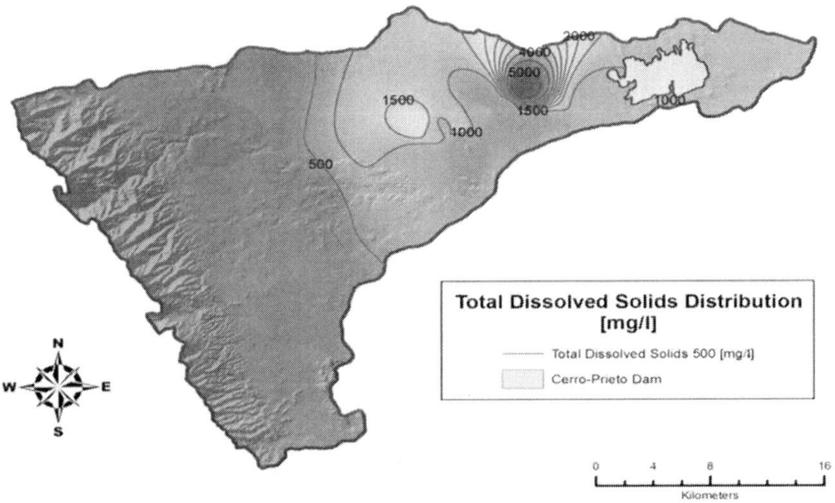

Figure 4. Total dissolved solids maps (TDS).

Sodium distribution is presented in **Figure 5**, the higher values are located in the central part and close to LC reaching ~2.75 mg/l and the lower values are placed in downstream direction near to CP water reservoir.

Potassium distribution is presented in **Figure 6** where the higher value (~23 mg/l) is located in the central portion of the study area.

In **Figure 7** the Calcium distribution is showed, the highest value (~511 mg/l) is near La Escondida (human settlement).

The general increment of Calcium is from west to east direction. **Figure 8** represents the spatial distribution of Magnesium, where the major concentration is located in La Escondida area reaching a value of ~143 mg/l.

The spatial distribution of anions like Chloride, Sulphate and Bicarbonate are showed in Figures 9-11 respectively, the highest values of Chloride, Sulphate and bicarbonate ~758, ~349 and ~522 mg/l respectively, as a general situation the values are increasing from west to east direction inside the study area.

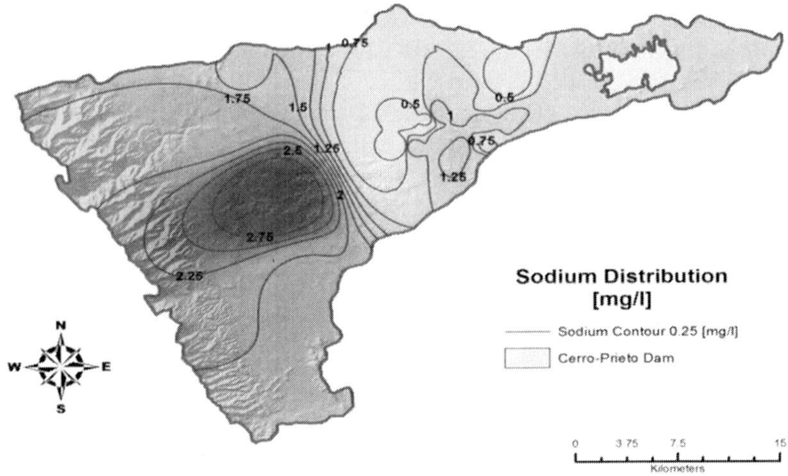

Figure 5. Sodium distribution.

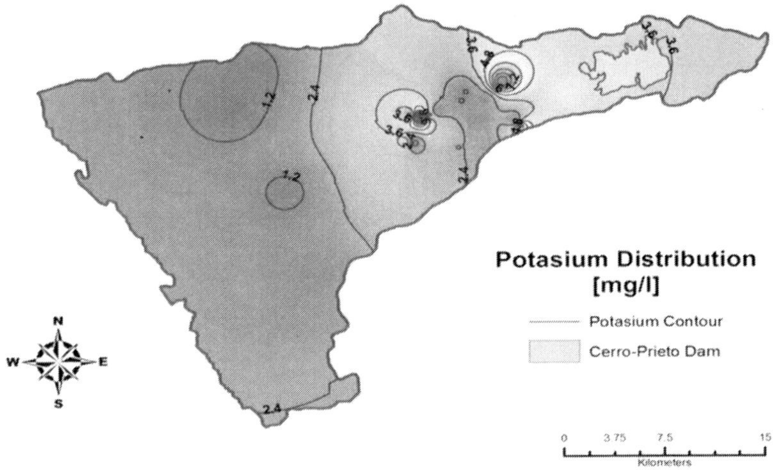

Figure 6. Potassium distribution.

The water types are categorized according to their content of major cations and anions by using the Piper diagram showed in the (**Figure 12**).

A correlation analysis of the major ions vs. TDS showed in **Figure 13** where, a) Mg^{2+}, b) Ca^{2+}, c) Cl^- and d) Na^+ have a coefficient of determination r^2 = 0.85, 0.80, 0.77 and 0.68, respectively. A possible reaction is input of Na^+ and Cl^- by dissolution of halite (NaCl), the possible sources of $Ca^{2+}$and Mg^{2+} could be the dissolution of dolomite, Calcite and Magnesite.

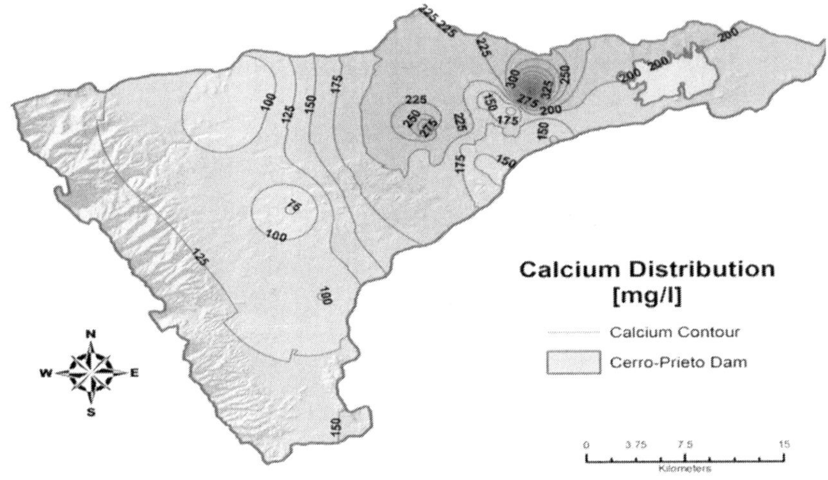

Figure 7. Calcium distribution.

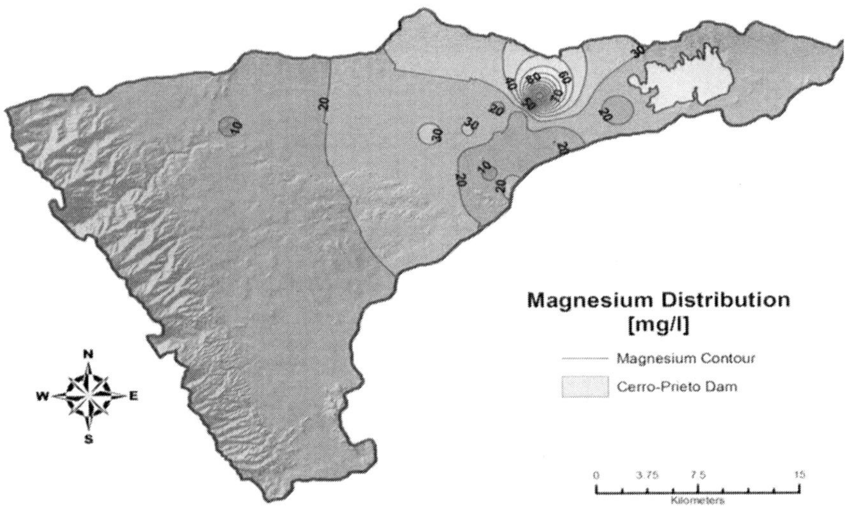

Figure 8. Magnesium distribution.

The relationship between Na^+ and Cl^- plotted in Figure 14(a), shows that the representative water samples of the aquifer are distributed close to the halite dissolution showing a weight ratio around 1, suggesting a similar origin of the water. The ratio Na^+/Cl^- (**Figure 14**(b)) is in general less than one expressing a possible recent recharge due to direct infiltration or river interconnection. The ratio of SO_4^{2-} /Cl⁻ plotted versus Chloride (**Figure 14**(c)), shows high ratios in the study area that means a high solubility of these ions suggesting a gypsum dissolution [20].

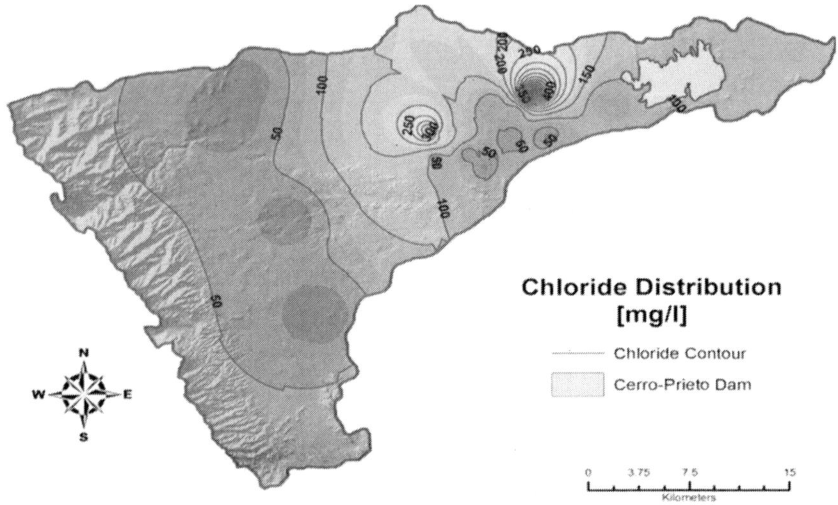

Figure 9. Chloride distribution.

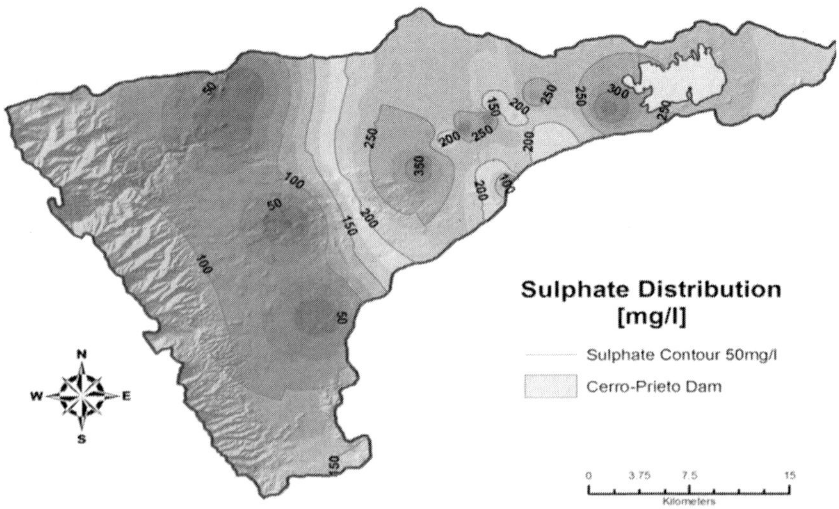

Figure 10. Sulfate distribution.

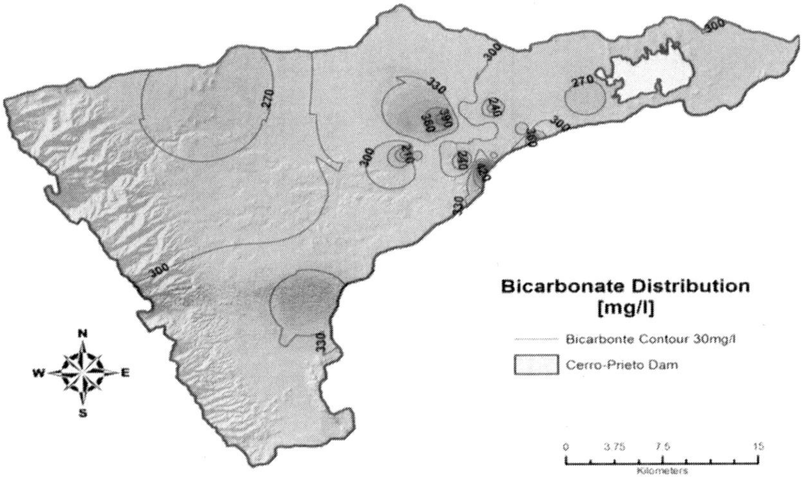

Figure 11. Bicarbonate distribution.

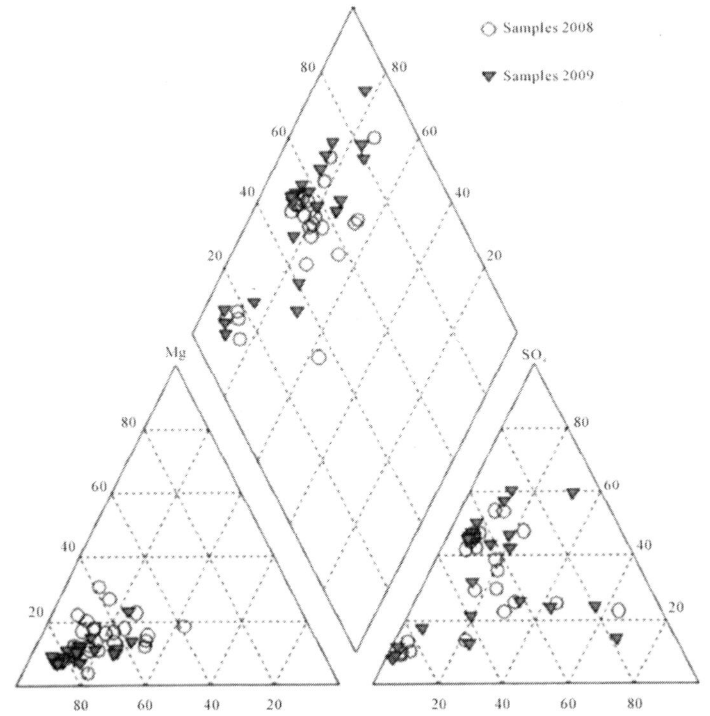

Figure 12. Piper diagram.

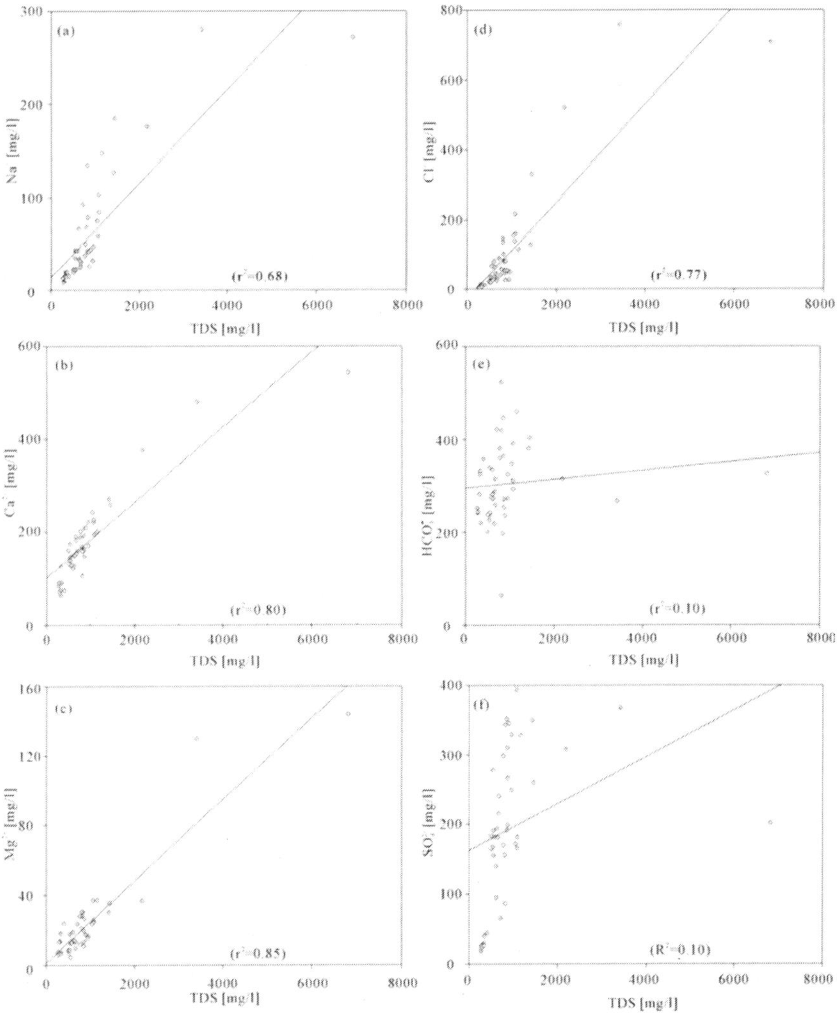

Figure 13. (a)-(f) Diagrams of variations between major ions and TDS in the groundwater.

6. CONCLUSIONS

Hydrochemistry data of water have been utilized to understand the origin, geochemical evolution and hydraulic interconnection of groundwater in Pablillo River Basin.

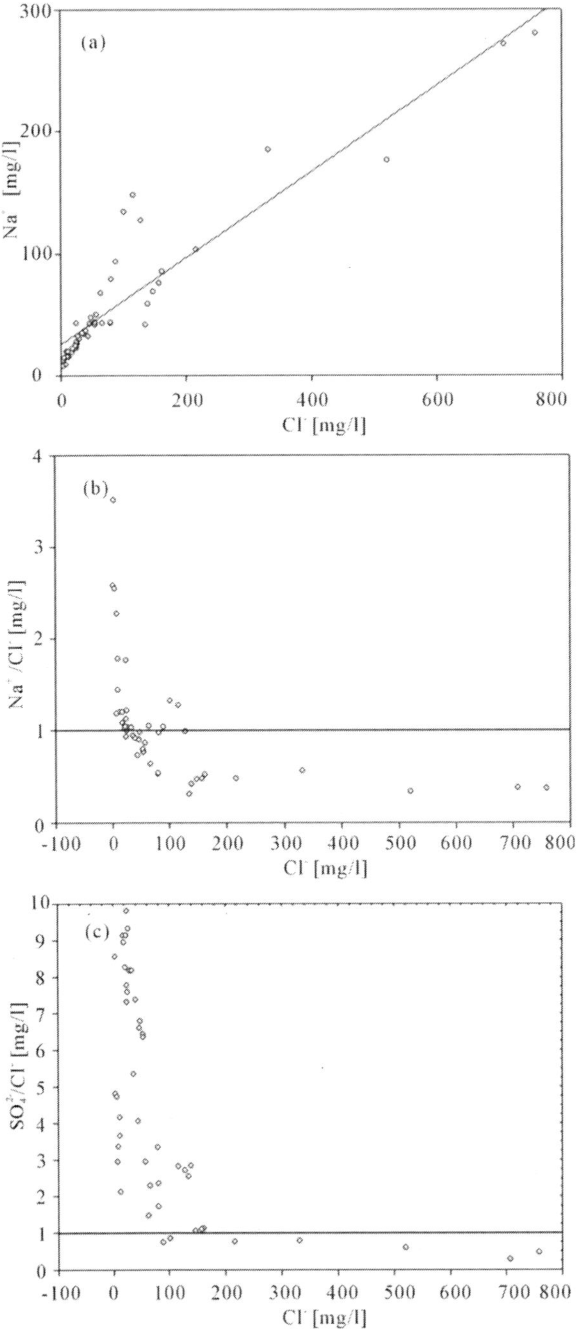

Figure 14. Relationship of Sodium, Na^+/Cl^- and SO_4^{2-} $/Cl^-$ ratios and chloride.

The main conclusions are as follows: the potentiometric map indicates the role of spatial recharge in the whole basin and in the region near to Sierra Madre Oriental that produce the alimentation of shallow water-table of the aquifers. Additionally, the hydrogeochemical data indicate that the fresh groundwater zones of the basin are mainly distributed close to the front pluvial fans of Sierra Madre Oriental.

The waters with more dissolved ions are predominant in the lower alluvial plains around Linares City, Hualahuises City and Cerro-Prieto dam. The concentration of TDS in groundwater increases along the groundwater flow from southwest to northeast and a very variable chemical composition of groundwater were discovered, e.g. EC varies in wide range between ~470 and 8240

μS/cm. The hydrochemical types in the area can be classified in three major types. 1) one type is Ca^{2+} - Na^{+} HCO_3^{-} predominance and located on the west side of the study area; 2) the second type is Ca^{2+} - Na^{+} - SO_4^{2-} predominance placed in the central portion close to the cities and 3) the third group has a Ca^{2+} - Mg^{2+} - Na^{+} present around La Escondida settlement in the northeast part.

The first and the second groundwater types maybe are affected by natural salinization due to: the dissolution of halite and gypsum and the third type is possible affected by the dissolution of dolomite and carbonates. More than a few mechanisms contribute to the later evolution of groundwater chemistry. In the upper topographic levels, a possible CO_2 gas dissolution and cation exchange predominates the water chemistry. These reactions can increase concentrations of $Na^{+}$and HCO_3^{-}. Moreover, in the lower alluvial plains, the geochemical evolution of groundwater is associated to mineral dissolution.

The SO_4^{2-} /Cl^{-} ratio is greater than 1 for the majority of samples in the all area, which suggests that gypsumhalite dissolution reactions take place under the condition of the rock weathering to some extent and evaporation is the dominant factor that determines the major ionic composition of groundwater. The overexploitation of groundwater in the citric region and cities intensifies the mixing of shallow groundwater (porous aquifer) and deep groundwater (fractured aquifer).

water affects its quality which can produce health problems in human beans. The groundwater pollution generated by anthropogenic activities should not be the only factor to be considerate in the quality of drinking water.

ACKNOWLEDGEMENTS

René Alberto Dávila Pórcel, Héctor de León-Gómez (DAAD-Germany) for support the Ph.D program. The authors would also like to acknowledge the associate editor and the anonymous reviewers for their valuable and constructive comments.

REFERENCES

1. CNA, "Determinación de la Disponibilidad de Agua en el Acuífero Citrícola sur, Estado de Nuevo León," Comisión Nacional del Agua, Monterrey, Nuevo León, 2002.
2. M. Jalali, "Geochemistry Characterization of Groundwater in an Agricultural Area of Razan, Hamadan, Iran," Environmental Geology, Vol. 56, No. 7, 2009, pp. 1479- 1488.
3. Q. Feng, G. Cheng and K. Masao, "Trends of Water Resources Development and Utilization in Arid North-West China," Environmental Geology, Vol. 39, No. 8, 2000, pp. 831-838.
4. L. Andre, M. Franceschi, P. Pouchan and O. Atteia, "Using Geochemical Data and Modelling to Enhance the Understanding of Groundwater Flow in a Regional Deep Aquifer, Aquitaine Basin South-West of France," Journal of Hydrology, Vol. 305, No. 1-4, 2005, pp. 40-65.
5. J. A. Ramos-Leal and R. Rodriguez-Castillo, "Aquifer Vulnerability Mapping in the Turbio River Valley, Mexico: A Validation Study," Geofísica Internacional, Vol. 42, No. 1, 2003, pp. 141-156.
6. INEGI, "Censo de Población y Vivienda 2010," 2013. http://www.inegi.org.mx/sistemas/consulta_resultados/iter2010.aspx?c=27329&s=est
7. Z. De Cserna, "Tectónica de la Sierra Madre Oriental de México, Entre Torreón y Monterrey, México," 20th Congreso Geológico Internacional, Mexico City, 1956, p. 87.
8. H. De León-Gómez, "Die Unterläufigkeit der Talsperre José López Portillo/Cerro Prieto Auf Einer Kalkstein— Mergelstein—Wechselfolge Bei Linares, Nuevo León/ México," Ph.D., Mittleilungen zur Ingenieurgeologie und Hydrogeologie, RWTH, Aachen Aachen, 1993.
9. N. Strahler, "Quantitative Analysis of Watershed Geomorphology," Transactions—American Geophysical Union, Vol. 38, No. 6, 1957, pp. 913-920.
10. S. I. De la Garza-González, "Estudio Geológico/Hidrogeológico de la Región Citrícola Linares-Hualahuises, Nuevo León, NE-México," Tesis de Licenciatura, Facultad de Ciencias de la Tierra, Universidad Autónoma de Nuevo León, Linares, 2000.
11. H. De León-Gómez, "Aspectos Hidrogeológicos GenRerales de la Cuenca del río Pablillo, Linares, N.L., Mé- xico," In: Primer Simposium Regional. Avances y Perspectivas de la Investigación de Recursos Ábioticos del Noreste de México: Agua, Clima, Suelo y Geología, Universidad Autónoma de Tamaulipas, Cd. Victoria, Tamaulipas, 2-4 de Abril de 1990, 1989.
12. J. M. Barbarín-Castillo, H. W. Hubberten, P. Meiburg and C. O. Rodríguez, "Hidrogeoquímica de las Aguas Termales del Baño San Ignacio, Linares, Nuevo León, Mé-xico," Actas Facultad Ciencias de la Tierra, Vol. 3, 1988, pp. 89-99.
13. S. M. Galván-Mancilla, "Cartografía Hidrogeológica de la Terraza Baja Entre Hualahuises y Linares, N.L.," Tesis de Licenciatura, Facultad de Ciencias de la Tierra, Universidad Autónoma de Nuevo León., Linares, 1996.
14. L. Lizárraga-Mendiola, H. De-León Gómez, F. Medina- Barrera and J. Návar, "Evaluation of the Impacted Aquifer by the Landfill of Linares, Mexico," Zbl. Geol. Paläont. Teil I, Vol. 23, 2004, pp. 225-244.

15. L. G. Lizárraga-Mendiola, "Análisis y Evaluación del Agua Subterránea del Área del Tiradero Municipal y La Petaca, Linares, N.L. México," Tesis de Maestria, Facultad de Ciencias de la Tierra, Universidad Autónoma de Nuevo León, Linares, 2003.
16. S. Schuster, "Ansätze für ein Hydrogeologisches Informationssystem für die Municipios de Linares und Hualahuises. Nuevo Leon," Diplomarbeit, Fakultät für Geowissenschaften, Universität Leipzig, Leipzig, 1999.
17. L. Moreno-Esparza, "Vulnerabilidad del Agua Subterránea a la Contaminación en la Región de Linares Mediante el Método DRASTIC," Engineering, Facultad de Ciencias de la Tierra, Universidad Autónoma de Nuevo León, Linares, 2009.
18. A. Aldama, J. Aparaicio, C. Gutiérrez-Ojeda, M. Martínez- Morales, L. González-Hita, G. Herrera-Zamarrón, et al., "Comportamiento Hidrogeológico de los Acuíferps Cuatrocienegas y El Hundido, Coahuila, México," Tecnología y Ciencias del Agua, Vol. 22, 2007, pp. 37-59.
19. M. Al-Shaibani, "Hydrogeology and Hydrochemistry of a Shallow Aluvial Aquifer, Western Saudi Arabia," Hydrogeology Journal, Vol. 16, No. 1, 2008, pp. 155- 165.
20. Y. Jiang, C. Zhang, D. Yuan, G. Zhang and R. He, "Impact of Land Use Change on Groundwater Quality in a Typical Karst Watershed of Southwest China: A Case Study of the Xiaojiang Watershed, Yunnan Province," Hydrogeology Journal, Vol. 16, No. 4, 2008, pp. 727- 735.

CHAPTER 10

Occurrence and Mobility of Mercury in Groundwater

Julia L. Barringer, Zoltan Szabo and Pamela A. Reilly

U.S. Geological Survey, USA

1. INTRODUCTION

1.1. Forms, toxicity, and health effects

Mercury (Hg) has long been identified as an element that is injurious, even lethal, to living organisms. Exposure to its inorganic form, mainly from elemental Hg (Hg(0)) vapor (Fitzgerald & Lamborg, 2007) can cause damage to respiratory, neural, and renal systems (Hutton, 1987; USEPA, 2012; WHO, 2012). The organic form, methylmercury (CH_3Hg^+; MeHg), is substantially more toxic than the inorganic form (Fitzgerald & Lamborg, 2007). Methylmercury attacks the nervous system and exposure can prove lethal, as demonstrated by well-known incidents such as those in 1956 in Minimata, Japan (Harada, 1995), and 1971 in rural Iraq (Bakir et al., 1973), where, in the former, industrial release of MeHg into coastal waters severely tainted the fish caught and eaten by the local population, and in the latter, grain seed treated with an organic mercurial fungicide was not planted, but eaten in bread instead. Resultant deaths are not known with certainty but have been estimated at about 100 and 500, respectively (Hutton, 1987). Absent such lethal accidents, human exposure to MeHg comes mainly from ingestion of piscivorous fish in which MeHg has accumulated, with potential fetal damage ascribed to high fish diets during their mothers' pregnancies (USEPA, 2001). Lesser human exposure occurs through ingestion of drinking water (USEPA, 2001), where concentrations of total Hg (THg; inorganic plus organic forms) typically are in the low nanograms-per-liter range[1] - , particularly from many groundwater sources, and concentrations at the microgram-per-liter level are rare.

1.2. Standards for mercury in water, and other regulations

For drinking water the World Health Organization (WHO) guideline for THg is 1,000 ng/L, a level (in some instances given as 0.001 mg/L or 1.0 μg/L) adopted by the European Union countries, the United Kingdom, Canada, and India (WHO, 1993; NIEA, 2011; Environment Canada, 2010; Srivastava, 2003). These standards are lower than the maximum contaminant level (MCL) adopted by the U.S. Environmental Protection Agency (USEPA), which is 2,000 ng/L (USEPA, 2001a). The USEPA has also adopted a reference dose (RfD) for MeHg in drinking water of 0.1 μg/kg bw/day (or 100 ng/kg bw/day, where bw = body weight) (USEPA 2001b).

Measures, such as regulations to restrict commerce in Hg, have been taken to reduce the amount of elemental Hg available globally. The amount mined in 2010, for example, was estimated at 2042 megagrams (Mg; or 2250 tons)(Brooks, 2010). In 2007, the European Union passed a ban on the export of Hg(0) and in 2008 this was expanded to include various Hg compounds. The USEPA, in 2008, passed a ban on the export of elemental Hg from the United States (USA), to take effect in 2013 (USEPA 2009). Because the USA is ranked as one of the world's top exporters of Hg (~417 Mg in 2010 (Brooks, 2010)), implementation of the act will remove a substantial amount from the global market, as it is exported to foreign countries where, among other uses it is employed in small-scale gold (artisanal) mining (USEPA, 2012). China, the world leader in Hg production in 2010, plans to lower production of some metals, including Hg, by 2015 (Brooks, 2010).

1.3. Previous studies of mercury in the environment

Given the potentially severe health effects of MeHg, the need to understand the production of MeHg and its bioaccumulation in the food web, as well as the role of atmospheric deposition in supplying Hg to soils and surface water, has led to numerous investigations of Hg inputs and transformations. Hence, the majority of studies and reviews of Hg in the environment focus on atmospheric inputs (e.g., Engstrom et al., 2007; Glass et al., 1991; Pacyna et al., 2006; Schuster et al., 2002) and the fate and transport of Hg in soils, sediments and surface-water settings (e.g., Bradley et al., 2011; Driscoll et al., 1994; Grigal., 2002; Porvari et al., 2003; Rudd, 1995; Shanley et al., 2002; Schuster et al., 2008; Selvendiran et al., 2008; Skyllberg, 2008; Ullrich et al., 2001; Yu et al., 2012). Many of these studies have their emphasis on the production, mobility, and bioaccumulation of MeHg. Far fewer studies have been performed and published on Hg in groundwater. Some of those have shown that Hg concentrations are low, in the nanograms-per-liter range (e.g. Krabbenhoft & Babiarz, 1992). However, other studies have shown that regional levels of Hg in groundwater can be high, in the micrograms-per-liter range (e.g. Barringer & Szabo, 2006), and some show that microgram-per-liter levels of Hg are being found in groundwaters previously not tested (e.g. Khatiwada et al., 2002).

As discussed below, not all contamination with Hg at the land surface causes contamination of groundwater. It is critical to know the complex biogeochemistry of Hg in order to understand how existing important research applies to Hg fate and transport in groundwater systems.

2. SOURCES AND FLUXES OF MERCURY

2.1. Mercury in natural materials

Mercury is one of the least abundant elements in crustal rocks. Concentrations in rocks of the upper continental crust typically range from 0.01 to about 2 mg/kg, although concentrations higher by more than two orders of magnitude are reported from igneous and sedimentary rocks of Crimea and the Donets Basin of Russia (Fleischer, 1970). A reasonable estimate of the average concentration of THg is 0.08 mg/kg (Fairbridge, 1972). The most abundant Hg mineral is cinnabar (HgS), which can be found in association with metacinnabar (β HgS—a metastable sulfide). Other Hg minerals, mainly sulfosalts that can also contain arsenic, are not common, and, when found, are generally associated with ore deposits.

Volcanic emissions have been estimated by some researchers to constitute the main natural source of Hg to the atmosphere where Hg is primarily present as gaseous elemental Hg (Martin et al., 2012). Kilauea, in Hawaii, was estimated to produce about 260,000 kg/yr (Siegel & Siegel, 1987). In contrast, Ferrara et al. (2000) suggest that volcanic emissions are not the main natural source of THg to the atmosphere, based on reported estimates of emissions from Mediterranean volcanoes: Vulcano, about 1 to 6 kg/yr; Stromboli, about 7 to 80 kg/yr; and Etna, about 60 to 500 kg/yr.

Volcanic emissions can cause local contamination of soils and surface waters, but, as shown at Mt. Etna in Sicily, concentrations of Hg in local groundwater were typically < 10 ng/L (Martin et al., 2012), indicating that Hg had not been easily mobilized from the land surface. Nevertheless, some of the Hg deposited to soils by volcanic emissions volatilizes and can be re-emitted to the atmosphere (Fig. 1). Further, evasion of Hg(0) from the ocean surface adds substantially to atmospheric Hg (Mason et al., 1994).

Characterization of atmospheric THg along the central USA Gulf Coast has shown inputs from sea spray increases atmospheric deposition of THg in coastal regions (Engle et al., 2008).

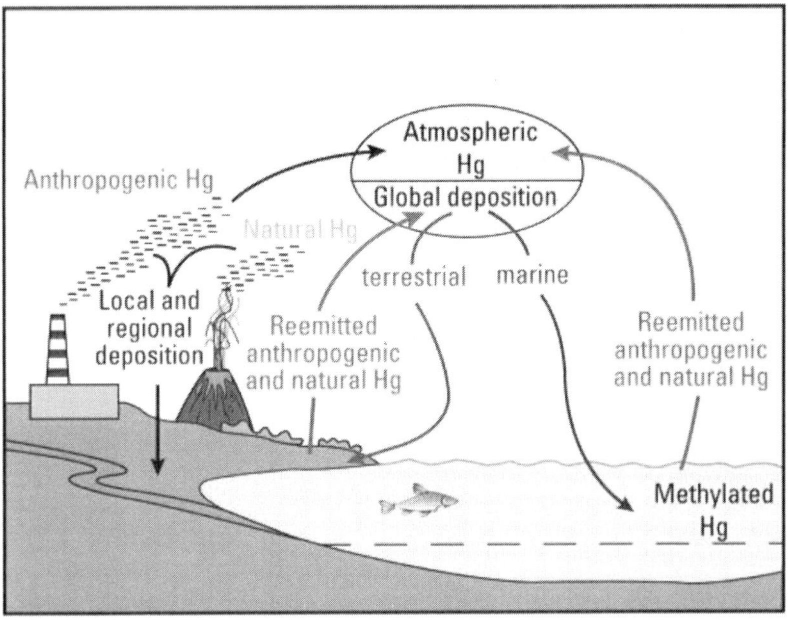

Figure 1. Sources and biogeochemical cycling of mercury between land surface, surface water, and the atmosphere (from Tewalt et al., 2001).

Mineralization associated with igneous activity has produced volcanic-hosted massive sulfide deposits, in which the ore mineral cinnabar forms (some mined since the third century B.C. (Navarro, 2008)). Such major geologic sources of Hg are found in Spain, Slovenia, China, and the western USA, in California. Volcanically derived sedimentary deposits containing other sulfides also can contain substantial Hg (Navarro 2008).

Associated with igneous activity, circulating geothermal fluids can contribute Hg to groundwater, surface water, and to solids that precipitate around mineral springs, geysers and fumaroles. In Russia, Yudovich and Ketris (2005) report that Hg contents in condensate from fumarole gases ranged up to 0.11 mg/kg at the Kamchatka Peninsula, and even higher (up to 0.40 mg/kg) at fumaroles in Japan and Guatemala. In Yellowstone National Park in the western USA, Hg is present at concentrations of 12 to 640 ng/L in waters of hot springs and geysers (Ball et al., 2006), although Hg minerals were not recognized in precipitates there (White et al., 1970).

Saline waters at Sulfur Bank and Wilbur Springs in the California Coast Range contained about 1,500 ng/L of Hg, and cinnabar and metacinnabar precipitate at Sulfur Bank (White et al., 1970).

Fleischer (1970) compiled literature values for Hg in crustal rocks, worldwide. Average values for mafic igneous rocks ranged from 0.001 to 0.240 mg/kg, whereas for silicic igneous rock, averages ranged from 0.005 to 0.190 mg/kg, except for igneous rocks from the Crimea and Donets Basin, where the average Hg contents ranged from 0.250 to 17.6 mg/kg. Average Hg contents of sandstones and limestones ranged from 0.018 to 5.70 mg/kg, and in shales from 0.05 to 2.3 mg/kg, with the higher contents typically found in the Russian shales. In metamorphic rocks, contents ranged from 0.060 to 2.50 mg/kg.

Many of the widely used sources of coal contain high concentrations of Hg; these include from China (Pirrone et al., 2010), Ukraine (Kolker et al., 2009), and Texas, USA (Tewalt et al., 2001). Coal burning is a major source of Hg release into the atmosphere (Wang et al., 2004); but potential effects on groundwater are mostly unknown. The Hg content of coals ranges from 0.01 to 1.85 mg/kg, with the highest content found in some Chinese coals (Pirrone et al., 2010 and references therein). Much of the Hg in coals is in associated pyrites, with contents that can be 10 mg/kg or greater (Yudovich & Ketris, 2005). Direct effects from coal seams on groundwater may not be obvious, however. Cravotta (2008) did not find detectable Hg concentrations in samples of waters discharging from abandoned coal mines in Pennsylvania, USA.

2.2. Mercury in atmospheric deposition and its effects

Emissions of Hg to the atmosphere have increased greatly since the beginning of the industrial era, although estimates of the increase in atmospheric levels vary widely—from about 3 to 24 times that of the pre-industrial period (Wang et al., 2004, and references therein). Recently, it has been estimated that total global emissions account for about 6500 megagrams (Mg) of Hg released annually to the environment. The uncertainty for current estimates is about a factor of two, according to Lohman et al., (2008). Nevertheless, as Gustin et al. (2008) point out, estimates of natural emissions vary widely (volcanic emissions are not constant, for example). Gustin et al. (2008) also note that the range in estimates of anthropogenic releases is small, relative to the range of estimates for natural sources. Coal burning and combustion of other fossil fuels were estimated to constitute about 60 % of the annual amount contributed to the atmosphere from anthropogenic sources (Swain et al., 2007). Total Hg contents of ice-core samples from the Upper Fremont Glacier in Wyoming, USA, when integrated over the past 270 years indicated that anthropogenic emissions accounted for 52%, volcanic events contributed 6%, and background sources supplied 42% (Schuster et al., 2002), and most of the anthropogenic contributions have occurred since about 1850. A current estimate of total annual emissions, both natural and anthropogenic, is about 7,300 Mg (Pirrone et al., 2010) (Table 1), higher than earlier estimates, and also with a higher uncertainty. In terms of regional emissions, those from Asia have increased from 38% to 64% of global emissions over the period of 1990-2007 (Pirrone et al., 2010).

Table 1. Estimates[a] of global mercury emissions from natural and anthropogenic sources, and year(s) for which estimate is made. [Mg, megagrams; Data from Pirrone et al., 2010 and references therein]

Source	**Mercury[e] (Mg/yr)**	**Year for estimate**
Natural		
Oceans	2,680	2008
Lakes	96	2008
Forests	342	2008
Tundra/grasslands[b]	448	2008
Desert/non vegetated areas[c]	546	2008
Volcanoes and geothermal areas	90	2008
Anthropogenic		
Non-ferrous metal production	310	2003-06
Pig iron and steel production	43	2003-06
Cement production	236	2003-06
Caustic soda production (chlor alkali process)	163	2003-06
Mercury production	50	2003-06
Gold production	400	2003-06
Waste disposal	187	2003-06
Stationary (fossil fuel) combustion	810	2003-06
Coal-bed fires	32	2003-06

Vinyl chloride monomer production	24	2003-06
Other	65	2003-06
Biomass burning[d]	675	2008
Agricultural areas[d]	128	2008

[i] - [a] the mean uncertainty associated with these estimates is + 25% (Pirrone et al., 2010)

[ii] - [b] Includes savannah, prairie, chaparral.

[iii] - [c] Includes metalliferrous areas.

[iv] - [d] Amounts may need reconfiguring from Pirrone et al.(2010) category of natural sources, p. 5953, because some burning is caused by humans, agricultural inputs of Hg are anthropogenic, and evasion is amplified by soil disturbance.

[v] - [e] Values have been rounded to the nearest whole number.Global production (mining and processing) of Hg has been reduced since the mid 20th century, although present-day production may still contribute about one third of emitted Hg from anthropogenic sources (Hylander & Meili, 2003). Amounts of Hg in atmospheric deposition are declining in some parts of the USA as there have been efforts to curb industrial and power-plant emissions. For example, in a study of Hg loading to Minnesota (USA) lakes, Engstrom et al. (2007) found that inputs, mainly from atmospheric deposition and subsequent soil erosion, peaked in the 1970s, and have declined substantially in recent years. Additionally, in the upper Great Lakes (Superior & Huron) in the USA and Canada, a decline in Hg in fish tissue is noted, although not in the lower Great Lakes (Erie & Ontario) (Bhavsar et al., 2010).

Atmospherically deposited Hg has affected mainly surface water and the organisms that live in water bodies. Hg deposited as Hg(0) may be oxidized to Hg(II), then transformed to MeHg by bacterial activity at and below the sediment/water interface or in algal mats. Low-trophic-level organisms (invertebrates) take up both THg and MeHg (Fig. 1). The concentration of MeHg in organisms increases with each step up the food chain—a process known as biomagnification (Alpers & Hunerlach, 2000; USGS, 2000). During the late 20th century, emissions from coal-fired plants in the Midwest of the USA, which are mainly in vapor (Hg(0)) form (Lindberg, 1987), deposited Hg on surface-water bodies. These emissions and other industrial emanations have resulted in fish consumption advisories because of high levels in the tissues of edible fish (Brooks, 2002). Advisories for non-commercial fish, as of 2006, now extend to freshwaters in 48 USA States, of which 23 are statewide advisories. In addition, 13 States have coastal and estuarine advisories (USEPA, 2010).

Results of two studies suggest that Hg from atmospheric deposition contributes to elevated Hg concentrations in groundwater. Bradley et al. (2012, p. 7507) found that strong hydraulic gradients toward a stream indicated deep groundwater discharge was the primary source of filtered Hg (FTHg) to a Coastal Plain stream, USA. Additionally, higher concentrations of FTHg in deeper wells near the stream (compared to shallower wells) and a lack of geologic Hg deposits in soils and sediments were "consistent with atmospheric Hg deposition on the terrestrial landscape as the distal source of FTHg to groundwater." Barringer and Szabo (2006) contend that Hg deposited from the atmosphere to soils of the Coastal Plain of southern New Jersey, USA, may be mobilized, along with pesticide residues, to groundwater where Hg concentrations in domestic well water are found to exceed the State and USEPA MCL of 2,000 ng/L.

2.3. Effects of mercury from manufacturing and agriculture on soils, surface water and groundwater

Mercury in consumer products and their manufacture constitutes a source of Hg to aquatic systems. Mercury is used in dry-cell batteries, fluorescent light bulbs and thermostats, and, in the mid-to late 20th century, as a fungicide in paints and in wood preservatives; Hg is also used in dental amalgams (Barringer et al., 1997; Brooks, 2000, 2002; USEPA, 2006). Use of Hg in dentistry has consumed large amounts of the metal (about 63,500 kg in Europe), resulting in emissions as well as constituting a substantial portion of the Hg in municipal waste streams (Hylander et al., 2006; Metro, 1991). Several studies, summarized by Morrison (1981), indicated that subsurface disposal of municipal sewage sludge in the USA resulted in several metals, including Hg, leaching to groundwater. Since the mid 20th century, industrial use of Hg has declined in the USA and elsewhere, however. There are now ongoing attempts in some countries, such as the USA and the United Kingdom, to reduce the amount of Hg used in products; particularly those that are then discarded (Bradley & Journey, 2012).

Groundwater contamination with THg from both inactive and active industrial sources is found in many countries where, in general, the contamination is relatively local. Chlor-alkali plants that once produced chlorine gas and sodium hydroxide using Hg-cell technology have been shown to contribute Hg to surface water, soils and groundwater—up to 22,900,000 ng/L in groundwater at a facility in Sydney, Australia (Orica, 2012). There are numerous such facilities; other examples are chlor-alkali plants in New York State, USA, and in Kazakhstan, which have also produced Hg contamination of surrounding soils and waters (Gbondo-Tugbawa & Driscoll, 1998; Ullrich et al., 2007). The use of Hg-cell technology is being voluntarily phased out in most countries; however, wastes from chlor alkali plants still remain at some locations (Hylander & Meili, 2003).

In some instances, Hg contamination of groundwater is more diffuse, perhaps coming from multiple (potentially small) point sources. Suspected industrial and wastewater discharges have introduced Hg contamination to the surficial alluvial aquifer in urban Madras, India (Somasundaram et al., 1993), where, along with other metals, Hg concentrations are reported to range from 1,000 to 18,000 ng/L in water from wells near the River Cooum in the Madras urban area, mainly exceeding the Indian drinking-water standard of 1,000 ng/L.

In some instances, Hg released from industrial operations results in contamination of soils at the site, but the soil characteristics are such that the Hg is sequestered or attenuated, and does not leach to groundwater, as found, for example, at a site in Trondheim, Norway (Saether et al., 1997). In other instances, Hg contamination of groundwater is noted locally near industrial sites. At a site in southern Germany where "kyanizing" (mercuric chloride ($HgCl_2$) used to preserve wood from decay) was performed, groundwater in a 1.3-km-long plume was found to be contaminated with Hg at concentrations that reached 230,000 ng/L (Bollen et al., 2008). $HgCl_2$ is extremely soluble in water, (7.4×10^{10} ng/L; O'Neil, 2006), and thus easily mobilized from inputs at the land surface to the water table.

Mercury has been used in munitions and for other purposes on military bases. This use has, in some cases, led to contamination of soils at the bases (Bricka et al., 1994); with the potential for further contamination of surface and groundwaters. Mercury contamination of soils, surface water and groundwater is found at a former military base in southern New Jersey, USA. The Hg content of soils ranged up to 555 mg/kg, and concentrations in some streamwater and groundwater samples exceeded 2,000 ng/L; determining whether the Hg derives solely from military activities is a part of ongoing investigations there (Barringer et al., 2012).

From the early-to-late 20th century, Hg was used in agricultural pesticides, but use of Hg compounds decreased substantially after 1970 in the USA (Murphy & Aucott, 1999). Mercurial compounds also were once used on golf courses in the USA, being most heavily applied in northern States to control snow mold. On a golf course in New Hampshire, USA, Estes et al. (1973) estimated that annual fungicide applications contributed 2.1 kg of Hg per hectare. In Australia, sugar-cane setts were treated with a fungicide containing methoxymethylmercury chloride before planting; concentrations of Hg from 30 to 670 ng/L were found in groundwater underlying cane cropland, but were interpreted as being within the range of naturally occurring Hg concentrations in area groundwater (Brodie et al., 1984). It is apparent that not all applications of Hg to soils result in groundwater contamination, but there are instances where Hg can be mobilized from soils to groundwater.

In the early to mid-20th century, phenylmercuric acetate was used in orchards; calomel and $HgCl_2$were used on row crops; these are known to be used on sandy agricultural soils of the Coastal Plain of southern New Jersey, USA (Murphy & Aucott, 1999). Inputs of Hg would have been high if the federally recommended application rates of 3.4 kg/hectare for highly soluble $HgCl_2$ were followed (Barringer et al., 1997 and references therein). Mobilization of Hg from applications of mercurial compounds, enhanced by subsequent disturbance from residential development of the land, may be the cause of elevated Hg concentrations on groundwater downgradient from such sources. Water from wells completed in the quartz sand aquifer beneath the former agricultural land contains Hg at concentrations exceeding 2,000 ng/L (Barringer et al., 2005; Barringer & Szabo, 2006).

In addition to pesticides, Hg also could be introduced to soils via fertilizers. A commercial 20-20-20 fertilizer solution prepared according to manufacturer's directions contained 280,000 ng/L of Hg (Barringer & MacLeod, 2001). Mercury was measured in several common fertilizers, with the highest concentration (5.1 mg/kg) found in calcium superphosphate, and a lower concentration (1.2 mg/kg) in 15-5-5 Nitrogen-Phosphorus-Potassium (NPK) fertilizer (Zhao & Wang, 2010). Some States in the USA are acting to regulate the amount of metals permissible in fertilizers and other agricultural chemicals (e.g. ODA, 2002).

2.4. Effects of human activities on naturally occurring mercury

Seawater intrusion along coasts typically occurs because withdrawals of freshwater resources reduce the freshwater hydraulic head, allowing seawater to enter aquifers on land. In southern Tuscany, Italy, the high chloride concentrations brought in by seawater intrusion may have been responsible for mobilizing Hg in the geologic materials to groundwater, perhaps as a chloride complex (Grassi & Netti, 2000; Protano et al., 2000). In Sardinia, dewatering of a lead-zinc (Pb-Zn) mine resulted in intrusion of deeper saline water that may have been responsible for increased Hg concentrations in shallow groundwater (Cidu et al., 2001).

Experiments support these interpretations; Behra (1986) showed that elevated chloride (Cl^-) concentrations mobilized Hg(II) in columns packed with quartz sand.

Contamination of soils, surface water and groundwater also arises from mining of cinnabar deposits. Mining of other metal ores has also resulted in Hg contamination, such as at the McLaughlin gold-mercury deposit in California (Sherlock, 2005), because of trace amounts of cinnabar or occurrence of Hg with other sulfide minerals. As a result of early 20th century copper mining activities along the Alaskan coast, USA, dissolved concentrations of Hg up to 4,100 ng/L were found in pore waters of sediments affected by mining waste (Koski et al., 2008).

In a Canadian gold and silver mine, accessory cinnabar in waste piles oxidized and leached to groundwater, resulting in concentrations that ranged up to 150,000 ng/L (Foucher et al., 2012). Not all mining activities and mining waste disposal procedures result in extremely high levels of Hg in groundwaters, however; concentrations in adit water samples from an abandoned mercury mine in Turkey were still high relative to most waters, but ranged from 250 to 274 ng/L (Gemici, 2008).

The process of amalgamation of silver and gold with Hg was developed in the 16th century and used on an industrial scale into the 19th century in Central and South Americas, primarily to extract silver, and into the 20th century in North America for gold extraction (Nriagu, 1994). Elemental Hg was lost to the environment during the process; Hg losses during gold and silver extraction in the Americas are estimated to be about 240,000 Mg (Nriagu, 1994). Similar extraction activities have taken place in the Philippines, Indonesia, Thailand, Vietnam, Tanzania, and China (Lacerda, 1997), though the scale is not as great as in the Americas. Although some of this mining-activity-related Hg has volatilized, adding to the atmospheric Hg burden, much of it apparently still remains in the areas where metal processing took place.

A few of these mining/extraction sites have been studied. In Tanzania and Zimbabwe, Hg sorption to iron-rich lateritic soils appears to have prevented groundwater contamination from Hg in tailings (van Straaten, 2000). Extreme surface-water loads have been documented from many of these mines including the Sierra Nevada gold mines in the USA (Domagalski, 1998), the Idrija Hg mine in Slovenia (Covelli et al., 2007), as well as Hg mining in Spain (Navarro, 2008) among others, with Hg sorption to iron-hydroxide-rich stream deposits a likely mechanism for Hg attenuation in some cases (Rytuba, 2000). Use of Hg in small-scale (artisanal) gold mining operations remains a substantial concern even today (Bradley & Journey, 2012).

3. BIOGEOCHEMISTRY OF MERCURY: FIELD AND EXPERIMENTAL STUDIES IN SOIL AND SURFACE-WATER ENVIRONMENTS

3.1. Impact of mercury reactivity on transport

Oxidation-reduction, precipitation-dissolution, aqueous complexation, and adsorption-desorption reactions will strongly influence the fate and transport of Hg in groundwater, and in the environment, generally. At the land surface, Hg participates in photochemical reactions (see the review of Zhang, 2006), but these reactions are not relevant to groundwater. Biogeochemical reactions in soils are of great importance to the fate and transport of Hg, however. Characteristics of soils, which include pH, carbon content, mineralogy, drainage properties, slope, and texture, all play a role in Hg retention or mobility and whether THg inputs to the land surface reach the water table. Concentrations of THg typically are higher in organic soil horizons than in the deeper mineral horizons because Hg typically is closely associated with organic matter (Amirbahman & Fernandez, 2012), and Andersson (1979) reports sorption to iron oxides (typical of some temperate-climate subsoils, and also tropical soils) at pH > 5.5. The reactions described below can occur in soils, in the surface-water environment, and, apparently, in groundwater as well.

3.2. Oxidation-reduction and sorption reactions

The three stable oxidations states of Hg in low-temperature aquatic systems are Hg(II), Hg(I) and Hg(0). The mercurous (Hg(I)) species is stable over a more limited range of conditions in sulfidic aqueous systems than it is when sulfur is absent (Hem, 1970). The Hg species vary in their solubility, complexation, adsorption (Stumm and Morgan, 1995) and their availability for microbial processes. Therefore, oxidation-reduction (redox) reactions will have a profound influence on Hg concentrations and mobility in groundwater. Both abiotic and biotic (primarily microbial) processes can drive Hg redox transformations.

Iron geochemistry is intimately associated with that of Hg. Anaerobic column experiments showed transport of Hg(II) retarded by sorption (as a Hg-Cl complex) to pyrite (FeS_2) (Bower, et al., 2008), and Hg(II)) has been shown to sorb to iron oxides at pH > 5.5 (Andersson, 1979). Given a positive association of Hg with iron (Fe) in iron-hydroxide-rich sub-soils in the New Jersey Coastal Plain, USA, sorption of Hg to Fe hydroxides appears to be a mechanism for attenuating Hg (Barringer & Szabo, 2006). The same mechanism appears present at some mining sites (Rytuba, 2000; van Staaten, 2000), although formation of aqueous and solid-phase sulfides controls Hg(II) concentrations in tailings-contaminated sediments from California, USA, mines (Rytuba et al., 2005). Fe(II) hydroxides can be reductively dissolved by sulfide, resulting in the release of sorbed Hg. Experiments showed that, in the presence of sulfide (S^{2-}), Fe (III) was reduced and concentrations of dissolved Hg increased (Slowey & Brown, 2007). It appears that these and the experiments of Bower et al. (2008) were not done in the dark, however. Consequently, applicability of results to a groundwater setting is not clear.

Field examples also demonstrate that oxygen-depleted conditions caused by septic-system-effluent releases led to reductive dissolution of Fe hydroxides, resulting in release of sorbed Hg(II) (Barringer & MacLeod, 2001). Further, the Fe(II) generated in such a reaction may adsorb to minerals where it can then reduce Hg(II) to Hg(0) (Charlet et al., 2002). Recent experiments show that Fe(II) in minerals also can reduce Hg(II) to Hg(0). For example, in sealed, dark bottles, magnetite was found to reduce Hg(II) to Hg(0) within minutes (Wiatrowski et al., 2009; Yee et al., 2010). Mercury (Hg(II)) was also rapidly reduced in anoxic solutions by Fe(II) under varying pH conditions, with aqueous $Fe(OH)^{+}$being the species that best described the electron transfer that occurred in the experiments (Amirbahman et al., 2012). Metals other than iron, such as tin, are also known to reduce Hg(II) to Hg(0) (e.g., Biester et al., 2000).

Natural organic matter has been shown to abiotically reduce Hg(II) to Hg(0) (Allard & Arsenie, 1991). Experiments under dark anoxic conditions by Gu et al. (2011) showed that dissolved organic matter (DOM) reduced Hg(II) to Hg(0) when low concentrations of DOM were present. At higher DOM concentrations, however, complexation with Hg inhibited Hg reduction reactions.

Microbially mediated redox reactions involving Hg also have been demonstrated. Hg(II) was reduced to gaseous Hg(0) by a *Pseudomonas* strain (Baldi et al., 1993). A newly isolated *mer*A-carrying *Bradyrhizobium* bacterium recently was found by Wang et al. (2012) to also reduce Hg(II) to Hg(0). The recent work by Wang et al. (2012) shows that Hg inhibits denitrification in groundwater in direct proportion to the concentration of Hg.

3.3. Organic and inorganic complexation

Mercury (Hg(II)) can be present not only as Hg^{2+}, but as $Hg(OH)_2$, $HgCl_2$ and other minor OH^- and Cl^- complexes, and in complexes with various organic anions, depending on pH, Eh, chloride concentrations and presence of DOM. Were DOM to be low or absent, then Hg could be present as hydroxide or chloride complexes in fresh waters (Reimers & Krenkel, 1974; Stumm & Morgan, 1995); at low to moderate pH and moderate to high chloride concentrations, chloride complexes would be most likely (Ravichandran, 2004). In the presence of dissolved sulfide, mercury-sulfide species may form (Benoit et al., 2001).

Mercury tends to form strong complexes with S^{2-} and, in DOM, Hg(II) binds preferentially to sulfur-containing functional groups such as thiols (Gabriel and Williamson, 2004; Ravichandran, 2004; Reimers & Krenkel, 1974). In anoxic environments, Hg can form complexes such as dissolved HgS, HgS_2^{2-}, $Hg(SH)_2$, $HgSH^+$, HgOHSH and HgClSH (Gabriel & Williamson, 2004). Although metals typically bind to acid sites (carboxyls, phenols, ammonium, alcohols, and thiols) in organic matter, Hg(II) binds preferentially with thiols and other reduced sulfur groups with which it forms strong covalent-like bonds. These sulfur-bearing groups are found in moderate abundance in organic matter in soils, in some surface water, and in wastewater (Hsu-Kim & Sedlack, 2003; Ravichandran, 2004). When the Hg/DOM ratio is high (> 10,000 ng Hg to 1 mg DOM), however, Hg also binds to the more abundant but less Hg-selective oxygen (ie., carboxyl) functional groups (Haitzer et al., 2002). Further, binding of DOM with Hg(II) is less strong at low pH than at high pH (Haitzer et al, 2003); this occurs because the extent of protonation of functional groups serving as Hg(II) ligands on DOM increases as pH decreases. Given the affinity of Hg(II) for thiol groups on DOM, it has been shown that DOM can dissolve cinnabar, inhibiting or preventing precipitation of metacinnabar and aggregation of HgS nanoparticles (Ravichandran et al., 1999; Reddy & Aiken, 2001; Slowey, 2010; Waples et al., 2005).

3.4. Microbial transformations

An important transformation of inorganic Hg involves its methylation to monomethyl- or dimethyl-mercury. Methylation of Hg(II) in soils and surface-water was found to be carried out under anoxic conditions by dissimilatory sulfate-reducing bacteria (DSRBs) (Gilmour et al., 1992). Dissimilatory iron-reducing bacteria (DIRBs) later were found to be able to methylate Hg(II) as well (e.g., Kerin et al., 2006). Populations of both DSRBs and DIRBs have been found to coexist in stream-bottom sediments where fine-grained sediments were "potential hot spots for both methylation and demethylation activities" (Yu et al., 2012).

Further, at low sulfate (SO_4^{2-}) concentrations, the methylating activity of SRBs is stimulated, but at high concentrations the methylating activity is inhibited because precipitated sulfides incorporate the Hg (Ullrich et al., 2001). Concentrations of SO_4^{2-} between 0.2 and 0.5 mM (about 19 to 48 mg/L) appeared to be optimum for promoting Hg methylation in freshwater (Gilmour and Henry, 1991). Barkay et al., (1997) discovered that high concentrations of DOM and salinity inhibited Hg(II) methylation because the Hg was complexed into forms that were not bioavailable to the methylating bacteria. The Hg in aqueous HgS complexes, which form in the presence of dissolved sulfide, was found to be bioavailable to the methylating bacteria, however (Benoit et al., 1999). Recent research shows that, although DOM can inhibit Hg bioavailability by complexing the Hg, DOM can also prevent HgS nanoparticles from aggregating, and thus the nanoparticles are bioavailable (Graham et al., 2012).

Mercury demethylation has also been studied in stream and lake sediments (e.g., Achá et al, 2011; Hintelmann et al., 2000; Pak & Bartha, 1998; Steffan et al., 1988). In experiments using sediments from southern New Jersey lakes, USA, Pak and Bartha (1998) showed that demethylation of MeHg is carried out by sulfidogenic and methanogenic bacteria, which are obligate anaerobes. Although the Hg methylation process was inhibited by low pH (4.4) conditions (which are common in southern New Jersey surface waters and groundwaters), demethylation of MeHg did not appear to be similarly affected for the pH range 4.4 to 8; inhibition occurred at pH < 4.4 (Steffan et al., 1988).

3.5. Colloids and particles

Extensive sorption of Hg(II) can limit concentration and mobility in groundwater unless the Hg(II) binds to colloidal solids under conditions where the colloids are stable and mobile. Colloids (particles < 1 μm in one dimension (Kretchmaar & Schafer, 2005)) in waters provide transport for various contaminants and generally are sufficiently small so as to pass the 0.45 μm pore-size filters used by most researchers in collected filtered water samples. Because of their large surface area relative to their volume, small particles and colloids can provide many sorption sites for strongly sorbing contaminants whose mobility would otherwise be minimal through soils and aquifers. Such movement can be triggered by chemical or physical disturbance of soils and sediments. For example, Hg sorbed to particles was released to runoff from boreal forest soils following clear cutting and scarification (Porvari et al., 2003).

Colloids can be formed by clay minerals; oxides and hydroxides of iron, aluminum, and manganese; silica; humic and fulvic acids; carbonates; phosphates (Ryan & Gschwend, 1994), also bacteria; and viruses (Kretchmaar & Schafer, 2005). Colloids are common in surface waters, soil and sediment porewaters. Colloids are found in groundwater as well. Changes in pH and redox reactions can cause dissolution or precipitation reactions that can form or release colloidal material. Examples are precipitation of colloid-sized minerals such as iron hydroxides when Fe(II) is oxidized, and oxides and carbonates that would precipitate at high pH.

In a southern New Jersey, USA, aquifer with Hg contamination, a study found colloids were more abundant in anoxic groundwater of an undeveloped area than in oxic groundwater (Ryan & Gschwend, 1990). The greater abundance likely occurred because iron hydroxide cements that bound clays to quartz-grain surfaces were being dissolved, liberating both Fe(II) and clay particles to solution (Ryan & Gschwend, 1994). With fluctuating water tables, some of the Fe(II) could be re-oxidized, forming colloidal precipitates.

In groundwater, colloids are subject to forces exerted by pumping. Sequential sampling of domestic wells in the unconfined aquifer system of southern New Jersey found that particulate Hg concentrations were commonly higher in first-draw water samples than in samples collected later during well purging (Szabo et al., 2010).

4. BACKGROUND CONCENTRATIONS OF THG IN GROUNDWATER AND GROUNDWATER/SURFACE-WATER INTERACTIONS

"Background" concentrations—that is—naturally occurring concentrations of THg (and MeHg) in groundwater probably depend upon ambient geochemical conditions, which would include pH that favors adsorption or desorption, presence and amount of Fe and DOM, and oxidizing or reducing conditions favorable to mineral precipitation or dissolution or microbial activities, including Hg(II) methylation. Absent known sources of contamination, background concentrations of THg in groundwater in several studies were found to be < 10 ng/L (Andren & Nriagu, 1979; Barringer & Szabo, 2006; Krabbenhoft & Babiarz, 1992; Kowalski et al., 2007). Total mercury concentrations in other groundwater studies, depending on filtration or lack thereof, ranged from <5 to 210 ng/L (Wiklander, 1969/1970; Reimann et al., 1999), with concentrations generally higher in unfiltered samples because of particulate material[7] - .

Relatively few studies have examined interactions between groundwater and surface water. In the Everglades swamp area in Florida, USA, groundwater pumping, dredging of canals, levee construction, and land subsidence have altered area hydrology (Harvey et al., 2002). In the surficial aquifer, THg, which is an element of concern because of severe MeHg impacts on Everglades biota, is recharged from surface water to groundwater, with higher concentrations (0.8 to 2.7 ng/L) tending to be in recharge from agricultural areas. Methylmercury (0.2 ng/L) was found only in groundwater recharged from agricultural areas and was not detected in groundwater elsewhere (Harvey et al. 2002). In a study of Hg inputs to Lake Superior, USA, however, groundwater was found to be an important source of MeHg to the lake, with concentrations as high as 12 ng/L in a hyporheic-zone sample (Stoor et al., 2006).

At a lake in glacial outwash (Wisconsin, USA), groundwater (sampled by piezometers and dug wells) both discharges to the lake and receives recharge from lake waters. Mercury enters from atmospheric deposition to the lake, and, apparently, through soils to groundwater. Mercury concentrations in groundwater discharge (mean 12 ng/L) to the lake was higher than that of water from nearby wells (mean 2.8 ng/L), showing the importance of reactions near and at the sediment/water interface (Krabbenhoft & Babiarz, 1992). In a New Jersey, USA, Coastal Plain watershed, groundwater discharging to a major river contained concentrations of THg in urban areas (some mainly in particulate form) that were higher than those in forested wetlands areas. Concentrations of THg in unfiltered water were 36 and 177 ng/L in discharge to the river at two sites in an urban area (Barringer et al., 2010a) and were not representative of background concentrations for that aquifer, which typically are < 10 ng/L (Barringer & Szabo, 2006). Bradley et al. (2012) also found Hg in groundwater discharge to be an important input to a southern USA Coastal Plain stream, although the THg concentrations in the groundwater were an order of magnitude lower than the concentrations in urban groundwater discharge in the study of Barringer et al. (2010a).

5. BIOGEOCHEMICAL PROCESSES AND MECHANISMS FOR MOBILIZING MERCURY IN GROUNDWATER — CASE STUDIES

5.1. New jersey coastal plain, usa

At more than 70 residential sites underlain by an areally extensive (7,770 km^2) unconfined non-calcareous quartz sand aquifer system in the Coastal Plain of New Jersey, USA, water from domestic wells has been found to contain Hg at concentrations that exceed the State MCL of 2,000 ng/L. In the same aquifer system, background levels of THg in the groundwater in neighboring forested areas and unaffected residential areas are <10 ng/L (Barringer et al., 1997; Barringer & Szabo, 2006). This highly permeable system is vulnerable to contamination, with a water-table depth commonly less than 7 m below land surface. Past agricultural use of mercurial pesticides and atmospheric deposition are two likely sources of THg to the aquifer. Currently, about 700 wells are known to have withdrawn water containing THg at concentrations ranging as high as 80,000 ng/L (Dr. Judith Louis, New Jersey Department of Environmental Protection, 2010, oral commun.); the New Jersey Department of Environmental Protection supplies treatment systems or connections to public supplies for the affected households. A small number of public-supply wells have also been affected in urban/suburban areas (Fischer et al., 2010). Groundwater contaminated with THg was found in a similar Coastal Plain aquifer in the neighboring State of Delaware (Koterba et al., 2006).

Redox reactions were found to be important in explaining the mobility of Hg in the New Jersey Coastal Plain aquifer system. Septic-system effluent discharges, which were ubiquitous in the unsewered residential areas with Hg-contaminated well water, apparently drove redox reactions that influenced THg mobility in the aquifer. Discharges with elevated concentrations of electron donors promoted reductive dissolution of Fe hydroxide coatings on subsoil and aquifer sediment grains and sorbed Hg (presumably as Hg(II)), was released to groundwater (Barringer & MacLeod, 2001; Barringer et al., 2006; Barringer & Szabo, 2006). Septage and leach-field effluent contained THg concentrations in the range <20 to 60 ng/L. Therefore the effluent is seen, not as a prominent source of Hg, but as a source of electron donors (or reduced solutes) capable of driving redox reactions, such as dissolution of Fe hydroxides and reduction of Hg(II) to Hg(0), that favor increased concentrations and mobility of Hg.

The THg in New Jersey groundwater is hypothesized to be mobilized from soils. The mobility of Hg in soils was tested during U.S. Geological Survey (USGS) studies by pulsed leaching of samples of disturbed subsoils with de-ionized water, with artificial road-salt solution and with 20-20-20 fertilizer solution. In all cases the Hg removed was particulate, and was associated with removal of particulate Fe, indicating that the most likely mode of transport for the THg from the mineral soil horizons was in association with Fe hydroxide particles (Reilly et al., 2012; Barringer et al., (2012)). In the Fe hydroxide-rich subsoils, the introduction of septic systems, and onset of reducing conditions in some instances provided the conditions for reductive dissolution of hydroxides and release of sorbed Hg(II) to the water table. From the water table, gradients toward numerous pumping domestic wells brought mobile Hg to levels in the aquifer where drinking-water supplies were impacted. The question then arises—in what form is the Hg mobile?

Although well-water samples were not analyzed for Hg(0) in the USGS studies in the New Jersey Coastal Plain, geochemical modeling results indicated that Hg in the groundwater could be reduced and that Hg(0) would be a likely species in the groundwater (Barringer & Szabo, 2006). The presence of Hg(0) in water from the Coastal Plain aquifer system has been shown by sampling at the military base mentioned previously, where reducing conditions are present in DOM-rich wetland areas (Barringer et al., 2012). As mentioned earlier, experimental studies have shown that reduction of Hg(II) to Hg(0) by sorbed Fe(II) is sufficiently fast to promote mobilization as Hg(0) (Amirbahman et al, 2012; Charlet et al., 2002). Because Hg(0) is slightly soluble in water (56,000 ng/L at 25° (O'Neil, 2006)), it is likely that, in parts of the aquifer system affected by effluent discharges, Hg(0) could form and be mobile as dissolved Hg within the aquifer.

Mercury in the New Jersey Coastal Plain aquifer also appears to be mobile in particulate and colloidal form. The differences between unfiltered Hg concentrations and filtered concentrations (passing 0.45- μm pore-size filters) is large for some groundwater samples; the magnitude of this difference is interpreted as the concentration of particulates greater than 0.45 μm in diameter (Barringer et al., 2006; Barringer & Szabo, 2006). The finding of Hg on particulates/colloids indicates presence of THg, either as oxidized Hg(II) or as reduced Hg(0) (Bouffard & Amyot (2009) sorbed to the mobile particles. Currently (2012) in New Jersey, the composition of the particulates is not known at the various residential Hg contamination sites—they could be Fe hydroxides (Ryan & Gschwend, 1990), clay minerals (Ryan & Gschwend, 1994), organic material (Hurley et al., 1998), mixed organo-oxide colloids (Chadwick et al., 2006), or sulfide particles (Slowey et al., 2005), depending on the geochemical environment in which they form. The presence of particulates indicates that the dissolved THg concentrations measured in the New Jersey Coastal Plain groundwater represent only some fraction of the total pool of THg present in parts of the aquifer system

5.2. Nepal

An incidence of Hg in groundwater in Nepal bears some resemblance to the New Jersey Coastal Plain contamination. In Nepal, public supply wells, deep private wells, and shallow dug wells, hand pumps and spouts completed in gravelly unconfined and confined aquifers were sampled. Twenty-three of the samples from 31 sampling sites contained Hg at concentrations that exceed the WHO Guideline of 1,000 ng/L; the highest concentration measured was about 300,000 ng/L in water from a dug well in an urban area. Concentrations of nitrogen species (NH_3 and NO_3^-) and dissolved organic carbon (DOC) were particularly high in several of the well water samples (ammonia (NH_3), up to 62 mg/L as N; nitrate (NO_3^-), 25 mg/L as N; and DOC, 63.6 mg/L), and reducing (anoxic) conditions were apparent in the confined aquifer (Khatiwada et al., 2002). No suggestions as to the source of the Hg were given, but the elevated nutrient and organic carbon levels are suggestive of anthropogenic inputs such as sewage that either include Hg, or that mobilize naturally occurring Hg, presumably under reducing conditions.

5.3. Cape cod, massachusetts, usa

The importance of redox conditions in Hg release and mobility also is demonstrated in recent research at the USGS research site downgradient from the Massachusetts Military Reservation on Western Cape Cod. Land disposal of treated wastewater to the unconsolidated sands and gravels of the shallow aquifer from the 1930s to December 1995 resulted in development of a plume of contaminated groundwater that has been investigated since the early 1980s (LeBlanc, 1984). In addition to standard monitoring wells, the area affected by the plume has been instrumented with numerous multi-level samplers (MLS) (LeBlanc et al, 1991). These MLS allow collection of point samples from distinct biogeochemical zones, which would otherwise be difficult owing to steep vertical gradients in groundwater chemistry in the plume. An extensive suboxic zone with elevated nitrate concentrations and denitrification, and an anoxic zone with dissimilatory iron reduction (DIR) have persisted for more than 15 years following cessation of inputs from the source (Repert et al. 2006; Savoie et al., 2012). A recent study that examined Hg fate and transport within the plume found that Hg(0) constituted > 50% of the Hg present in the DIR zone near the source. About 1-2 km downgradient from the source, the anoxic zone had essentially no dissolved iron, but concentrations of both ammonium (NH_4^+) and NO_3^- were high. Methylmercury comprised nearly 100% of the dissolved THg present in some samples from this region of the plume. Under the original infiltration beds, THg concentrations ranged up to about 200 ng/L (1,000 pM), but concentrations were rapidly attenuated with distance downgradient from the beds. Concentrations of dissolved THg in the oxic, uncontaminated groundwater, where pH values ranged from 5.0 to 5.6, were about 0.2 ng/L. The study shows there are at least two distinct redox environments and two different microbial regimes operating within the plume. However, the distribution of dissolved THg suggests that downgradient from the source, dissolved THg has been mainly mobilized from the non-calcareous, quartzitic aquifer sediments rather than being transported long distances in the wastewater stream (Carl Lamborg, Woods Hole Oceanographic Institution & Douglas Kent, USGS, 2012, written commun.).

5.4. Maine, USA

Crystalline rocks can contain Hg that may become mobile. Mercury in water from domestic wells completed in fractured rock aquifers of the Waldoboro pluton complex in Maine, USA, exceeded the USEPA drinking water standard of 2,000 ng/L (Sidle, 1993). Most high Hg concentrations were found in water from granitoid rock aquifers that contain several joint sets as well as cataclasite-fault breccias zones and shear-mylonite zones. Mineral-rich pegmatite dikes follow one of the joint sets in the Waldoboro pluton. A few anomalously high concentrations occurred in water from wells completed in associated metamorphic rock aquifers (gneisses and amphibolites) of the Bucksport Formation.

The Hg content of the rocks varied substantially, from 0.005 to about 500 mg/kg, with a median content of 78 mg/kg. Mercury concentrations in well-water samples ranged from 40 to 6200 ng/L. Mercury in surficial sediments, which contained glacially-derived deposits, ranged up to about 0.100 mg/kg. The presumption is that the Hg derived from the geologic materials, and that the fracturing aided in transporting Hg-rich water through the aquifers. There were no other chemical data reported for the domestic wells, and therefore it is not clear whether natural weathering of the Hg-enriched rocks was the cause of the high THg levels measured in well water, or whether inputs of anthropogenic chemicals could have mobilized Hg from the geologic substrate.

5.5. Groundwater discharge to coastal areas

Largely within the last decade, researchers have turned to an examination of the impacts of fresh groundwater inputs on coastal waters. Groundwater that discharges to bays and estuaries has been found to contribute nutrients to coastal waters (e.g., Slomp & van Cappellen, 2004), and recent research has expanded to examine contributions of trace elements from this source. Within the past 5 years, at the volcanic island of Jeju, south of Korea, on the south coast of the English Channel at Caux, France, and along both east and west coasts of the USA, submarine groundwater discharge (SGD) of Hg and MeHg to estuaries and bays has been investigated (Black et al., 2009; Bone et al, 2007; Ganguli et al., 2012; Laurier et al., 2007; Lee et al., 2011).

Mass balance indicated that Hg in SGD at Jeju Island constituted 34 and 67% of the total Hg annually in waters of two bays, whereas atmospheric deposition contributed from 23 to 25% of the Hg. Methylmercury in SGD at the island also constituted the majority of MeHg in the coastal waters (Lee et al., 2011).

At Waquoit Bay in Massachusetts, USA, sampling of SGD showed that Hg concentrations in groundwater ranged from <0.64 ng/L to 52.4 ng/L, and that daily discharge of Hg was from 94 to 380 ng/m^2 (Bone et al., 2007). Further, total dissolved Hg and DOC were not correlated, a lack of relation that was also observed in some of the New Jersey, USA, Coastal Plain studies, particularly at THg concentrations much higher than those found in waters discharging to Waquoit Bay.

At Caux, France, the Hg concentrations in tissues of blue mussels (*Mytilus edulis*) were higher than elsewhere along the coast, prompting an investigation of Hg in SGD. Although the total Hg concentrations were not greatly elevated at Caux (a mean of 0.73 ng/L), they were higher than in the nearby industrialized Seine estuary (mean of 0.32 ng/L) and the dissolved Hg likely represented a source of bioavailable Hg to the mussel population. Dissolved MeHg concentrations at Caux constituted from 2.4 to 7.8% of the total Hg present in SGD (Laurier, et al., 2007).

On the west coast of the USA, north of San Francisco, SGD contributed from 0.24 ng/L to 5.7 ng/L to California coastal waters; the Hg concentrations were significantly correlated with NH_4^+ and silica (SiO_2) (Black et al., 2009). Farther south, at Malibu Lagoon, SGD also transported Hg and MeHg to coastal water. Mixing between groundwater and seawater was inferred. MeHg concentrations in seawater increased at low tide, as did filtered Hg concentrations; inputs from SGD were thought to change Hg partitioning and solubility in the seawater (Ganguli et al., 2012).

6. CONCEPTUAL MODEL FOR PROCESSES INFLUENCING MERCURY FATE AND TRANSPORT IN GROUNDWATER

Overall, the studies of the Hg-contaminated groundwater in New Jersey and Cape Cod, in particular, have provided opportunities to investigate mechanisms for releasing Hg from the land surface to the water table, and to suggest further avenues to explore biogeochemical reactions that mobilize Hg from the subsurface and within aquifers. A conceptual model for Hg mobilization has been developed by New Jersey researchers (Reilly et al., 2011) and is shown in figure 2.

Mercury, either naturally occurring, of anthropogenic origin, or both, is released to the water table as Hg(II) from surface soils and subsoils by weathering or by inputs of anthropogenic chemicals such as road salt or fertilizers, or by subsurface inputs of septic-system effluent. Under oxidizing conditions, dissolved THg is mobile as a complex with DOM, and sorbed Hg(II) is mobile on Fe hydroxide particles. Effluent discharges provide electron donors and sorbed Hg(II) is released as Fe hydroxides reductively dissolve, and Hg(II) may be reduced either by DOM or by Fe(II). Where anoxic conditions are present, sulfate reduction is an important terminal electron accepting process, and methylation may take place. Additionally, sulfides may precipitate, removing Hg from the aqueous phase. Hg(0) may be re-oxidized to Hg(II) should groundwater become more enriched in oxygen farther down a flowpath.

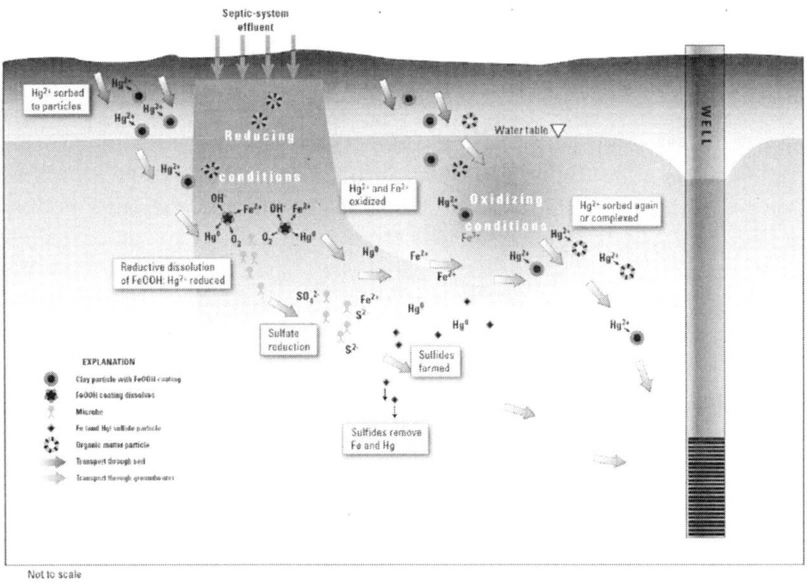

Figure 2. Mobilization of mercury from land surface to groundwater and biogeochemical transformations along flow paths in an unconsolidated sandy, acidic aquifer.

7. CONCLUSIONS

Mercury is relatively rare compared to most other elements, but owing to its toxicity at low concentrations, Hg is an important potential contaminant. There is a large reservoir of inorganic Hg in the environment—much of it derived from human activities, most associated with industrialization. Some of that Hg enters freshwater supplies where conditions may be conducive to methylation and thus the production of MeHg that readily bioaccumulates. Relatively less Hg is mobilized to groundwater than to surface water, in part because Hg can be attenuated by sorption to clays, iron oxides, and residual soil organic matter. Studies of the fate and transport of Hg in the subsurface are beginning to reveal how transport from land surface to groundwater might occur and how Hg remains mobile within aquifers.

In none of the above studies of Hg discharge to coastal waters have the sources of the Hg in SGD been identified, nor have the mechanisms for maintaining Hg mobility in groundwater discharging to the coasts been discerned. Given the recent discoveries of Hg and MeHg inputs from SGD, there clearly are avenues for further investigations into this phenomenon. Uptake into estuarine biota in the biodiverse and biomass rich estuarine and coastal waters is of key concern for these sensitive ecosystems and for human health, given the great potential for biomagnification. The importance of groundwater inputs thus cannot be ignored.

Studies in New Jersey and Cape Cod, USA, have investigated processes leading to Hg mobility in groundwater in settings that involve inputs from sewage to the subsurface. Given that sewage effluent contains materials that can fundamentally alter biogeochemical environments, mobilization of metals such as Hg, whatever their origin, may be an ever increasing process as humans continue to develop their surroundings. It is hoped that the research, past and ongoing, that is discussed herein will be of use to readers who seek to understand, to prevent, or to mitigate Hg contamination of groundwater supplies.

(Any use of trade, product, or firm names is for descriptive purposes only and does not imply endorsement by the U.S. Government.)

REFERENCES

1. Achá, D., Hintelmann, H., & Yee, J. (2011). Importance of sulfate reducing bacteria inmercury methylation and demethylation in periphyton from Bolivian Amazon region. *Chemosphere,* 82, 911-916.
2. Allard, B. & Arsenie, I. (1991). Abiotic reduction of mercury by humic substances inaquatic systems—an important process for the mercury cycle. *Water, Air, and Soil Pollution,*56, 457-464.
3. Alpers, C.N. & Hunerlach, M.P. (2000). Mercury contamination from historic goldmining in California. U.S. Geological Survey Fact Sheet FS-061-00. Amirbahman, A. & Fernandez, I.J. (2012). Chapter 7. The role of soils in storage and cycling of mercury. In. Bank, M.S. (ed.) Mercury in the environment: Pattern and Process. Berkely, University of California Press, 97-116.
4. Amirbahman, A., Kent, D.B., & Curtis, G.P. (2012). Kinetics of abiotic mercury (II) reductionby iron (II). (Abstract), Program: BIOGEOMON, the 7th International Symposium on Ecosystem Behavior, July 15-20. 2012, Northport, Maine.
5. Andersson, A. (1979). Mercury in soils. 79-112. In: Nriagu, J.O., (ed). The biogeochemistryof mercury in the environment. Amsterdam:, The Netherlands: Elsevier/North Holland Biomedical Press.

6. Andren, A.W. & Nriago, J.O. (1979). The global cycle of mercury. 1-21. In. Nriagu,J.O., (ed.), The biogeochemistry of mercury in the environment. Amsterdam, The Netherlands: Elsevier/North Holland Biomedical Press.
7. Bakir, F., Damluji, S.F., Amin-Zaki, L., Murtada, M., Khalidi, A., al-Rawi, N.Y., Tikriti,S., Dahahir, H.I., Clarkson, T.W., Smith, J.C., & Doherty, R.A. (1973). Methylmercurypoisoning in Iraq. *Science*, 181, 230-241.
8. Baldi, F., Parati, F., Semplici, F. & Tandoi, V. (1993). Biological removal of inorganicHg(II) as gaseous elemental Hg(0) by continuous culture of a Hg-resistant *Pseudomonas-Putida* strain FB-1, *World Journal of Microbiology and Biotechnology*, 9, 275.
9. Ball, J.W., McCleskey, R.B., Nordstrom, D.K., & Holloway, J.M. (2006). Water-chemistrydata for selected springs, geysers, and streams in Yellowstone National Park,Wyoming, 2003-2005. *U.S. Geological Survey Open-File Report 2006-1339*. 137 p.
10. Barkay, T., Gillman, M. & Turner, R.R. (1997). Effects of dissolved organic carbon andsalinity on bioavailability of mercury. *Applied and Environmental Microbiology*, 63,4267-4271.
11. Barringer, J.L., & MacLeod, C.L. (2001). Relation of mercury to other chemical constituents in ground water in the Kirkwood-Cohansey aquifer system, New Jersey CoastalPlain, and mechanisms for mobilization of mercury from sediments to ground water. *U.S. Geological Survey Water-Resources Investigations Report*, 00-4230.
12. Barringer,.J.L., MacLeod, C.L., & Gallagher, R.A. (1997). Mercury in ground water,soils, and sediments of the Kirkwood-Cohansey aquifer system in the New Jersey Coastal Plain. *U.S. Geological Survey Open-File Report 95-475.*
13. Barringer, J.L., Riskin, M.L., Szabo, Z., Reilly, P.A., Rosman, R., Bonin, J.L., Fischer,J.M. & Heckathorn, H.A. (2010a). Mercury & methylmercury dynamics in a CoastalPlain watershed, New Jersey, USA. *Water, Air, and Soil Pollution*, 212, 251-273.
14. Barringer, J.L., & Szabo, Z. (2006). Overview of investigations into mercury inground water, soils, and septage, New Jersey Coastal Plain. *Water Air, and Soil Pollution*,175, 193-221.
15. Barringer, J.L., Szabo, Z., Kauffmann, L.J., Barringer, T.H., Stackelberg, P.E. Ivahnenko,T., Rajagopalan, S. & Krabbenhoft, D.P. (2005). Mercury concentrations in waterfrom an unconfined aquifer system, New Jersey Coastal Plain. *Science of the Total Environment*,346, 169-183.
16. Barringer, J.L., Szabo, Z., Reilly, P.A. & Riskin, M.L. (2010b) Mobilization of mercury to a first-order Coastal Plain stream in New Jersey: Geological Society of America Abstracts, v. 42, no. 1, p. 154, Abstract 63-6.
17. Barringer, J.L., Szabo, Z., Schneider, D., Atkinson, W.D. & Gallagher, R.A. (2006).Mercury in ground water septage, leach-field effluent, and soils in residential areas,New Jersey Coastal Plain. *Science of the Total Environment*, 361, 144-162.
18. Barringer, J.L., Szabo, Z. & Reilly, P.A. (2012). Mercury in waters, soils and sediments of the New Jersey Coastal Plain: A comparison of regional distribution and mobility to the mercury contamination at the William J. Hughes Technical Center, AtlanticCounty, New Jersey. *U.S. Geological Survey Scientific Investigations Report 2012-5115*
19. Behra, P. (1986). Evidences for the existence of a retention phenomenon during themigration of a mercurial solution through a saturated porous medium. *Geoderma*, 38,209-222.
20. Benoit, J.M., Gilmour, C.C., Mason, R.P. & Heyes, A. (1999). Sulfide controls on mercury speciation and bioavailability to methylating bacteria in sediment pore waters. *Environmental Science and Technology*, 33, 951-957.
21. Benoit, J.M., Mason, R.P., Gilmour, C.C. & Aiken, G.R. (2001). Constants for mercurybinding by dissolved organic matter isolated from the Florida Everglades. *Geochimicaet Cosmochimica Acta*, 65, 4445-4451.
22. Bhavsar, S.P., Gewurtz, S.B., McGoldrick, D.J., Keir, M.J. & Backus, S.M., (2010).Changes in mercury levels in Great Lakes fish between 1970s and 2007. *EnvironmentalScience and Technology*, 44, 3273-3279.

23. Biester, H., Schihmacher, P. & Müller, G. (2000). Effectiveness of mossy tin filters toremove mercury from aqueous solution by Hg(II) reduction and Hg(0) amalgamation.*Water Research*, 34, 2031-2036.
24. Black, F.J., Paytan, A., Knee, K.L., de Sieyes, N.R., Ganguli, P.M., Gray, E. & Flegal, R.(2009). Submarine groundwater discharge of total mercury and monomethylmercuryto Central California coastal waters. *Environmental Science and Technology*, 43,5652-5659.
25. Bollen, A., Wenke, A. & Biester, H. (2008). Mercury speciation analyses in HgCl2-contaminatedsoils and groundwater—implications for risk assessment and remediationstrategies. *Water Research*, 42, 91-100.
26. Bone, S.E., Charette, M.A., Lamborg, C.H. & Gonneea, M.E. (2007). Has submarinegroundwater discharge been overlooked as a source of mercury to coastal waters?*Environmental Science and Technology*, 41, 3090-3095.
27. Bouffard & Amyot (2009). Importance of elemental mercury in lake sediments. *Chemosphere*,74, 1098-1103.
28. Bower, J., Savage, K.S., Weinman, B., Barnett, W.P. & Harper, W.F. (2008). Immobilizationof mercury by pyrite (FeS2), *Environmental Pollution*, 156, 504-514.
29. Bradley, P., Burns, D., Murray, K., Brigham, M., Button, D., Chasar, L., Marvin-Di-Pasquale, M., Lowery, M. & Journey, C.A. (2011). Spatial and seasonal variability ofdissolved methylmercury in two stream basins in the eastern United States. *Environmental*
30. Bradley, P. & Journey, C. (2012). Hydrology and methylmercury availability inCoastal Plain streams. Chapter 8, 169-190. In: Water Resources Management and Modeling: Nayak, P., (ed.). InTech Open Access Publishing. Available at http://www.intechopen.com/books/ water-resources-management-and-modeling.
31. Bradley, P.M., Journey, C.A., Lowery, M.A., Brigham, M.E., Burns, D.A., Button,D.T., Chapelle, F.H., Luz, M.A., Marvin-DiPasquale, M.C. & Riva-Murray, K. (2012).Shallow groundwater mercury supply in a Coastal Plain stream. *Environmental Scienceand Technology*, 46, 7503-7511.
32. Bricka, R.M., Williford, C.W. & Jones, L.W. (1994). Heavy metal soil contamination atU.S. Army installations: Proposed research and strategy for technology development.Technical Report IRRP-94-1, U.S. Army Corps of Engineers, Waterways ExperimentStation, Vicksburg, MS.
33. Brodie, J.E., Hicks, W.S., Richards, G.N. & Thomas, F.G. (1984). Residues related to agricultural chemicals in the groundwaters of the Burdekin River Delta, North Queensland. *Environmental Pollution (Series B)* 8, 187-215.
34. Brooks S.C. & Southworth, G.R. (2011). History of mercury use and environ-mentalcontamination at the Oak Ridge Y-12 Plant. *Environmental Pollution*, 159, 219-228.
35. Brooks, W.E., (2000). Mercury. U.S. Geological Survey 2000 Minerals Yearbook.
36. Brooks, W.E., (2002). Mercury. U.S. Geological Survey 2002 Minerals Yearbook.
37. Chadwick, J.P., Babiarz, C.L., Hurley, J.P. & Armstrong, D.E. (2006). Influences ofiron, manganese, and dissolved organic carbon on the hypolimnetic cycling ofamended mercury. *Science of the Total Environment*, 306, 177-188.
38. Charlet, L., Bosbach, D. & Peretyashko, T. (2002). Natural attenuation of TCE, As, Hglinked to oxidation of Fe(II): an AFM study. *Chemical Geology*, 190, 303-319.
39. Cidu, R., Biagini, C., Fanfani, L., La Ruffa, G. & Marras, I. (2001) Mine closure at Monteponi (Italy): effect of the cessation of dewatering on the quality of shallow groundwater. *Applied Geochemistry*, 16, 489-502.
40. Covelli, S., Piani, R., Acquavita, A., Predonzani, S. & Frageneli, J., (2007). Transport and dispersion of particulate Hg associated with a river plume in coastal Northern Adriatic environments. *Marine Pollution Bulletin*, 55, 436-450.
41. Cravotta, C.A., III. (2008). Dissolved metals and associated constituents in abandoned coal-mine discharges, Pennsylvania, USA, Part I: Constituent quantities and correlations. *Applied Geochemistry*, 23, 166-202.
42. Davidson, B. & Fisher, R.S. (2005). Groundwater quality in Kentucky: Mercury. InformationCircular 8, Series XII, Kentucky Geological Survey. Available atwww.uky.edu/kgs.

43. Domagalski, J.L. (1998). Occurrence and transport of total mercury and methyl mercuryin the Sacramento River Basin, California. *J. Geochemical Exploration*, 64, 277-291.
44. Driscoll, C.T., Yan, C., Schofield, C.L., Munson, R. & Holsapple, J. (1994). The mercurycycle and fish in the Adirondack lakes. *Environmental Science and Technology*, 28,136A-143A.
45. Engle, M. A., Tate, M. T., Krabbenhoft, D. P., Kolker, A., Olson, M. L., Edgerton, E. S.,DeWild, J. F. & McPherson, A. K., 2008, Characterization and cycling of atmosphericmercury along the central U.S. Gulf Coast: *Applied Geochemistry*, v. 23, #3, p. 419-437,doi:1016/j.apgeochem.2007.12.024.
46. Engstrom, D.R., Balogh, S.J. & Swain, E.B. (2007) History of mercury inputs to Minnesotalakes: Influences of watershed disturbance and localized atmospheric deposition.*Limnology and Oceanography*, 52, 2467-2483.
47. Environment Canada (2010). Environment Canada Risk Management Strategy forHg. Available at www.ec.gc.ac/doc/mercury-mercury/1241/index_e.htm#gotoC00
48. Estes, G.O., Knoop, W.E. & Houghton, F.D., 10973. Soil-plant response to surface-appliedmercury. *Journal of Environmental Quality*, 2, 451-452.
49. Fairbridge, R.W. (1972). The encyclopedia of geochemistry, and environmental sciences.*Encyclopedia of Earth Sciences Series, vol. IV.*, New York: A. Van Nostrand ReinholdCo.
50. Ferrara, R., Mazzolai, B., Lanzillotta, E., Nucaro, E. & Pirrone, N. (2000). Volcanoes asemission sources of atmospheric mercury in the Mediterranean basin. *Science of theTotal Environment*, 259, 115-121.
51. Fischer, J.M., Szabo, Z., Barringer, J.L. & Jacobsen, E. (2010). Mercury in a North AtlanticCoastal Plain aquifer system: A flowpath study: *Geological Society of AmericaAbstracts, v. 42*, no. 1, p. 178, Abstract 78-4.
52. Fitzgerald, W. & Lamborg, C.H. (2007). Geochemistry of mercury in the environment.107-148. In: *Environmental Geochemistry, V. 9;* Lollar, B.S. (ed.), Oxford, Elsevier.
53. Fleischer, M. (1970) Summary of the literature on the inorganic geochemistry of mercury.6-13. In. Mercury in the Environment, *U.S. Geological Survey Professional Paper713.*
54. Foucher, D., Hintelmann, H., Al, T.A. & MacQuarrie, K.T. (2012) Mercury isotopefractionation in waters and sediments of the Murray Brook mine watershed (NewBrunswick, Canada): Tracing mercury contamination and transformation. *ChemicalGeology*, (in press).
55. Gabriel, M.C. & Williamson, D.G. (2004). Principal biogeochemical factors affectingthe speciation and transport of mercury through the terrestrial environment. *EnvironmentalGeochemistry and Health*, 26, 421-434.
56. Ganguli, P.M., Conaway, C.H., Swarzenski, P.W., Izbicki, J.A. & Flegal, A.R. (2012)Mercury speciation and transport via submarine groundwater discharge at a southern California coastal lagoon system. *Environmental Science and Technology*, 46,1480-1488.
57. Gbondo-Tugbawa, S. & Driscoll, C.T. (1998) Application of the regional mercury cyclingmodel (RMCM) to predict the fate and remediation of mercury in Onondaga Lake, New York. *Water, Air, and Soil Pollution*, 105, 417-426.
58. Gemici, U. (2008) Evaluation of the water quality related to the acid mine drainage ofan abandoned mercury mine (Alasehir, Turkey). *Environmental Monitoring and Assessment*, 147, 93-106.
59. Gilmour, C.C. & Henry, E.A. (1991) Mercury methylation in aquatic systems affectedby acid deposition. *Environmental Pollution*, 71, 131-169.
60. Gilmour, C.C., Henry, E.A. and Mitchell, R. (1992). Sulfate stimulation of mercurymethylation in freshwater sediments. *Environmental Science and Technology,* 26, 2281.
61. Glass, G.E., Sorensen, J.A., Schmidt, K.W., Rapp. G.R., Yap, D. & Fraser, D. (1991).Mercury deposition and sources for the upper Great Lakes region. *Water, Air, and SoilPollution*, 56, 235-249.
62. Graham, A.M., Aiken, G.R. & Gilmour, C.G. (2012). Dissolved organic matter enhancesmicrobial mercury methylation under sulfidic conditions. *Environmental Scienceand Technology,* 46, 2715-2723.

63. Gu, B., Bian, Y., Miller, C.L., Dong, W., Jiang, X. & Liang, L. (2011). Mercury reductionand complexation by natural organic matter in anoxic environments. *Proceedingsof the National Academy of Sciences*, 108, 1479-1483.
64. Grassi, S. & Netti, R., (2000). Sea water intrusion and mercury pollution of somecoastal aquifers in the province of Grosseto (Southern Tuscany, Italy). *Journal of Hydrology*,237, 198-211.
65. Grigal, D.F. (2002). Inputs and outputs of mercury from terrestrial watersheds: A review.*Environmental Reviews*, 10, 1-39.
66. Gustin, M.S., Lindberg, S.E. & Weusberg, P.J. (2008). An update on the natural sourcesand sinks of atmospheric mercury. *Applied Geochemistry*, 23, 482-493.
67. Haitzer, M., Aiken, G.R. & Ryan, J.N. (2002). Binding of mercury (II) to dissolved organicmatter: the role of mercury-to-DOM concentration ratio. *Environmental Scienceand Technology*, 36, 3564-3570.
68. Haitzer, M., Aiken, G.R. & Ryan, J.N. (2003). Binding of mercury (II) to aquatic humicsubstances: Influence of pH and source of humic substances. *Environmental Scienceand Technology*, 37, 2436-2441.
69. Harada, M. (1995). Minimata disease: Methylmercury poisoning in Japan caused byenvironmental pollution. *Critical Reviews in Toxicology*, 25, 1-24.
70. Harvey, J.W., Krupa, S.L., Gefvert, C., Mooney, R.H., Choi, J., King, S.A. & Giddings,J.B. (2002). Interactions between surface water and ground water and effects on mercury transport in the North-central Everglades. *U.S Geological Survey Water-Resources Investigations Report 02-4050.*
71. Hem, J.D. (1970). Study and interpretation of the chemical characteristics of naturalwater. *U.S. Geological Survey Water-Supply Paper 2254.*
72. Hintelmann, H., Keppel-Jones, K. & Evans, R.D. (2000). Constants of mercury methylationand demethylation rates in sediments and comparison of tracer and ambientmercury availability. *Environmental Toxicology and Chemistry*, 19, 2204-2211.
73. Hsu-Kim, H. & Sedlak, D.L. (2005). Similarities between inorganic sulfide and thestrong Hg(II)-complexing ligands in municipal wastewater effluent. *EnvironmentalScience and Technology*, 39, 4035-4041.
74. Hurley, J.P. Cowell, S.E., Shafer, M.M. & Hughes, P.E. (1998). Tributary loading ofmercury to Lake Michigan: Importance of seasonal events and phase partitioning.*Science of the Total Environment 213*, 129-137.
75. Hutton, M. (1987). Human health concerns of lead, mercury, cadmium and arsenic,Chapter 6. 53-68. In Hutchinson, T.C. and Meema, K.M. (eds.) Lead, Mercury, Cadmium and Arsenic in the Environment, SCOPE, London, John Wiley and Sons, Ltd.
76. Hylander, L.D., Lindvall, A. & Gahnberg, L. (2006). High mercury emissions fromdental clinics despite amalgam separators. *Science of the Total Environment*, 362, 74-84.
77. Hylander L.D. & Meili, M. (2003). 500 years of mercury production: global annual inventoryby region until 2000 and associated emissions. *The Science of the Total Environment*,304, 13-27.
78. Kerin, E.J., Gilmour, C.C., Roden, E., Suzuki, M.T., Coates, J.D. & Mason, R.P. (2006).Mercury methylation by dissimilatory iron-reducing bacteria. *Applied and Environmental Microbiology*, 72, 7919-7921.
79. Khatiwada, N.R., Takizawa, S., Tran, T.V.N. & Inoue, M. (2002). Groundwater contaminationassessment for sustainable water supply in Kathmandu Valley, Nepal.*Water Science and Technology*, 46, 147-154.
80. Kolker, A., Panov, B. S., Panov,Y. B., Landa, E. R., Korchemagin, V. A., Conko, K. M.& Shendrik, T., 2009, Geochemistry of Donbas Coals and associated mine water inthe vicinity of Donetsk, Ukraine: *International Journal of Coal Geology*, v. 79, p. 83-91.
81. Koski, R.A., Munk, L., Foster, A.L., Shanks, W.C., III & Stillings, L.L. (2008). Sulfideoxidation and distribution of metals near abandoned copper mines in coastal environments,Prince William Sound, Alaska, USA. *Applied Geochemistry*, 23, 227-254
82. oterba, M.T., Andres, A.S.,Vrabel, J., Crilley, D.M., Szabo, Z. & DeWild, J.T.. (2006).Occurrence and distribution of mercury in the surficial aquifer, Long Neck

peninsula,Sussex County, Delaware, 2003-04. *U.S. Geological Survey Scientific InvestigationsReport 2006-5011.*

83. Kowalski, A., Siepak, M. & Boszke, L. (2007). Mercury contamination of surface and ground waters of Poland. *Polish Journal of Environmental Studies*, 16, 67-74.
84. Krabbenhoft, D.P. & Babiarz, C.L. (1992). The role of groundwater transport in aquatic mercury cycling. *Water Resources Research*, 28, 3119-3128.
85. Kretzschmar, R. & Schäfer, T. (2005). Metal retention and transport on colloidal particles in the environment. *Elements,* 1, 205-210.
86. Lacerda, L.D. (1997). Global mercury emissions from gold and silver mining. *Water,Air, and Soil Pollution,* 97, 209-221.
87. Laurier, F., J.G., Cossa, D., Beucherm C. & Brévière, E. (2007). The impact of groundwaterdischarges on mercury partitioning, speciation, and bioavailability to musselsin a coastal zone. *Marine Chemistry*, 104, 143-155.
88. LeBlanc, D.R. (1984). Sewage plume in a sand and gravel aquifer, Cape Cod, Massachusetts.*U.S. Geological Survey Water-Supply Paper* 2218.
89. LeBlanc, D.R., Garabedian, S.P., Hess, K.M., Gelhar, L.W., Quadri, R.D., Stollenwerk,K.G. & Wood, W.W. (1991). Large-scale natural-gradient tracer test in sand and gravel,Cape Cod, Massachusetts: 1. Experimental design and observed tracer movement.*Water Resources Research*, 27, 895-910.
90. Lee, Y-g., Moklesur Rahman, M.D., Kim, G. & Han, S. (2011) Mass balance of totalmercury and monomethylmercury in coastal embayments of a volcanic island: significanceof submarine groundwater discharge. *Environmental Science and Technology,* 45,9891-9900.
91. Lindberg, S.E., (1987). Emission and deposition of atmospheric mercury vapor,Chapter 8, 89-106. In; Lead, Mercury, Cadmium and Arsenic in the Environment,Hutchinson, T.C. and Meema, K.M., (eds.) SCOPE, London, John Wiley and Sons,Ltd.
92. Lohman, K., Seigneur, C, Gustin, M. &Lindberg, S. (2008). Sensitivity of the global atmosphericcycle of mercury to emissions. *Applied Geochemistry*, 23, 454-466.
93. Martin, R.S., Witt, M.L.I., Sawyer, G.M., Thomas, H.E., Watt, S.F.L., Bagnato, E., Calabrese,S., Aiuppa, A., Delmelle, P., Pyle, D.M. &Mather, T.A. (2012). Bioindication ofvolcanic mercury (Hg) deposition around Mt. Etna (Sicily). *Chemical Geology*, 310-311,12-22.
94. Mason, R.P., Fitzgerald, W.F. & Morel, F.M.M. (1994). The biogeochemical cycling ofelemental mercury: Anthropogenic influences. *Geochimica et Cosmochimica Acta,* 58,3191-3198.
95. Metro. (1991). Dental office waste stream characterization study (prepared by CynthiaWelland). *Seattle, WA, Municipality of Metropolitan Seattle*, September 1991.
96. Morrison, R.D. (1981). Potential health impacts of subsurface sewage sludge disposalupon groundwater resources. In: Quality of groundwater; Proceedings of an Internationalsymposium, Noordwijkerhout, Netherlands, 23-27 March, 1981. Van Duijvenbooden,W., Glasbergen, P. & van Lelyveld, H. (eds.) Studies in EnvironmentalScience, 17. Netherlands, Elsevier.
97. Murphy, E.A. & Aucott, M. (1999). A methodology to assess the amounts of pesticidalmercury used historically in New Jersey. *Water, Air, and Soil Pollution*, 78, 61-72.
98. Navarro, A. (2008). Review of characteristics of mercury speciation and mobilityfrom areas of mercury mining in semi-arid environments. *Reviews of EnvironmentalScience and Biotechnology,* 7, 287-306.
99. NIEA. (2011). European and National Drinking Water Qualty Standards. NorthernIreland Environmental Agency (NIEA), October, 2011. Available at www.doeni.gov.uk/niea.
100. Nriagu, J.O. (1994). Mercury pollution from the past mining of gold and silver in theAmericas. *The Science of the Total Environment*, 149, 167-181.
101. O'Neil, M.J., 2006, (Ed.) Merck Index—14th Edition: Whitehouse Station, NJ, Merckand Co., Inc.,. p.1017-1018.
102. Orica (2012). Former chlor alkali plant mercury contamination—Investigation overview.Available from http://www.oricabotanytransformation.com/index?page=107 andproject=103; accessed 7/12/12.

103. Pacyna, E.G. Pacyna, J.M., Fudala, J., Strzelecka-Jastrzab, E., Hlawiczka, S., and Panasiuk,D. (2006). Mercury emissions to the atmosphere from anthropogenic sources in Europe in 2000 and their scenarios until 2020. *Science of the Total Environment*, 370, 147-156.
104. Pak, K-R, and Bartha, R. (1998). Mercury methylation and demethylation in anoxiclake sediments and by strictly anaerobic bacteria. *Applied and Environmental Microbiology*,64, 1013-1017.
105. Pirrone, N., Cinnirella, S., Feng, X., Finkelman, R.B., Friedli, H.R. Leaqner, J., Nason,R., Mukherjee, A.B., Stracher, G.B., Streets, D.G. & Telmer, K. (2010). Global mercuryemissions to the atmosphere from anthropogenic and natural sources. *Atmospheric Chemistry and Physics*, 10, 5951-5964.
106. ODA (2002). Administrative rule developed to declare limits of certain metals in fertilizer,agricultural amendments, agricultural mineral and lime products distributed in Oregon. *Oregon Department of Agriculture; ODA White Paper 08/2002*.
107. Porvari, P, Verta, M., Munthe, J. & Haapanen, M. (2003). Forestry practices increasemercury and methyl mercury output from boreal forest catchments. *EnvironmentalScience and Technology*, 37, 2389-2393.
108. Protano, G., Riccobono, F. & Sabatini, G. (2000). Does salt water intrusion constitute amercury contamination risk for coastal fresh water aquifers? *Environmental Pollution*,110, 451-458.
109. Ravichandran, M. (2004). Interactions between mercury and dissolved organic matter—a review. *Chemosphere*, 55, 319-331.
110. Ravichandran, M., Aiken, G.R., Ryan, J.N. & Reddy, M.M., (1999). Inhibition of precipitationand aggregation of metacinnabar (mercuric sulfide) by dissolved organicmatter isolated from the Florida Everglades. *Environmental Science and Technology*, 33,1418-1423.
111. Reddy, M.M. & Aiken, G.R. (2001). Fulvic acid-sulfide ion competition for mercuryion binding in the Florida Everglades. *Water, Air, and Soil Pollution, 132,* 89-104.
112. Reilly, P.A., Barringer, J.L. & Szabo, Z. (2012). Mobility of mercury from sandy soilsinto acidic groundwater, New Jersey Coastal Plain: Geological Society of AmericaAbstracts, v. 44, no. 2, p. 118, Abstract 53-2.
113. Reimann, C., Siewers, U., Skarphagen, H. & Banks, D. (1999). Influence of filtrationon concentrations of 62 elements analyzed on crystalline bedrock groundwater samplesby ICP-MS. *Science of the Total Environment*, 234, 155-173.
114. Reimers, R.S. & Krenkel, P.A. (1974). Kinetics of mercury adsorption and desorptionin sediments. *Journal of the Water Pollution Control Federation*, 46, 3520365.
115. Repert, D.A, Barber, L.B., Hess, K.M., Keefe, S.H., Kent, D.B., LeBlanc, D.R. & Smith,R.L. (2006). Long-term natural attenuation of carbon and nitrogen within a groundwaterplume after removal of the treated wastewater source. *Environmental Scienceand Technology*, 40, 1154-1162.
116. Rudd, J.W.M. (1995). Sources of methyl mercury to freshwater ecosystems—a review.*Water, Air, and Soil Pollution*, 80, 697-713.
117. Ryan, J.N. & Gschwend, P.M. (1990). Colloid mobilization in two Atlantic CoastalPlain aquifers: field studies. *Water Resources Research,* 26, 307-322.
118. Ryan, J.N. & Gschwend, P.M. (1994). Effect of solution chemistry on clay colloid releasefrom an iron oxide-coated aquifer sand. *Environmental Science and Technology*,28, 1717-1726.
119. Rytuba, J.J. (2000). Mercury mine drainage and processes that control its environmentalimpact. *Science of the Total Environment* 260, 57-71.
120. Rytuba, J.J., Ashley, R., Slowey, A.,J. Brown, G.E. & Foster, A. (2005). Mercury speciationand transport from historic placer gold and mercury mines, California. (Abstract).*Geological Society of America, Abstracts with Programs*, 37, 104.
121. Saether, O.M., Storroe, G., Segar, D. & Krog, R. (1997). Contamination of soil andgroundwater at a former industrial site, Trondheim, Norway. *Applied Geochemistry*,12, 327-332.
122. Savoie, J.G., LeBlanc, D.R., Fairchild, G.M., Smith, R.L., Kent, D.B., Barber, L.B., Repert,D.A., Hart, C.P., Keefe, S.H. & Parsons, L.A. (2012). Groundwater-quality data for a treated-

wastewater plume near the Massachusetts Military Reservation, Ashumet Valley, Cape Cod, Massachusetts, 2006-08. *U.S. Geological Survey Data Series* 648,11 p.

123. Schuster, P.F., Krabbenhoft, D.P., Naftz, D.L., Cecil, L.D., Olson, M.L., DeWild, J.F.,Susong, D.D., Green, J.R. & Abott, M.L. (2002). Atmospheric mercury deposition during the last 270 years: a glacial ice core record of natural and anthropogenic sources. *Environmental Science and Technology*, 36, 2303-2310.
124. Schuster, P.F., Shanley, J.B., Marvin-diPasquale, M, Reddy, M.M., Aiken, G.R., Roth,D.A., Taylor, H.E, Krabbenhoft, D.P. & DeWild, J.F. (2008). Mercury and organic carbondynamics during runoff episodes from a northeastern USA watershed. *Water, Air, and Soil Pollution*, 187, 89-108.
125. Selvendiran, P., Driscoll, C.T., Bushey, J.T. & Montesdeoca, M.R. (2008). Wetland influenceon mercury fate and transport in a temperate forested watershed. *EnvironmentalPollution*, 154, 46-55.
126. Shanley, J.B., Schuster. P.F., Reddy, M.M., Roth, D.A., Taylor, H.E. & Aiken, G.R.(2002). Mercury on the move during snowmelt in Vermont. *EOS*, 83, 45-48.
127. Sherlock, R.L. (2005). The relationship between the McLaughlin gold-mercury depositand active hydrothermal systems in the Geysers-Clear Lake area, Coast Ranges,California. *Ore Geology Reviews*, 26, 349-382.
128. Sidle, W.C. (1993). Naturally occurring mercury in a pristine environment? *EnvironmentalGeology*, 21, 42-50.
129. Siegel, B.Z. & Siegel, S.M. (1987). Hawaiian volcanoes and the biogeology of mercury.In: Decker, R.W., Wright, T.L. & Stauffer, P.H. (eds.) Volcanism in Hawaii. *U.S.Geological Survey Professional Paper* 1350, 1, 827-839.
130. Skyllberg, U. (2008) Competition among thiols and inorganic sulfides and polysulfidesfor Hg and MeHg in wetland soils and sediments under suboxic conditions: illuminationof controversies and implications for MeHg net production. *Journal ofGeophysical Research: Biogeology*, 113, G00C03.
131. Slomp, C.P. & van Cappellen, P. (2004). Nutrient inputs to the coastal ocean throughsubmarine groundwater discharge. Controls and potential impact. *Journal of Hydrology*,295, 64-86.
132. Slowey, A.J. (2010). Rate of formation and dissolution of mercury sulfide nanoparticles:the dual role of natural organic matter. *Geochimica et Cosmochimica Acta*, 74,4693-4708.
133. Slowey, A.J. & Brown, G.E., Jr. (2007). Transformations of mercury, iron, and sulfurduring reductive dissolution of iron hydroxides by sulfide. *Geochimica et CosmochimicaActa*, 71, 877-894.
134. Slowey, A.J., Johnson, S.B., Rytuba, J.J. & Brown, G.E., Jr. (2005). Role of organic acidsin promoting colloidal transport of mercury from mine tailings. *Environmental Scienceand Technology*, 39, 7869-7874.
135. Smith, R.L., Harvey, R.W. & LeBlanc, D.R. (1991). Importance of closely spaced verticalsampling in delineating chemical and microbiological gradients in groundwater studies. *Journal of Contaminant Hydrology*, 7, 285-300.
136. Somasundaram, M.V., Ravindran, G. & Tellam, J.H. (1993). Ground-water pollutionof the Madras urban aquifer, India. *Ground Water*, 31, 4-11.
137. Srivastava, R.C. (2003). Guidance and awareness raising materials under new UNEPMercury Programs (Indian Scenario). Center for Environmental Pollution Monitoringand Mitigation. Available at http://www.chem.unep.ch/mercury/2003-gov-sub/Indiasubmission.pdf..
138. Steffan, R.J., Korthals, E.T., and Winfrey, M.R. (1988). Effects of acidification on mercurymethylation, demethylation, and volatilization in sediments from an acid-susceptiblelake. *Applied and Environmental Microbiology*, 54, 2003-2009.
139. Stoor, R.W., Hurley, J.P, Babiarz, C.L. & Armstrong, D.E. (2006). Subsurface sourcesof methyl mercury to Lake Superior from a wetland-forested watershed. *Science of theTotal Environment*. 368, 99-110.
140. Stumm, W. & Morgan, J.J. (1995). Aquatic chemistry, third ed. New York, John Wileyand Sons.

141. Swain, E.B., Jakus, P.M., Rice, G., Lupi, F., Maxson, P.A., Pacyna, J.M., Penn, A., Spiegel,S.J. & Veiga, M.M. (2007). Socioeconomic consequences of mercury use and pollution.*Ambio,* 36, 45-61.
142. Szabo, Z., Barringer, J.L., Jacobsen, E., Smith, N.P., Gallagher, R.A. & Sites, A. (2010)Variability of mercury concentrations in domestic well water, New Jersey CoastalPlain: Geological Society of America Abstracts, v. 42, no. 1, p. 178, Abstract 78-5.
143. Tewalt, S.J., Bragg, L.J. & Finkelman, R.B., (2001). Mercury in U.S. coal – Abundance,distribution, and modes of occurrence. *U.S. Geological Survey Fact Sheet FS-095-01.*
144. Ullrich, S.M., Ilyushchenko, M.A., Kamberov, I.M. & Tanton, T.W. (2007). Mercurycontamination in the vicinity of a derelict chlor-alkali plant. Part I: Sediment and watercontamination of Lake Balkyldak and the River Irtysh. *Science of the Total Environment,*381, 1-16.
145. Ullrich, S.M., Tanton, T.W. & Abdrashitova, S.A. (2001). Mercury in the aquatic environment:a review of factors affecting methylation. *Critical Reviews in EnvironmentalScience and Technology,* 31, 241-293.
146. USEPA (2001a). National primary drinking water standards. EPA 816-F-01-007
147. USEPA (2001b). Water quality criterion for the protection of human health: Methylmercury.EPA-823-R-01-001.
148. USEPA (2009). Report to Congress: Potential export of mercury compounds from theUnited States for conversion to elemental mercury. U.S. Environmental ProtectionAgency, Office of Pollution Prevention and Toxic Substances. Available at http://www.epa.gov/mercury/pdfs/mercury-rpt-to-congress.pdf.
149. USEPA (2010). Mercury in fish: Data. America's Children and the Environment.Available at ttp://www.epa.gov/ace/emerging_issues/fish-table.html#tableei1
150. USEPA (2012). Mercury: Laws and Regulations. Available from http://www.epa.gov/hg/regs.htm 2012, Accessed 8/13/2012.
151. USGS, (2000). Mercury in the Environment. *U.S. Geological Survey Fact Sheet 146-00.*Available at http://www.usgs.gov/themes/factsheet/146-00/index.html.
152. Van Straaten, P. (2000). Mercury contamination associated with small-scale gold miningin Tanzania and Zimbabwe. *The Science of the Total Environment,* 259, 105-113.
153. Wang, Q, Kim, D., Dionysiou, D.D., Sorial, G.A. & Timberlake, D. (2004). Sources andremediation for mercury contamination in aquatic systems—a literature review. *EnvironmentalPollution,* 131, 323-336.
154. Wang, Y., Wiatrowski, H.A., John, R., Lin, C-C, Young, L.Y., Kerkhof, L.J., Yee, N. &Barkay, T. (2012). Impact of mercury on denitrification and denitrifying microbialcommunities in nitrate enrichments of subsurface sediments. *Biodegradation,* in press,DOI 10.1007 s10532-012-9555-8.
155. Waples J.C., Nagy,.L., Aiken, G.R. & Ryan, J.N. (2005). Dissolution of cinnabar (HgS)in the presence of organic matter. *Geochimica et Cosmochimica Acta,* 69, 1575-1588.
156. White, D.E., Hinckle, M.E. & Barnes, I. (1970). Mercury contents of natural thermaland mineral fluids, 25-28. In. Mercury in the Environment. U.S. Geological SurveyProfessional Paper 713.
157. WHO (1993). Guidelines for drinking water quality, Vol, 2, Health critera and othersupporting nformation.
158. WHO (2012). Mercury and health. Fact Sheet No. 361. Available from http://www.who.int/medicentre/factsheets/fs361/en/index.html; accessed 7/12/12.
159. Wiatrowski, H.A., Das, S., Kukkadupu, R., Ilton, E.S., Barkay, T. & Yee, N. (2009). Reductionof Hg(II) to Hg(0) by magnetite. *Environmental Science and Technology,* 43,5307-5313.
160. Wiklander, L. (1969). The content of mercury in Swedish ground and river water. *Geoderma,*3, 75-79.
161. Yee, N., Barkay, T., Parikh, M., Lin, C-C., Wiatrowski, H. & Das, S. 2010. Biotic/abioticpathways of Hg(II) reduction by dissimilatory iron reducing bacteria. GeologicalSociety of America Abstracts, v. 42, no. 1, p. 178, Abstract 78-1.

162. Yu, R-Q., Flanders, J.R., Mack, E.E., Turner, R., Mirza, M.B. & Barkay, T. (2012). Contributionof coexisting sulfate and iron reducing bacteria to methylmercury productionin freshwater river sediments. *Environmental Science and Technology,* 46,2691-2684.
163. Yudovich. Ya E. & Ketris, M.P.(2005). Mercury in coal: a review. Part I. Geochemistry.*International Journal of Coal Geology*, 62, 107-134.
164. Zhang, H. (2006). Photochemical redox reactions of mercury. 37-79. In: *Recent developmentsin mercury science. Structure and bonding*, 120. DOI: 10.1007/430_015.
165. Zhao, X. & Wang, D. (2010). Mercury in some chemical fertilizers and the effect ofcalcium superphosphates on mercury uptake by corn seedlings (*Zea mays. L*). *Journalof Environmental Science,* 22, 1184-1188.

NOTES

[1] - Because many studies report Hg concentrations in units of nanograms per liter, results reported for aqueous samples in other units herein will be converted to nanograms per liter. Contents of solid materials will be reported in megagrams, kilograms and (or) milligrams per kilogram.

[7] - An issue with studies of low concentrations of Hg historically has been sample contamination. Studies since the early 1990s have used sampling protocols that typically obtain reliable samples; e.g. Krabbenhoft and Babiarz, (1992).

CHAPTER 11

Mobility and Transformation of Inorganic Contaminants in Mining-impacted Groundwater

Anita Etale[1*] and Sabelo Mhlanga[1]

Nanotechnology and Water Sustainability Research Unit, College of Engineering, Science and Technology, University of South Africa, Florida Campus, Johannesburg, South Africa

ABSTRACT

Mining often results in the contamination of groundwater by metal, sulphate and radionuclide ions following their percolation from tailings impoundments. This chapter discusses the processes by which elements within tailings are transformed and translocated to groundwater and the role of aquifer characteristics and colloids in these processes. The prevention and remediation of contaminated groundwater is also discussed, with particular attention given to the use of permeable reactive barriers and sulphate reducing bacteria.

Keywords: colloidal transport, acid mine drainage, inorganic contaminants, groundwater remediation, permeable reactive barriers, sulphate reducing bacteria

1. INTRODUCTION

Mining has been the mainstay of many economies across the globe for centuries. There is evidence, for example, of copper mining in Cyprus from as early as 4000 BC and from the Rio Tinto deposits in Spain from 1200–1500 BC [1]. However, along with the economic growth spurred by mining came unprecedented environmental pollution. The contamination of water resources by high concentrations of metals, non-metals and radionuclide elements has been reported from Spain and Portugal [2], to Australia [3] and South Africa [4]. Groundwater may be contaminated by direct infiltration of leachate from mine tailings and other mine wastes or following underground disposal of mine wastes [5]. Contaminated groundwater then recharges surface water with acidic metal-laden water (acid mine drainage) (**Figure 1**).

Figure 1. A river in the west of Johannesburg, South Africa, contaminated by acid mine drainage from disused underground mines. Note the orange colour which is due to deposition of iron flocs on the river bed.

Acid mine drainage (AMD) is formed via a cascade of reactions (Equations 1–4) when sulphide minerals are exposed to oxygen by mining [6]. The process begins when oxygenated water percolates through the finely divided tailings and pyrite is oxidised to ferrous iron (Equation 1) and then to ferric iron (Equation 2). Ferric iron which is soluble at pH below 3.5 then acts as an additional oxidising agent for pyrite (Equation 3). Above pH 3.5, ferric iron precipitates as $Fe(OH)_3$ (Equation 4); a reaction that is able to buffer the pH of AMD at pH 2.5–3.5 [7].

$$FeS_2 + {}^7/_2O_2 + H_2O \rightarrow Fe^{2+} + 2SO_4^{2+} + 2H^+ \qquad (1)$$

$$Fe^{2+} + {}^1/_4O_2 + H^+ \rightarrow Fe^{3+} + {}^1/_2H_2O \qquad (2)$$

$$FeS_2 + Fe^{3+} + 12H_2O \rightarrow 12Fe^{2+} + 2SO_4^{2+} + 16H^+ \qquad (3)$$

$$Fe^{3+} + 3H_2O \rightarrow Fe(OH)_3 + 3H^+ \qquad (4)$$

This sustained acidity leads to the dissolution of other sulphide ores hence the presence of ions including Ag, Au, Cd, Co, Mn, Ni, Hg, Mo, Se, U, Th and Zn in mine drainage [4] and metal-laden water percolates through the tailings heaps to recharge groundwater. A conceptual model of this process was supplied by Tutu et al. [4] (**Figure 2**). In this model, ingress through the dump by oxygenated water results in the oxidation of tailings and the dissolution of elements followed by a downward movement of these dissolved ions into groundwater. This chapter focuses on the transformation and mobility in groundwater, of inorganic contaminants originating from mining activity. Here, groundwater encompasses water in aquifers below tailings dumps as well as that in pores within tailings (pore water).

2. TRANSFORMATION OF INORGANIC CONTAMINANTS IN GROUNDWATER

The transformation of inorganic contaminants in groundwater is influenced by a number of chemical and physical characteristics of aquifers such as pH, redox potential (Eh) and organic matter [8]. These properties have an influence over contaminant transformation reactions including precipitation/co-precipitation-dissolution, oxidation-reduction and acid–base/sorption–desorption [9].

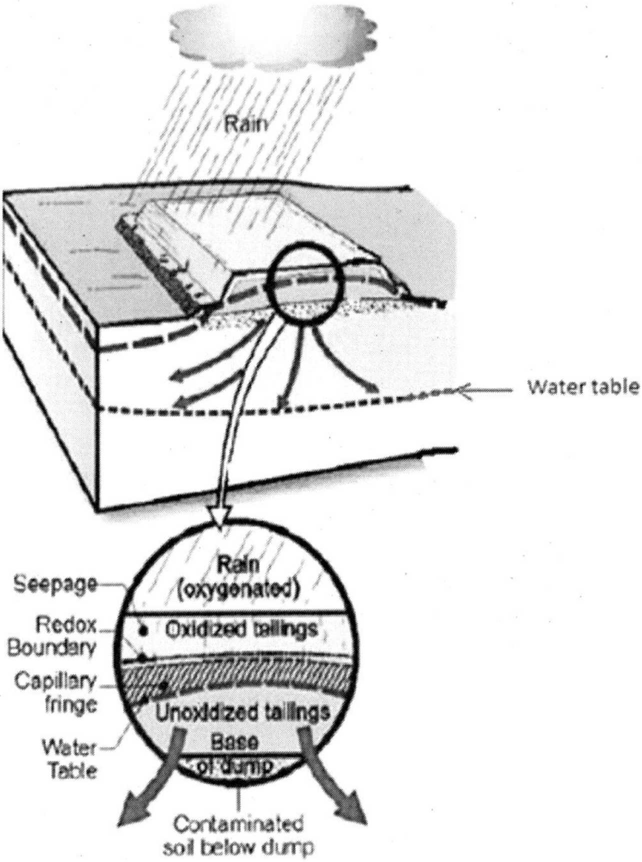

Figure 2. A conceptual model of the downward movement of elements through tailings dumps into groundwater (After [4]).

2.1. Precipitation/dissolution

Precipitation/co-precipitation and dissolution reactions play an important role in determining elemental concentrations in groundwater. Jean and Bancroft [10] showed, using pyrite (FeS), sphalerite (ZnS), galena (PbS) and pyrrhotite (Fe*1*-*x*S), that sulphide minerals are excellent scavengers for dissolved Hg^{2+}, Pb^{2+}, Zn^{2+} and Cd^{2+} ions, acting as adsorbents for these metals. However, Özverdi et al. [11] reported that under acidic conditions (< pH 3), metal removal by pyrite was by precipitation of metal sulphides due to the presence of H_2S. In contrast, metal hydroxides were precipitated under basic conditions. These observations were supported by the field studies of Muller et al. [12] in the Kristineberg mines in Sweden and Al et al. [13] in the Kidd Creek tailings impoundment in Ontario, Canada, which that both adsorption and precipitation reactions were responsible for metal attenuation.

Ferric iron is ubiquitous in AMD-contaminated environments. Its precipitation via various pathways (Equations 5–7) is a significant process in the attenuation of metal on concentrations in mine drainage via co-precipitation reactions [14]. Indeed, Fe- and Al-compounds are commonly used for the chemical precipitation of As(III) and As(V) in water treatment plants [15].

$$Fe^{3+} + 3H_2O \rightarrow Fe(OH)_3 + 3H^+ \tag{5}$$

$$Fe^{3+} + 3H_2O \rightarrow Fe(OH)_3 + 3H^+ \tag{6}$$

$$K^+ + Fe^{3+} + 2SO_4^{2-} + 6H_2O \rightarrow KFe_3(SO_4)_2(OH)_6 + 6H^+ \tag{7}$$

Ferrous iron, on the other hand, is controlled by the precipitation of siderite ($FeCO_3$) (Equation 8), a reaction occurring mainly in shallow tailings.

$$Fe^{2+} + HCO_3 \rightarrow FeCO_3 + H^+ \tag{8}$$

Elemental concentrations are also controlled by simple precipitation reactions in response to solution pH through the formation of insoluble hydroxides such as $Pb(OH)_2$ and $Cu(OH)_2$ or following reaction with sulphides and carbonates to form insoluble compounds [11].

With respect to dissolution, McGregor et al. [16] found that Ca and Mg concentrations in the Copper Cliff tailings in Ontario, Canada, were controlled by the dissolution of carbonate and aluminosilicate minerals during pH buffering reactions. Similarly, Mn was derived from the dissolution of pyroxene, chlorite, amphibole or carbonates, Al and Si from weathering of biotite, orthopyroxene and feldspars and Ni and Zn concentrations from the oxidation and dissolution of pentlandite ($(Fe,Ni)_9S_8$) and sphalerite (ZnS). K and Na on the other hand were controlled by dissolution of aluminosilicate minerals although their concentrations were limited by equilibrium with respect to jarosite ($KFe_3(SO_4)_2(OH)_6$) and natrojarosite ($NaFe_3(SO_4)_2(OH)_6$). Dissolution may also be microbially-driven. Cummings et al. [17] reported the release of arsenate following the dissolution of scorodite ($FeAsO_4.2H_2O$) by an iron-reducing bacterium, *Shewanella alga*.

2.2. Oxidation–reduction

Oxidation–reduction reactions may be chemically- or biologically-driven [18]. Selenate (SeO_4^{2-}) may be reduced to elemental selenium (Se^0) by ferrous hydroxide [19] and zero-valent iron (ZVI) has been used for the reductive precipitation of As from contaminated water [20]. Microbial oxidation-reduction in some cases can be many times faster than abiotic reactions [21]. The oxidation of As(III) to As(V) by a *Thermus* species, for example, was found to be approximately 100 times faster than abiotic rates [22].

Redox states of As have environmental implications because of their effect on As speciation. As(V), the predominant form in aerobic environments, is more strongly sorbed to mineral surfaces and thus less mobile than As(III) which sorbs less strongly and is thus more mobile [23]. Routh et al. [24] conducted microcosm experiments to investigate As behaviour in mine tailings near the Adak mine in northern Sweden. They found that microbial reduction of As(V) to As(III) increased the concentrations of the latter in aqueous media and as such, enhanced As remobilisation from sediments. In contrast, As(V) concentrations increased in sediments and aqueous media of control experiments treated with formaldehyde and $HgCl_2$.

Treatments applied to mine tailings may also have an effect on microbial activity and As behaviour in mining-impacted environments. Macur et al. [25] reported that the addition of lime (CaO), a common treatment applied to mine tailings to immobilise metal ions, stimulated As-reducing microorganisms (*Caulobacter*-, *Sphingomonas*- and *Rhizobium*-like populations) and in turn, enhanced As(V) reduction and mobilisation in tailings.

Fe, an important variable in mining-impacted environments, also influences As mobilisation. Han et al. [26] reported that Fe(II) significantly inhibited the removal of As(III) by MnO_2 in acidic environments (pH 3). They postulated that ferric iron compounds formed a coating on MnO_2 surfaces which inhibited access of As(III) ions to oxidation sites on MnO_2. The inhibition was however dependent on how Fe(II) ions were introduced into the system. Where the MnO_2 was pre-treated with Fe(II), As(III) diffused through the schwertmannite coatings that formed on the MnO_2 and its oxidation to As(V) was possible. However, where Fe(II) and As(III) were introduced simultaneously, competitive oxidation of the two ions prevented the complete oxidation of As(III) due to the formation of FeOHAs or $FeAsO_4$ coatings on the MnO_2 surface.

Cr exists mainly as Cr(III) and Cr(VI). Cr(VI) is highly soluble and therefore more mobile while Cr(III), tends to precipitate as amorphous hydroxides e.g. $Cr(OH)_3$ and ((Fe,Cr)OH_3) in slightly acidic and alkaline environments. Cr(III) is commonly oxidised to Cr(VI) by manganese oxides. In fact, manganese oxides are the only naturally-occurring inorganic phases capable of this reaction [27, 28] which Weaver and co-workers [29] found proceeded in multiple stages. However, Eary and Rai [28] reported that Cr(III) oxidation by pyrolusite (β-MnO_2) was slow in both acidic and slightly acidic solutions. In an acidic solution, slow oxidation was likely the result of the strong sorption of the oxidation product, Cr(VI), to the pyrolusite surface. Such sorption limited contact of unoxidised species with the pyrolusite surface, inhibiting additional oxidation [30]. In slightly acidic to basic media, slow oxidation was the result of the low solubility of $Cr(OH)_3$. Clearly, the equilibrium favoured the trivalent ion. Nevertheless, as with As, a suite of reducing microorganisms exist for reduction of Cr(VI) to Cr(III). Native isolates of *Acinetobacter sp.* from the Sukinda Valley in Jaipur, India were able to reduce initial Cr(VI) concentrations of 5 mg L^{-1} by 80% in 7 hours [31]. Similar results were reported by Dhal et al. [32] using a *Bacillus sp.* bacterium from chromite mine soils in Boula-Nuasahi mine in Orissa, India. The strain reduced > 90% of 100 mg L^{-1} Cr(VI) in 144 hours at pH 7 and 35°C.

2.3. Acid–base/sorption–desorption

Variable charge/amphoteric minerals such as crystalline and short-range ordered Fe-, Al-, and Mn- significantly influence the concentrations of elements in groundwater through adsorption/ion exchange reactions. This is because of (i) their large surface areas (ii) the acid–base surface hydroxyl groups resulting from the dissociative chemisorption of water molecules on their surfaces [33, 34]. Surface functional groups on mineral oxides undergo protonation (Equation 9) and deprotonation (Equation 10) reactions depending on solution pH [35]. As such, oxide surfaces are positively charged and primed for anion adsorption at lower pH values and negatively charged and primed for cation sorption at higher pH values [35, 36].

$$FeOH + H^+ \rightarrow FeOH_2^+ \quad (9)$$

$$FeOH \rightarrow FeO^- + H^+ \quad (10)$$

Adsorption of ions from solution (**Figure 3**) is therefore a bid to maintain electric neutrality both on the oxide surface as well as in the solution (ion exchange) [36].

The adsorption of divalent ions to oxide surfaces including goethite, hydrous iron and manganese oxides has been reported in several studies. Borah and Senapati [37] investigated the factors influencing the adsorption of Cd^{2+} to natural pyrite. They found that metal uptake increased with decreased pyrite particle size and was maximal at 30°C and pH 6. At this pH, Cd^{2+} ions were the main ions in solution and metal uptake was thus an exchange between the H^+ and Cd^{2+} ions on the pyrite surface. Similar findings were reported by Forbes et al. [38] in the adsorption of Cd^{2+}, Co^{2+}, Pb^{2+}, and Zn^{2+} on goethite and by Gadde and Laitinen [39] in the adsorption of Pb^{2+}, Cd^{2+}, Zn^{2+} and Tl^+ onto hydrous iron and manganese oxides. Ion exchange may also occur with fixed charge minerals such as zeolites where contaminated water comes into contact with clay minerals [40–42]. Adsorption is not always accompanied by proton loss. In the adsorption of arsenate to goethite, for example, $FeAsO_4H_2^0$ and $FeAsO_4H^-$ were the dominant species at pH < 5 and pH 5–8 respectively, in reactions that were not accompanied by the loss of protons from the goethite surface [35].

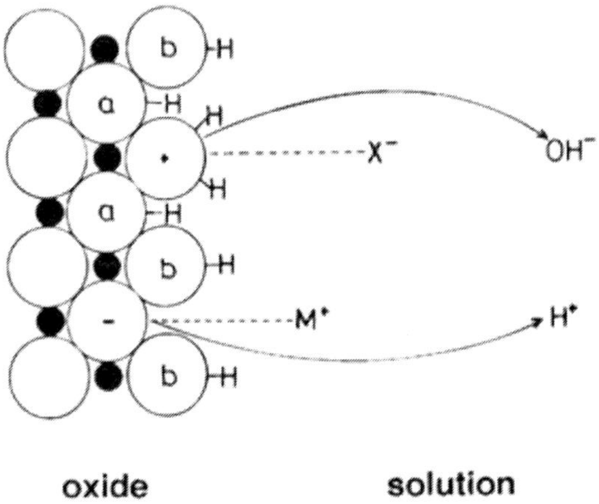

Figure 3. Acid and base hydroxyl sites on metal oxide surfaces and ion exchange reactions at the oxide-solution interface: ●=metal ions, ○=oxide ions, a=acid hydroxyl sites and b=base hydroxyl sites [36].

Determining the mode of contaminant binding is essential to predicting their behaviour in groundwater. Contaminants may be sorbed by electrostatic, hydrogen or covalent bonds. Electrostatic bonds are formed between charged hydrated species and oppositely charged mineral surfaces forming weak outer-sphere complexes. Contaminants sorbed in this way are easily desorbed by perturbations in solution parameters e.g. pH, ionic strength. Hydrogen bonds have intermediate strength while covalent bonding results in strong sorption of contaminants. Contaminants sorbed this way are harder to desorb and colloidal transport may play an important role in their transport through groundwater.

3. MOBILITY OF INORGANIC CONTAMINANTS IN GROUNDWATER

The mobility of contaminants in groundwater has been the subject of many investigations. Early models of contaminant mobility divided contaminants between only two phases: the dissolved phase (mobile) and a sorbed phase (immobile). However, after contaminants were detected at distances further than was predicted by this model, a third phase, the mobile colloidal phase, was applied (**Figure 4**) [16, 33, 43–47]. Buddemeier and Hunt [48] found that colloids were responsible for the transport of Mn, Co, Ce and Eu from the Nevada ammunitions test site to a location 300 m away from the disposal site. Similar findings were made by Kersting et al. [47] for the transport of Pu from the same site and by Hochella et al. [49] for the transfer of As, Cu, Pb and Zn from the largest Superfund site in the USA.

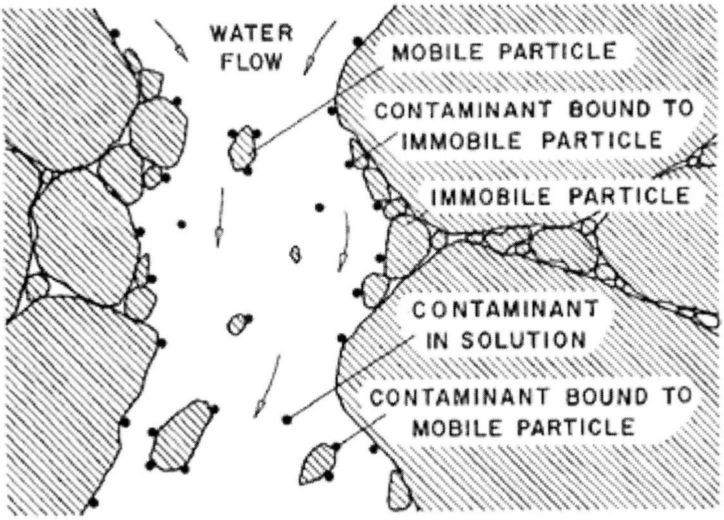

Figure 4. Schematic representation of colloid facilitated transport in a subsurface water-saturated medium. Contaminants (•) are either dissolved in solution, adsorbed to mobile phases (colloids) or to stationary phases [50].

For colloid-facilitated contaminant transport to be efficient, three criteria must be met: (i) colloids must be generated, (ii) a strong association must be formed between contaminants and the colloids, (iii) the colloid-contaminant composites must be transported through groundwater [50, 51]. Colloids are particles in the 1 nm to 1 μm size range [52].

They may be organic e.g. humic acids and microbes or inorganic e.g. metal oxy(hydr)oxides, carbonates, silicates and phosphates [52–54]. Inorganic colloids are particularly important in mining-contaminated groundwater. In these environments, colloids may be formed biogenically, as a result of ore processing or by precipitation from supersaturated solutions [53, 55].

Webster et al. [54] suggested that colloids formed in AMD were more effective sorbents than pure minerals due to the presence of sulphates and the influence of bacterial activity on their synthesis. Colloids may also be mobilised by perturbations to groundwater properties including pH, ionic strength and flow velocity (e.g. flow through fractures or variable infiltration following rainfall events). pH shifts are especially important in AMD-impacted environments as pH influences the formation of Fe and Al colloids [53, 56]. It also influences colloidal surface charges, the affinity of contaminants for colloid surfaces and the suspension or precipitation of colloids.

Solution ionic strength also influences colloid mobilisation due to its effect on the electric double layer of ions as put forward in the DVLO (Derjaguin-Landau-Verwey-Overbeek) theory. According to this theory, colloid mobilisation increases with decreasing ionic strength because the electrostatic double layer around colloids expands resulting in greater repulsion between like-charged colloids. Thus, ^{137}Cs by kaolinite through quartz found that transport was substantially increased at low ionic strengths because kaolinite colloids were more mobile and bound more ^{137}Cs. [57]. Increases in ionic strength on the other hand, lead to compression of the double layers, hence a decrease in repulsive forces and colloid aggregation/coalescence [58, 59].

Thus, contaminant transport may be retarded due to colloidal sedimentation. Kimball et al. [56] found that while Fe colloids aided the transport of As, Cd, Cu, Mn, Pb and Zn from mining flows, the colloidal load decreased by half after the first 50 km due to aggregation and sedimentation of colloids in the stream bed. Retardation may also be due to colloidal plugging/blockage of pores [60] within transport matrices.

4. PREVENTION AND REMEDIATION OF GROUNDWATER CONTAMINATION BY AMD

Conventional methods of AMD treatment involve the hydroxide precipitation using quicklime (CaO), hydrated lime ($Ca(OH)_2$), caustic soda (NaOH) or soda ash (Na_2CO_3) [9] or sulphide precipitation agents such as NaS, NaHS, BaS, FeS or H_2S [61]. This latter approach is superior because metal sulphides are generally less soluble that hydroxide counterparts, and as such, allows more complete precipitation (although the objectionable odour and toxicity of H_2S has to be considered).

Iron, in its reduced form, either as ferrous salts or zero valent form, has also been used for the chemical treatment of contaminated water. The primary advantages associated with ZVI use include low cost, simplicity in handling and the formation of strong complexes between contaminants and the iron oxides [62]. Farrell et al. [20] and Melitas et al. [63] applied ZVI for the removal of arsenate ions from solution. Adsorption was resolved as the removal mechanism because no As(III) was detected in solution. ZVI was also effective for the removal of CrO_4^{2-}, TcO_4^-, MoO_2^{2-} and UO_2^{2+} ions by reductive precipitation [26, 62, 64–67]. The rate of removal followed the order CrO_4^{2-} >TcO_4^- >UO_2^{2+}>>MoO_2^{2-} with CrO_4^{2-} concentrations decreasing from 10,000 μg L^{-1} to < 5 μg L^{-1} in 1 hour [65]. Cationic contaminants e.g. Ag^+ and Hg^+ may also be reduced to their zero valent metallic forms [68]. The use of ZVI exploits redox reactions and the strong reducing properties of iron to convert contaminants into less soluble immobile forms [63]. In this process, ZVI is first oxidised by water to Fe(II) which then acts as an electron donor for the reduction of dissolved ions (Equation 11) [20, 69].

$$Fe^0 + 2H_2O \rightarrow Fe^{2+} + H_2 + 2OH^- \qquad (11)$$

With the advent of nanotechnology, nanoscale ZVI (nZVI) use has also been attempted in laboratory and field studies [70, 71]. Li et al. [72] reported that for Cr(IV) degradation, reaction rates for nZVI were at least 25–30 times faster and the sorption capacity was much higher compared with granular iron. In another study, 25% of As(V) was reduced to As(III) after 90 days [73]. Despite this, the environmental and human health risks associated with nanoparticles have meant that larger scale application of these materials has been approached with caution. Tratnyek and Johnson [74], however, pointed out that the mobility of nanoscale ZVI was less than a few metres under almost all relevant environmental conditions and thus, human exposure was likely to be minimal.

Sulphate reducing bacteria (SRB) have been increasingly investigated for mine drainage remediation since Tuttle et al. [75] documented sulphate reduction in an AMD-contaminated stream. Samuel et al. [31] reported the remediation of Cr(VI) by indigenous isolates of *Bacillus, Acinetobacter* and *Escherichia spp.* from chromite mines in the Sukinda Valley of Orissa, India. SRB utilize organic carbon or hydrogen to reduce sulphates to sulphides (Equation 12) which then facilitate the precipitation of metal sulphides (Equation 13). This reaction also increases the alkalinity and pH of solutions, further promoting metal precipitation [76]. As such, the use of SRB for AMD treatment reduces sulphate concentrations, precipitates metal ions from solution and raises solution pH.

$$2CH_2O + SO_4^{2-} + 2H^+ \rightarrow H_2S + 2CO_2 + 2H_2O \quad (12)$$

$$Fe^{2+} + H_2S \rightarrow FeS + 2H^+ \quad (13)$$

Cardenas et al. [77] reported the biological in situ remediation of uranium contaminated groundwater. The growth of denitrifying, Fe(III)-reducing and SRB including *Desulfovibrio, Geobacter, Anaero- myxobacter, Desulfosporosinus, Acidovorax, Ferribacterium* and *Geothrix spp* through weekly injections of ethanol into the subsurface. After 2 years, U concentrations were reduced from 60 mg $L^{-1}$to < 30 μg L^{-1}. Sulphate concentrations also decreased when ethanol was injected and rebounded when injection stopped, indicating SRB activity in the subsurface.

Such in situ approaches for the treatment of mine drainage contamination have attracted much attention in recent times. Designed to intercept contaminants in the subsurface with reactive materials, in situ treatment has the advantage of treating contaminated groundwater prior to the oxidation of Fe^{2+}(Reaction 2), thus preventing the generation of additional acidity and mobilisation of additional metal ions [64]. Sulphate reduction is also optimised in the near-neutral pH characteristic of many aquifers and the process is less costly because the volumes of water to be treated are lower than in pump-and-treat systems [78]. Two possible approaches exist for *in situ* remediation. The first involves injecting reactants into the subsurface to form a reactive treatment zone in which reactants are adsorbed onto aquifer materials (**Figure 5a**). The procedure by Cardenas et al. [77] described above falls in this category. Alternatively, permeable reactive barriers may be keyed in to underlying bedrock in the flow path of contaminants (**Figure 5b**).

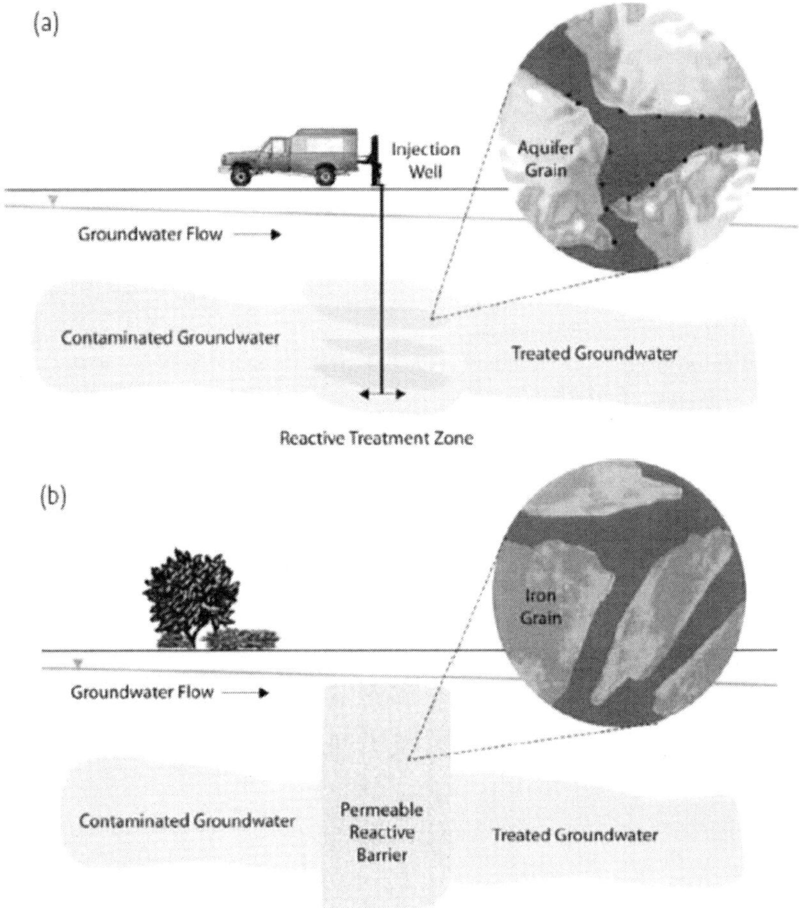

Figure 5. In situ treatment of groundwater may be by a reactive treatment zone (a) or a permeable reactive barrier (b) (After [74]).

A number of studies have documented the use of PRBs for the remediation of AMD-contaminated groundwater [64, 78, 79]. The reactive materials within PRBs may be chemical or biological [80]. Baker et al. [81] used a mixture containing 50 wt% silica sand, 45 wt% crushed limestone and 5 wt% metal oxide for the removal of phosphates while ZVI, $FeCO_3$, FeS were investigated for Cr(VI) removal [82]. Biologically-driven PRBs (biobarriers) mostly comprise of SRB [61, 83] and require (i) an anaerobic environment (redox potential of ~200 mV), (ii) pH values greater than 5, (iii) a sulphate species to be reduced and (iv) an energy source (electron donor), mostly short chain organic substrates e.g. ethanol although a variety of natural substrates including leaf mulch, vegetal compost and sawdust [84] have been tested. Benner et al. [85] installed a PRB 20 m long, 4 m thick and 3.5 m tall into the Nickel Rim aquifer downstream of a tailings impoundment. They recorded, after 22 months, dramatic changes in the concentrations of several contaminants. Concentrations of sulphates, Fe and Ni decreased by 2000–3000 mg L^{-1}, 270–1300 mg L^{-1} and 30 mg L^{-1}, respectively. In addition, alkalinity increased by 800–2700 mg L^{-1} and the populations of SRB were 10, 000 times greater than before the installation of the PRB. Column experiments by Waybrant et al. [86] showed similar results: iron concentrations decreased from 300–1200 mg/L to <0.01–220 mg/L while Zn and Ni decreased from 0.6–1.2 mg L^{-1} to 0.01–0.15 mg L^{-1} and 0.8–12.8 mg L^{-1} to <0.01 mg L^{-1}, respectively. The pH increased slightly from 5.5–6.0 to 6.5–7.0 and alkalinity from <50 mg/L to 300–1300 mg L^{-1}. Biobarriers are also effective for the attenuation of Cr(VI) [87] radionuclide ions [88] and sulphates [89]. Sulphate concentrations were decreased from 1800 to < 250 mg L^{-1} and the mine waters neutralised using only bacterially-mediated alkalinity. Natural treatment as well as permeable reactive barriers therefore both hold promise for the treatment of groundwater contamination. Future research should look into the incorporation of nanomaterials e.g. embedded in polymers, into PRBs to facilitate faster reaction times and more efficient removal of contaminants.

5. CONCLUSIONS AND RECOMMENDATIONS

This chapter explored the behaviour of inorganic contaminants from mining activity, in groundwater. The processes influencing their transformation i.e. acid/base, redox and (co)precipitation/dissolution reactions have been discussed in detail. The factors influencing their mobility, specifically, the role of colloids as well as the prevention and remediation of contamination have also been reviewed. Because of their low costs and preclusion of the need for pumping out aquifers, PRBs remain a viable option for the prevention and treatment of contaminated groundwater. Future research should look into the use of nanomaterials in PRBs. With their large surface areas and faster reaction rates, the use of nanomaterials in PRBs will likely result in greater treatment efficacy. Although the current high cost of some nanomaterials may hinder their application, this will be achievable as production technologies improve and prices decrease. On the other hand, the use of SRB in PRBs may provide a more natural alternative devoid of the environmental concerns associated with the introduction of engineered nanomaterials to the environment. It also has the double advantages of removing both metal and sulphate ions. Research into the use of SRB in PRBs will involve identification and isolation of bacterial strains suited for each contaminated site as studies have shown bacteria to be site specific and not transferable.

REFERENCES

1. Leblanc M. 4,500-year-old mining pollution in southwestern Spain: Long-term implications for modern mining pollution. Econ. Geol. 2000;95:655–662.
2. Nieto JM, Sarmiento AM, Olías M, Canovas CR, Riba I, Kalman J, Delvalls TA. Acid mine drainage pollution in the Tinto and Odiel rivers (Iberian Pyrite Belt, SW Spain) and bioavailability of the transported metals to the Huelva Estuary. Environ. Int. 2007;33:445–455.
3. Hogsden KL, Harding JS. Anthropogenic and natural sources of acidity and metals and their influence on the structure of stream food webs. Environ. Pollut. 2012;162:466–474.
4. Tutu H, McCarthy TS, Cukrowska E. The chemical characteristics of acid mine drainage with particular reference to sources, distribution and remediation: The Witwatersrand Basin, South Africa as a case study. Appl. Geochemistry. Elsevier Ltd; 2008 Dec;23:3666–3684.
5. Roychoudhury AN, Petersen J. Geochemical evaluation of soils and groundwater affected by infiltrating effluent from evaporation ponds of a heavy mineral processing facility, West Coast, South Africa. J. Geochemical Explor. 2014;144:478–491.
6. Singer PC, Stumm W. Acidic mine drainage: The rate-determining step. Science. 1970;167:1121–1123.
7. España JS, Pamo EL, Santofimia E, Aduvire O, Reyes J, Barettino D. Acid mine drainage in the Iberian Pyrite Belt (Odiel river watershed, Huelva, SW Spain): Geochemistry, mineralogy and environmental implications. Appl. Geochemistry. 2005;20:1320–1356.
8. Roy SB, Dzombak DA. Chemical factors influencing colloid-facilitated transport of contaminants in porous media. Environ. Sci. Technol. 1997;31:656–664.
9. Evanko CR, Dzombak DA. Remediation of Metals-Contaminated Soils and Groundwater. 1997;1–61.
10. Jean GE, Bancroft GM. Heavy metal adsorption by sulphide mineral surfaces. Geochim. Cosmochim. Acta. 1986;50:1455–1463.
11. Özverdi A, Erdem M. Cu2+, Cd2+ and Pb2+ adsorption from aqueous solutions by pyrite and synthetic iron sulphide. J. Hazard. Mater. 2006;137:626–632.
12. Muller B, Axelsson MD, Öhlander B. Analyses of trace elements on quartz surfaces in sulfidic mine tailings from Kristineberg (Sweden) a few years after remediation. Environ. Geol. 2003;45:98–105.
13. Al TA, Blowes DW, Martin CJ, Cabri LJ, Jambor JL. Aqueous geochemistry and analysis of pyrite surfaces in sulfide-rich mine tailings. Geochim. Cosmochim. Acta. 1997;61:2353–2366.
14. Banks D, Younger PL, Egil RA, Sheila RI. Mine-water chemistry: The good, the bad and the ugly. Environ. Geol. 1997;32:157–174.
15. Manceau A, Charlet L, Boisset MC, Didier B, Spadini L. Sorption and speciation of heavy metals on hydrous Fe and Mn oxides. From microscopic to macroscopic. Appl. Clay Sci. 1992;7:201–223.
16. McGregor R, Blowes D, Jambor J, Robertson W. The solid-phase controls on the mobility of heavy metals at the Copper Cliff tailings area, Sudbury, Ontario, Canada. J. Contam. Hydrol. 1998; 33:247–271.
17. Cummings DE, Caccavo F, Fendorf S, Rosenzweig RF. Arsenic mobilization by the dissimilatory Fe(III)-reducing bacterium *Shewanella alga BrY*. Environ. Sci. Technol. 1999;33:723–729.

18. Dhal B, Thatoi HN, Das NN, Pandey BD. Chemical and microbial remediation of hexavalent chromium from contaminated soil and mining/metallurgical solid waste: A review. J. Hazard. Mater. 2013;250–251:272–291.
19. Murphy AP. Removal of selenate from water by chemical reduction. Ind. Eng. Chem. Res. 1988;27:187–191.
20. Farrell J, Wang J, O'Day P, Conklin M. Electrochemical and spectroscopic study of arsenate removal from water using zero-valent iron media. Environ. Sci. Technol. 2001;35:2026–2032.
21. Jones CA, Langner HW, Anderson KT, McDermott R, Inskeep WP. Rates of microbially mediated arsenate reduction and solubilization. Soil Sci. Soc. Am. J. 1999;64:600–608.
22. Gihring TM, Banfield JF. Arsenite oxidation and arsenate respiration by a new *Thermus* isolate. FEMS Microbiol. Lett. 2001;204:335–340.
23. Oremland RS, Stolz JF. The Ecology of Arsenic. Science. 2003;300:939–944.
24. Routh J, Bhattacharya A, Saraswathy A, Jacks G, Bhattacharya P. Arsenic remobilization from sediments contaminated with mine tailings near the Adak mine in Västerbotten district (northern Sweden). J. Geochemical Explor. 2007;92:43–54.
25. Macur RE, Wheeler JT, McDermott TR, Inskeep WP. Microbial populations associated with the reduction and enhanced mobilization of arsenic in mine tailings. Environ. Sci. Technol. 2001;35:3676–3682.
26. Han X, Li YL, Gu JD. Oxidation of As(III) by MnO2 in the absence and presence of Fe(II) under acidic conditions. Geochim. Cosmochim. Acta. 2011;75:368–379.
27. Rai D, Eary LE, Zachara JM. Environmental chemistry of chromium. Sci. Total Environ. 1989;86:15–23.
28. Eary LE, Rai D. Kinetics of Chromium(III) oxidation to Chromium(VI) by reaction with manganese dioxide. Environ. Sci. Technol. 1987;21:1187–1193.
29. Weaver RM, Hochella MF, Ilton ES. Dynamic processes occurring at the CrIIIaq-manganite (ɣ-MnOOH) interface: Simultaneous adsorption, microprecipitation, oxidation/reduction, and dissolution. Geochim. Cosmochim. Acta. 2002;66:4119–4132.
30. Fendorf SE, Fendorf M, Sparks DL, Gronsky R. Inhibitory mechanisms of Cr(III) oxidation by δ-MnO2. J. Colloid Interface Sci. 1992;153:37–54.
31. Samuel J, Paul ML, Pulimi M, Nirmala MJ, Chandrasekaran N, Mukherjee A. Hexavalent chromium bioremoval through adaptation and consortia development from Sukinda chromite mine isolates. Ind. Eng. Chem. Res. 2012;51:3740–3749.
32. Dhal B, Thatoi H, Das N, Pandey BD. Reduction of hexavalent chromium by *Bacillus sp.* isolated from chromite mine soils and characterization of reduced product. J. Chem. Technol. Biotechnol. 2010;85:1471–1479.
33. Brown Jr GE, Parks GA. Sorption of trace elements on mineral surfaces: Modern perspectives from spectroscopic studies, and comments on sorption in the marine environment. Int. Geol. Rev. 2001;43:963–1073.
34. Kinniburgh DG, Jackson ML, Syers JK. Adsorption of alkaline earth, transition, and heavy metal cations by hydrous oxide gels of iron and aluminum. Soil Sci. Soc. Am. J. 1976;40:796–799.
35. Gao Y, Mucci A. Acid base reactions, phosphate and arsenate complexation, and their competitive adsorption at the surface of goethite in 0.7 M NaCl solution. Geochim. Cosmochim. Acta. 2001;65:2361–2378.
36. Tamura H, Katayama N, Furuichi R. Modeling of ion-exchange reactions on metal oxides with the frumkin isotherm. 1. acid–base and charge characteristics of MnO2, TiO2, Fe3O4 and Al2O3 surfaces and adsorption affinity of alkali metal ions. Environ. Sci. Technol. 1996;30:1198–1204.
37. Borah D, Senapati K. Adsorption of Cd(II) from aqueous solution onto pyrite. Fuel. 2006;85:1929–1934.
38. Forbes EA, Posner AM, Quirk JP. The specific adsorption of divalent Cd, Co, Cu, Pb, and Zn on goethite. J. Soil Sci. 1976;27:154–166.
39. Gadde RR, Laitinen HA. Studies of heavy metal adsorption by hydrous iron and manganese oxides. Anal. Chem. 1974;46:2022–2026.

40. Nagy NM, József Kónya, Urbin Z. The competitive exchange of hydrogen and cobalt ions on calcium-montmorillonite. Colloids and Surfaces A Physicochem. Eng. Asp. 1997;121:117–124.
41. Nagy NM, Kónya J. The exchange of Mn and Ca ions on montmorillonite. React. Polym. 1992;17:9–13.
42. Das NC, Bandyopadhyay M. Removal of copper using vermiculite. Water Environ. Res. 1992;64:852–857.
43. Cauwet G, Dai M, Martin J. The significant role of colloids in the transport and transformation of organic carbon and associated trace metals (Cd, Cu and Ni) in the Rhone delta (France). Mar. Chem. 1995;51:159–175.
44. Kaplan DI, Bertsch PM, Adriano DC. Facilitated transport of contaminant metals through an aquifer. Groundwater. 1995;33:708–717.
45. McCarthy JF, Zachara JM. Subsurface transport of contaminants. Environ. Sci. Technol. 1989;23:496–502.
46. Kaplan DI, Hunter DB, Bertsch PM, Bajt S, Adriano DC. Application of synchrotron X-ray fluorescence spectroscopy and energy dispersive X-ray analysis to identify contaminant metals on groundwater colloids. Environ. Sci. Technol. 1994;28:6–9.
47. Kersting AB, Smith DK. Migration of plutonium in ground water at the Nevada Test Site. Nature. 1999;397:56–59.
48. Buddemeier RW, Hunt JR. Transport of colloidal contaminants in groundwater: Radionuclide migration at the Nevada Test Site. Appl. Geochemistry. 1988;3:535–548.
49. Hochella Jr MF, Kasama T, Putnis A, Putnis C V., Moore JN. Environmentally important, poorly crystalline Fe/Mn hydrous oxides: Ferrihydrite and a possibly new vernadite-like mineral from the Clark Fork River Superfund Complex. Am. Minerol. 2005;90:718–724.
50. Grolimund D, Borkovec M, Barmettler K, Sticher H. Colloid-facilitated transport of strongly sorbing contaminants in natural porous media: A laboratory column study. Environ. Sci. Technol. 1996;30:3118–3123.
51. Ryan JN, Elimelech M. Colloid mobilization and transport in groundwater. Colloids Surfaces A Physicochem. Eng. Asp. 1996;107:1–56.
52. Hassellov M, von der Kammer F. Iron Oxides as Geochemical Nanovectors for Metal Transport in Soil-River Systems. Elements. 2008;4:401–406.
53. Schemel LE, Kimball BA, Bencala KE. Colloid formation and metal transport through two mixing zones affected by acid mine drainage near Silverton, Colorado. Appl. Geochemistry. 2000;15:1003–1018.
54. Webster JG, Swedlund PJ, Webster KS. Trace metal adsorption onto an acid mine drainage Iron(III) oxy hydroxy sulfate. Environ. Sci. Technol. 1998;32:1361–1368.
55. Chapman BM, Jones DR, Jung RF. Processes controlling metal ion attenuation in acid mine drainage streams. Geochim. Cosmochim. Acta. 1983;47:1957–1973.
56. Kimball A, Axtmann V. Effects of colloids on metal transport in a river receiving acid mine drainage, Upper Arkansas River, Colorado, USA. 1995;10:285–306.
57. Saiers JE, Hornberger GM. The influence of ionic strength on the facilitated transport of Cesium by kaolinite colloids. Water Resour. Res. 1999;35:1713–1727.
58. Kanti Sen T, Khilar KC. Review on subsurface colloids and colloid-associated contaminant transport in saturated porous media. Adv. Colloid Interface Sci. 2006;119:71–96.
59. Nightingale HI, Bianchi WC. Groundwater turbidity resulting from artificial recharge. Groundwater. 1977;15:146–152.
60. Torkzaban S, Bradford SA, Vanderzalm JL, Patterson BM, Harris B, Prommer H. Colloid release and clogging in porous media: Effects of solution ionic strength and flow velocity. J. Contam. Hydrol. 2015;181:161–171.
61. Hammack RW, Edenborn HM, Dvorak DH. Treatment of water from an open-pit copper mine using biogenic sulfide and limestone: A feasibility study. Water Res. 1994;28:2321–2329.
62. Manning BA, Hunt ML, Amrhein C, Yarmoff JA. Arsenic(III) and arsenic(V) reactions with zerovalent iron corrosion products. Environ. Sci. Technol. 2002;36:5455–5461.
63. Melitas N, Wang J, Conklin M, O'Day P, Farrell J. Understanding soluble arsenate removal kinetics by zerovalent iron media. Environ. Sci. Technol. 2002;36:2074–2081.

64. Blowes DW, Ptacek CJ, Jambor JL. In-situ remediation of Cr(VI)-contaminated groundwater using permeable reactive walls: Laboratory studies. Environ. Sci. Technol. 1997;31:3348–3357.
65. Cantrell KJ, Kaplan DI, Wietsma TW. Zero-valent iron for the in situ remediation of selected metals in groundwater. J. Hazard. Mater. 1995;42:201–212.
66. Gu B, Liang L, Dickey MJ, Yin X, Dai S. Reductive precipitation of uranium(VI) by zero-valent iron. Environ. Sci. Technol. 1998;32:3366–3373.
67. Su C, Puls RW. Arsenate and arsenite removal by zerovalent iron: Kinetics, redox transformation, and implications for in situ groundwater remediation. Environ. Sci. Technol. 2001;35:1487–1492.
68. Gould JP, Masingale MY, Miller M. Recovery of silver and mercury from COD samples by iron cementation. J. Water Pollut. Control Fed. 1984;56:280–286.
69. Kanel SR, Manning B, Charlet L, Choi H. Removal of arsenic (III) from groundwater by nanoscale zero-valent iron. Environ. Sci. Technol. 2005;39:1291–1298.
70. Nowack B. Pollution prevention and treatment using nanotechnology. In: Krug H, editor. Nanotechnology. Weinheim: WILEY-VCH Verlag GmbH & Co. KGaA, Weinheim; 2010. 1–15.
71. Crane RA, Scott TB. Nanoscale zero-valent iron: Future prospects for an emerging water treatment technology. J. Hazard. Mater. 2012;211–212:112–125.
72. Li X, Elliott DW, Zhang W. Zero-valent iron nanoparticles for abatement of environmental pollutants: Materials and engineering aspects. Crit. Rev. Solid State Mater. Sci. 2006;31:111–122.
73. Kanel SR, Greneche JM, Choi H. Arsenic(V) removal from groundwater using nano scale zero-valent iron as a colloidal reactive barrier material. Environ. Sci. Technol. 2006;40:2045–2050.
74. Tratnyek PG, Johnson RL. Nanotechnologies for environmental cleanup. Nano Today. 2006;1:44–48.
75. Tuttle JH, Dugan PR, Randles CI. Microbial sulfate reduction and its potential utility as an acid mine water pollution abatement procedure. Appl. Microbiol. 1969;17:297–302.
76. Christensen B, Laake M, Lien T. Treatment of acid mine water by sulfate-reducing bacteria; Results from a bench scale experiment. Water Res. 1996;30:1617–1624.
77. Cardenas E, Wu WM, Leigh MB, Carley J, Carroll S, Gentry T, Luo J, Watson D, Gu B, Ginder-Vogel M, Kitanidis PK, Jardine PM, Zhou J, Criddle CS, Marsh TL, Tiedje JM. Microbial communities in contaminated sediments, associated with bioremediation of uranium to submicromolar levels. Appl. Environ. Microbiol. 2008;74:3718–3729.
78. Benner SG, Blowes DW, Ptacek CJ. A full-scale porous reactive wall for prevention of acid mine drainage. Gr. Water Monit. Remediat. 1997. 99–107.
79. Fruchter J. In Situ Treatment of chromium contaminated groundwater. Environ. Sci. Technol. 2002;36:464A – 472A.
80. Scherer MM, Richter S, Valentine RL, Alvarez PJJ. Chemistry and microbiology of permeable reactive barriers for in situ groundwater clean up. Crit. Rev. Microbiol. 2000;26:221–264.
81. Baker MJ, Blowes DW, Ptacek CJ. Phosphorous adsorption and precipitation in a permeable reactive wall: Applications for wastewater disposal systems. L. Contam. Reclam. 1997;5:189–193.
82. Puls RW, Paul CJ, Powell RM. The application of in situ permeable reactive (zero-valent iron) barrier technology for the remediation of chromate-contaminated groundwater: A field test. Appl. Geochemistry. 1999;14:989–1000.
83. Gibert O, de Pablo J, Cortina JL, Ayora C. Treatment of acid mine drainage by sulphate-reducing bacteria using permeable reactive barriers: A review from laboratory to full-scale experiments. Rev. Environ. Sci. Biotechnol. 2002;1:327–333.
84. Waybrant KR, Blowes DW, Ptacek CJ. Selection of reactive mixtures for use in permeable reactive walls for treatment of mine drainage. Environ. Sci. Technol. 1998;32:1972–1979.
85. Benner SG, Blowes DW, Gould WD, Herbert RB, Ptacek CJ. Geochemistry of a permeable reactive barrier for metals and acid mine drainage. Environ. Sci. Technol. 1999;33:2793–2799.

86. Waybrant KR, Ptacek CJ, Blowes DW. Treatment of mine drainage using permeable reactive barriers: Column experiments. Environ. Sci. Technol. 2002;36:1349–1356.
87. Jeyasingh J, Somasundaram V, Philip L, Bhallamudi SM. Pilot scale studies on the remediation of chromium contaminated aquifer using bio-barrier and reactive zone technologies. Chem. Eng. J. 2011;167:206–214.
88. Michalsen MM, Peacock AD, Smithgal AN, White DC, Spain AM, Sanchez-Rosario Y, Krumholz LR, Kelly SD, Kemner KM, McKinley J, Heald SM, Bogle MA, Watson DB, Istok JD. Treatment of nitric acid-, U(VI)-, and Tc(VII)-contaminated groundwater in intermediate-scale physical models of an in situ biobarrier. Environ. Sci. Technol. 2009;43:1952–1961.
89. Hammack RW, de Vegt AL, Schoeneman AL. The removal of sulfate and metals from mine waters using bacterial sulfate reduction: Pilot plant results. Mine Water Environ. 1998;17:8–27.

CHAPTER 12

A Columbia River Basalt Group Aquifer in Sustained Drought: Insight from Geophysical Methods

Mark W. Piersol and Kenneth F. Sprenke *

Department of Geological Sciences, University of Idaho, Moscow, ID 83844-3022, USA

ABSTRACT

Aquifers within the Columbia River Basalt Group (CRBG) provide a critical water supply throughout much of the Pacific Northwest of the United States. Increased pumping has resulted in water level declines in this region. Recharge into this aquifer system is generally not well understood. Recent suggestions of probable decades-long droughts in the 21st century add to this problem. We show that geophysical methods can provide useful parameters for improved modeling of aquifers in a primary CRBG aquifer located on the eastern edge of the Columbia Plateau. Groundwater models depend in part on the area, thickness, porosity, storativity, and nature of confinement of this aquifer, most of which are poorly constrained by existing well information and previous stress tests. We have made use of surface gravity measurements, borehole gravity measurements, barometric efficiency estimates, earth tidal response, and earthquake seismology observations to constrain these parameters. We show that the aquifer, despite its persistent drawdown, receives a great deal of recharge annually. Much of the recharge to the aquifer is due to leakage from overlying flows, ultimately tied to precipitation, an important result for future aquifer management in times of sustained drought.

Keywords: aquifer; recharge; geophysics; hydrogeophysics; Columbia River Basalt Group

1. INTRODUCTION

The Columbia River Basalt Group (CRBG) in the Pacific Northwest of the United States hosts a regional aquifer system that extends from the foothills of the northern Rocky Mountains in west-central Idaho into the Columbia Basin of Washington and Oregon (Figure 1). In many locations within the region, it is the only water source for communities, residences, industry, agriculture, and aquaculture [1]. The CRBG aquifer system is a significant concern to water resources managers because of long-term water level decline and a general lack of knowledge about natural recharge.

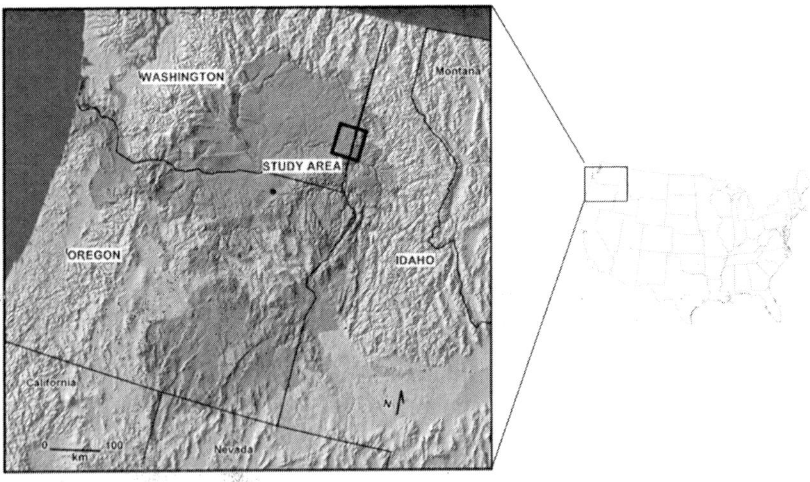

Figure 1. The darkly shaded area is the CRBG regional aquifer system in the Pacific Northwest of the contiguous United States. The rectangle on the left map shows the location of the general study area, the Palouse basin.

A recent study has suggested that in the western states, drought risk in the 21st century will likely exceed the driest centuries of the past millennia, leading to unprecedented drought conditions [2]. Much of the inland Pacific Northwest is characterized by an arid to semiarid climate. The majority of surface water in this region originates as upland precipitation. The causes of past and future drought will not be identical but the paleoclimatic record demonstrates the plausibility of extensive, severe droughts [3]. In the past half century, the up-dip areas of the Columbia Basin abutting the northern Rocky Mountains region have experienced a substantial decline in peak snow water equivalent [4]. A recent USGS study of the Columbia Basin suggests that climate change will result in a drop in water levels across the region with resultant degradation and disappearance of aquatic ecosystems [5]. Water managers in the Columbia Basin need to understand the nature of recharge into the aquifer system and the effect a sustained drought will have on aquifer system productivity.

Recharge, if any, of the deeper CRBG aquifers is generally inferred to occur (1) from groundwater movement originating around the edge of the Columbia Basin in up-dip areas where units pinch out; and (2) from vertical leakage through younger more shallow flows via open faults and tectonic fractures. Both of these sources are intimately related to precipitation and ultimately climate change. The general area of this study (the Palouse basin), lies at the eastern up-dip edge of the CRBG in good position to study the nature of CRBG recharge (Figure 1). Furthermore, excellent records of pumping volumes and historic records of annual water level declines are maintained by the Palouse Basin Aquifer Committee [6].

The key to understanding the nature of recharge in the study area is to have an adequate groundwater model that explains the dynamics of the aquifer system. However, previous groundwater models [7,8] for the study area have neither predicted the annual drop in water levels within the aquifer system nor have they provided any useful insight into the nature of recharge. We believe the lack of success of these models was at least in part the result of incorrect parameters employed, particularly with respect to the area, thickness, porosity, storativity, and nature of confinement of the aquifer system. The purpose of this study is to show that geophysical methods including surface gravity measurements, borehole gravity measurements, barometric efficiency estimates, earth tidal response, and earthquake seismology observations can be useful to better constrain these parameters for future groundwater models.

1.1. CRBG Aquifers

Numerous studies of CRBG aquifers have been conducted within the Columbia Basin to better understand their hydrogeology [9,10,11,12,13,14,15,16,17]. One of the most significant findings of these studies is the similarity of the hydrogeological characteristics, properties, and behavior of the CRBG aquifers across the region. These general characteristics include: (1) simple sheet flows of basalt 10 to >100 m thick of great lateral extent that act as aquitards; (2) intraflow structures such as vesicular flow-tops, rubble zones, and sedimentary interbeds that act as aquifers; (3) control of lateral continuity of the aquifers by erosion, flow pinchouts, folds, faults and dikes; (4) flow pathways dominantly sub-horizontal down-dip; (5) limited vertical groundwater movement. These similar characteristics allow for the application of the knowledge of the general hydraulic characteristics and behavior of the CRBG aquifers in one area to be applied to CRBG aquifers in other areas [18].

1.2. Study Area

Our general study area (Figure 2) is the Palouse basin of western Idaho and eastern Washington abutting the crystalline rocks of the northern Rockies foothills. The Palouse is an important dryland agricultural region, with a semi-arid climate and topography dominated by rolling hills composed predominantly of windblown loess. A larger amount of precipitation falls in the mountainous terrain which bounds the study area to the east, southeast, and north [19]. The Palouse River, its tributaries, and other local streams drain the area and flow towards the northwest to eventually join the Snake River. Some 800 water wells have been cataloged in the Palouse basin [20] which covers an area of 1280 km^2. Most are shallow, very-low productivity domestic wells hosted in the loess soil cover or in the crystalline rocks of the surrounding highlands.

1.3. The Moscow-Pullman Basin

Based on the very similar piezometric elevations and historic water level declines, the primary aquifer system that serves the cities of Pullman and Moscow appears to be restricted to an area referred to as the Moscow-Pullman basin (Figure 2) which is partially hydraulically isolated from the larger Palouse basin and the main body of the CRBG regional ground-water flow system. This is a lava embayment of the CRBG Grande Ronde Formation which flowed into the area during Miocene time from the southwest following a pre-basalt stream system cut into the western slope of the northern Rockies. The Moscow-Pullman basin has an area of 420–660 km^2 and encompasses the cities of Pullman, Washington and Moscow, Idaho plus multiple smaller communities and rural areas, as well as two large universities. This basin is part of the much larger Palouse basin, also shown on Figure 2.

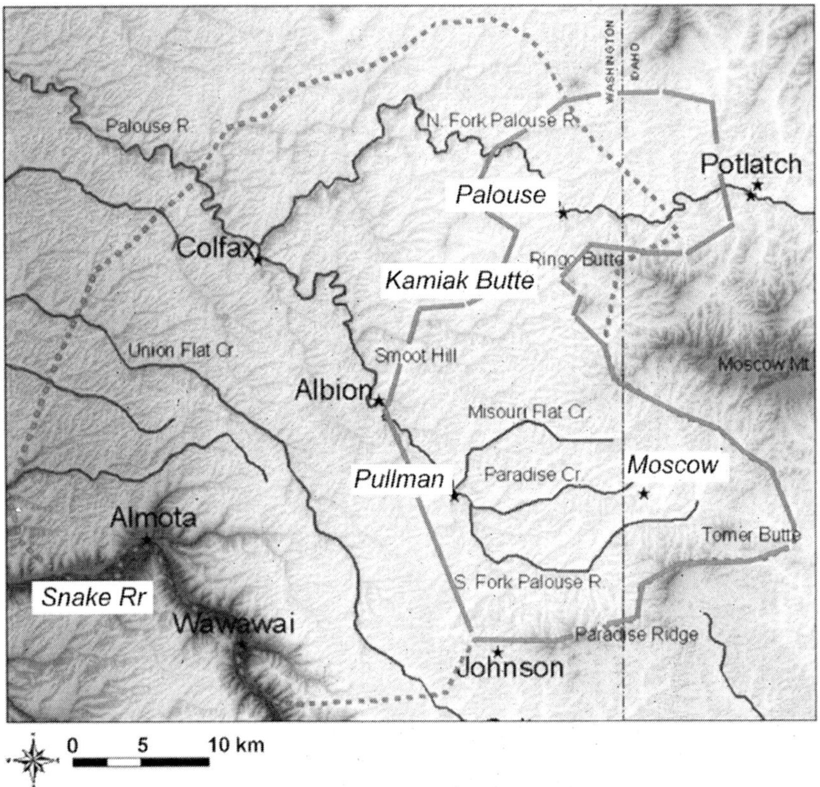

Figure 2. The Palouse basin general study area. See Figure 1 for location. The specific study area, the Moscow-Pullman basin is indicated by the solid purple line. The dashed line is the boundary of the larger area under the political jurisdiction of the Palouse Basin Aquifer Committee.

A geological section (Figure 3) through Moscow and Pullman illustrates the general stratigraphy in the Moscow-Pullman basin. Beneath a loess soil cover are the Saddle Mountain, Wanapum, Grande Ronde, and Imnaha basalt formations of the Miocene CRBG, each of which is composed of multiple flows. Interspersed between flows are sedimentary interbeds collectively called the Latah Formation. The pre-basalt rocks are Cretaceous granites and Precambrian crystalline metasediments.

As shown in Figure 4, hydrographs are available for about 30 wells in the Moscow-Pullman basin. Fairly common are wells 30–40 m deep, hosted in the Wanapum Formation. These wells account for 10% of the water produced in the basin. The primary aquifer system (known locally as the "lower" or "Grande Ronde" aquifer system) is completely contained within the Grande Ronde Formation. The high yield municipal and university wells tap the Grande Ronde Formation typically at depths of 90–120 m. The Grande Ronde Formation accounts for 90% of the 2.45 billion gallons (62.4×10^6 m^3) pumped annually in the Palouse basin [6].

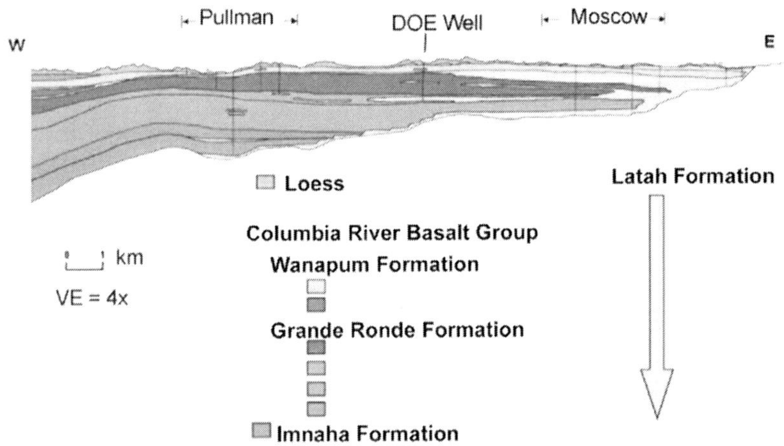

Figure 3. Diagrammatic west to east geological cross section through Pullman to Moscow showing the basalt stratigraphy of the study area.

Except at the entrance to the basin, the aquifer system is largely bounded by crystalline rocks of the surrounding highlands. The entrance to the Moscow-Pullman basin near Pullman coincides with a ground water divide apparent on the piezometric surface of the primary aquifer system (Figure 5).

To the southwest of Pullman, a poorly understood barrier to groundwater flow [7] prevents groundwater from flowing down the steep gradient to the elevation of the Snake River, lying in a deep canyon some 500 m in elevation below the Moscow-Pullman basin (Figure 5). Geochemical observations [21] have shown that groundwater does not seep into the Snake Canyon. It has also been noted that there is a conspicuous lack of large springs on the walls of the Snake River canyon [22]. Generally to the west of the Moscow-Pullman basin piezometric elevations fall off more moderately toward Colfax, dropping some 135 m. As shown in the stratigraphic and hydrogeological cross-section (Figure 3 and Figure 6), the top of the Grande Ronde drops in elevation east towards Moscow and west and northwest away from Pullman.

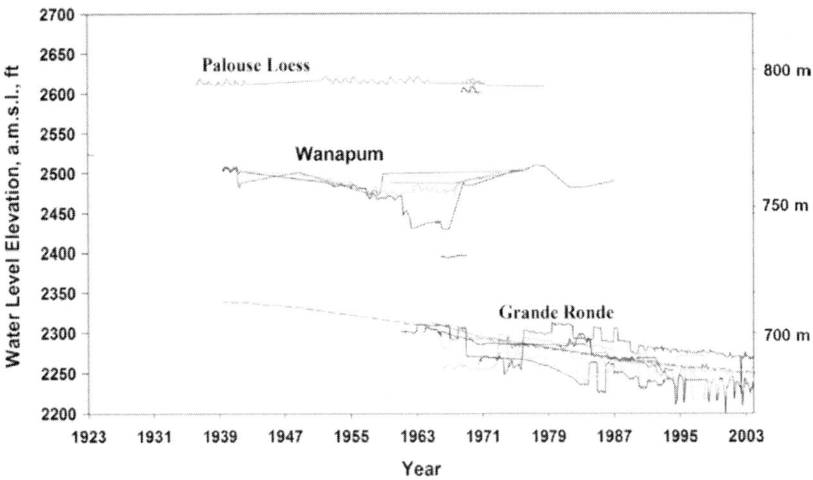

Figure 4. Composite hydrograph of 30 water wells in the Palouse basin. Modified from [20].

1.4. The Primary Aquifer System

At Pullman, the top of the Grande Ronde Formation is at a relatively high elevation; the depth to the Grande Ronde Formation top is less than 15 m in places (Figure 6). Here, water-level elevations are essentially the same in wells completed deep in the aquifer system and in more shallow open wells. At Moscow and Palouse, however, the primary aquifer system is definitely confined. The Grande Ronde Formation is more than 130 m deep at Moscow, yet the piezometric surface rises to the same level as at Pullman. Wells completed in the overlying Wanapum aquifer system have higher ground-water levels as compared to wells completed in the primary aquifer system (Figure 4 and Figure 6). This strong downward hydraulic gradient suggests that the Wanapum aquifer system, might leak into the primary aquifer system. However, no hydraulic connection has ever been observed between the Wanapum aquifer system and the Grande Ronde aquifer system in the Moscow area.

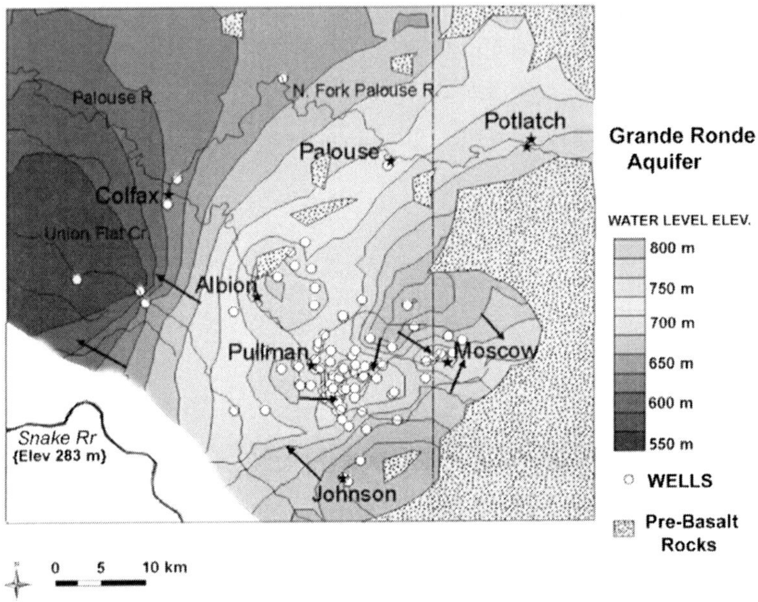

Figure 5. Piezometric map of the primary aquifer. Arrows indicate postulated flow directions of groundwater.

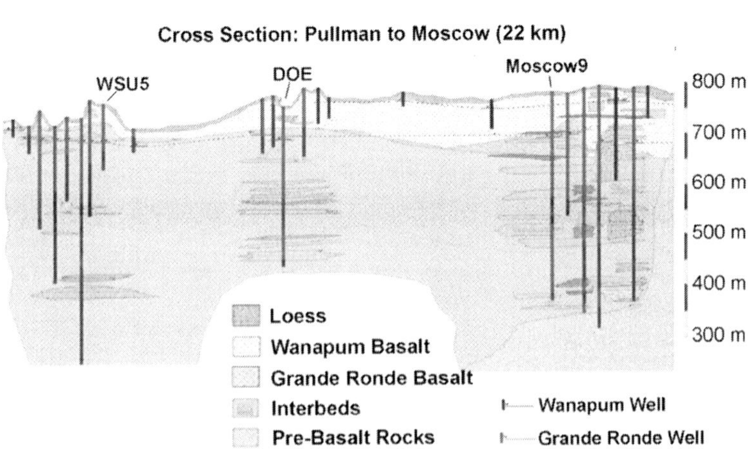

Figure 6. Diagrammatic hydrogeological cross-section from the Pullman wells on the left (west) to the Moscow wells on the right (east).

Four predominant studies using isotopic concentrations in groundwater have been completed in the Moscow-Pullman basin [23,24,25,26]. The apparent radiocarbon ages of water in the Pullman-Moscow area range from 11,000 to 26,000 year in the primary aquifer and from modern to 10,000 year in the overlying units [25,26]. For stable oxygen isotopes, the $\delta^{18}O$ values range from −16% to −17.5% in the primary aquifer and from −16% to −14.8% in the overlying units [24,26]. These numbers suggest that the majority of the water in the Grande Ronde was recharged during ice age conditions in the Pleistocene. On the other hand, tritium concentrations, no older than 60 years, have been detected in the upper part of primary aquifer indicating that modern recharge is mixing with the Pleistocene water in the Grande Ronde [26].

In the Moscow-Pullman basin, recharge, if any, to the primary aquifer system comes from either the mountains on edges of the basin where the basalt flows pinch out against the crystalline rocks of the highlands or from vertical leakage from overlying more shallow aquifers perched between flows above the regional aquifer system [27]. It is of critical importance for future management of the aquifer system to determine if there is any recharge to the primary aquifer system and whether it is generated in the surrounding highlands at flow pinchouts or by vertical leakage throughout the basin. If there is no recharge, then new sources of water need to be found for future development, such as a pipeline to the Snake River, some 20 km south and 500 m in elevation below the study area. If there is recharge, then the optimum locations of recharge galleries need to be determined.

The rational approach to understanding the nature of recharge is to develop a ground water model based on a sound hydrogeological conceptual model consistent with pumpage records and the history of water level declines. Previous groundwater models [7,8] of the study area have failed in all these regards. Below we show how geophysical methods have provided better parameters for future groundwater models.

2. METHODS

Groundwater models depend in part on the porosity, area, thickness, storativity, and nature of confinement of the aquifer. These parameters, in the Moscow-Pullman basin, are poorly constrained by existing well information and previous stress tests. We have made use of geophysical measurements, including surface gravity measurements, borehole gravity measurements, barometric efficiency estimates, earth tidal response, and earthquake seismology observations to constrain these parameters.

2.1. Porosity of the Aquifer System

The porosity of the CRBG strata exhibits a bimodal distribution due to the difference between the dense, massive interiors of flows (aquitards) and the intraflow structures (aquifers). To better quantify this porosity difference, we made use of extraordinarily detailed borehole gravity data that have been carried out in CRBG strata in three wells at the Hanford site in central Washington State as part of the design process for a new waste treatment plant [28]. The Hanford Site is a mostly decommissioned nuclear production complex sitting on CRBG bedrock operated by the United States federal government on the Columbia River in the U.S. state of Washington. This borehole gravity survey is stated to include the longest section (approximately 1280 m in the three wells) of measurements data taken at such a small station spacing (3 m).

The borehole gravity meter tool has a very large horizontal depth of investigation so it is not influenced by washouts and borehole construction. The survey goal at Hanford was to provide CRBG rock densities with an estimated error of ±0.05 g/cm^3 or less. We plotted the 302 CRBG basalt density determinations from the Hanford survey as a histogram in Figure 7. The distribution is clearly bimodal indicating the dichotomy between the massive flows and the porous intraflow zones encountered in the section of CRBG strata surveyed. As also shown in Figure 7, we fit the histogram with two Gaussian curves [29]. The higher peak is at a density 2.80 ± 0.12 mg/cm^3 which we take as the mean density of the massive flows. The lower peak is at density 2.02 ± 0.17 g/cm^3 which we interpret as the density of the intraflow zones. The density contrast between the massive flow zones and the intraflow units is 0.78 ± 0.20 g/cm^3. If we treat the massive flows as having no effective porosity, but with an unconnected vesicle volume typical of basalt (7.8%) [30], we find the mean porosity of the intraflow zones is 0.355 ± 0.04.

2.2. Area of the Aquifer System

The area of the main Grande Ronde aquifer system that should be included in a groundwater model of the Moscow-Pullman basin is quite controversial. By area in this regard, we mean the areal extent to which aquifer parameters determined from well tests in the deep city wells should be applied. In some studies the southwest boundary of the Moscow-Pullman aquifer system is thought to be bounded by the Snake River [8,31], however, geochemical studies have shown that Moscow-

Pullman basin ground water does not seep into the Snake River Canyon [21]. It appears likely that northwest trending folds paralleling the Union Flat Creek restrict groundwater movement towards the Snake River Canyon, instead channeling it northwest [7,20,32,33,34,35,36,37].

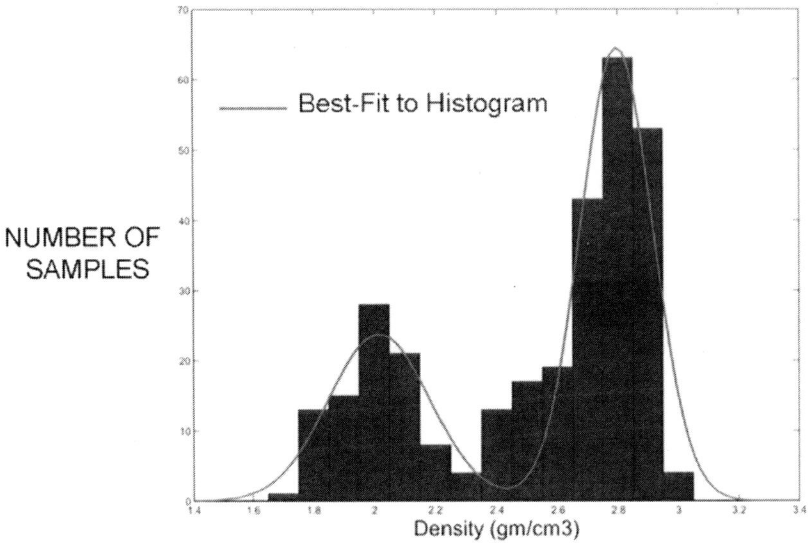

Figure 7. Histogram of basalt densities found from a borehole gravity survey [28]. The survey densities were binned in 0.1 g/cm^3 increments and then fitted to a bimodal curve using the EzyFit toolbox [29]. The lower density peak is modeled as 2.02 ± 0.17 g/cm^3 and the higher density peak as 2.80 ± 0.12 g/cm^3.

Another controversy concerns whether the Grande Ronde aquifer system hydrologically connects the southern cities of Pullman and Moscow to the northern city of Palouse. A break in the crystalline rocks within the basin, called the Kamiak Gap, occurs between Pullman and Palouse immediately east of Kamiak Butte (Figure 2). Hydraulic connection between Palouse and the rest of the basin through the Kamiak Gap has been investigated on a geophysical basis by several researchers [38,39]. A magnetotelluric study [38] suggested continuity of the Grande Ronde Formation through the gap. On the other hand, gravity methods [39] initially suggested that the Grande Ronde was interrupted by a saddle of crystalline basement rock obstructing aquifer conductivity south of Palouse. However, new test well data [40], providing much needed control on the elevations of formation tops, has allowed us to reinterpret these gravity data. Gravity readings were taken along four lines in the Kamiak Gap (Figure 8) [39].

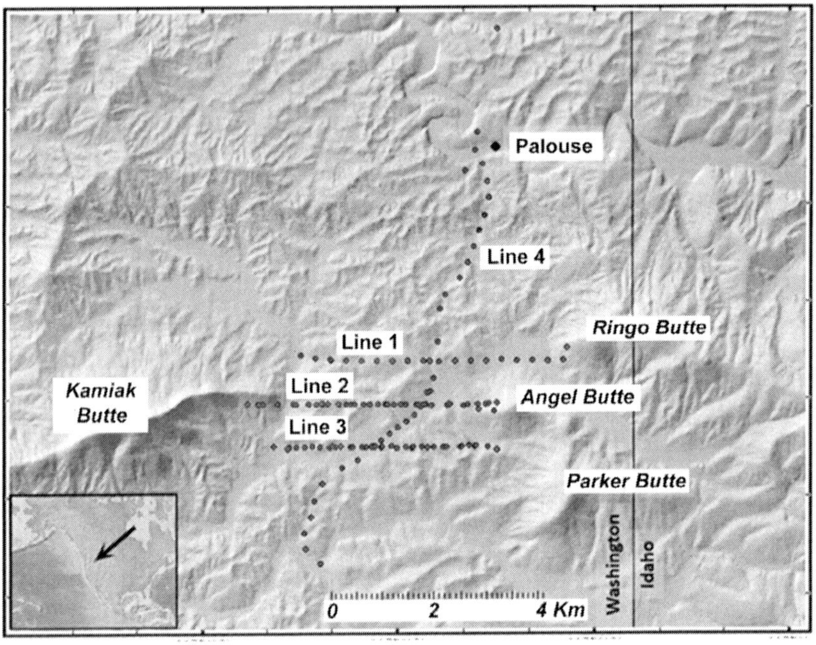

Figure 8. Gravity survey area in the Kamiak Gap [39]. See Figure 2 for location of Kamiak Butte within the study area. Lines 1, 2, and 3 are east-west transects of gravity stations across the gap and Line 4 meanders approximately north-south crossing the other three lines. Figure generated with GeoMapApp [41].

The stratigraphy of the DOE Butte Gap well, drilled at the latitude of the southernmost east-west transect, is remarkably similar to the stratigraphy of the Palouse City #3 well drilled to the north of the Kamiak Gap [40]. Using these constraints we propose a new geophysical model (Figure 9) suggesting that the Grande Ronde basalts are indeed continuous through the Kamiak Gap with a net thickness in excess of 100 m. Thus, Kamiak Gap should not create a hydraulic barrier to north-south groundwater flow.

Thus, for improved ground-water models, the total area of the primary aquifer system that directly affects the Moscow-Pullman basin should be about 620 km^2, with 420 km^2 south of Kamiak gap, and 200 km^2 north of the gap.

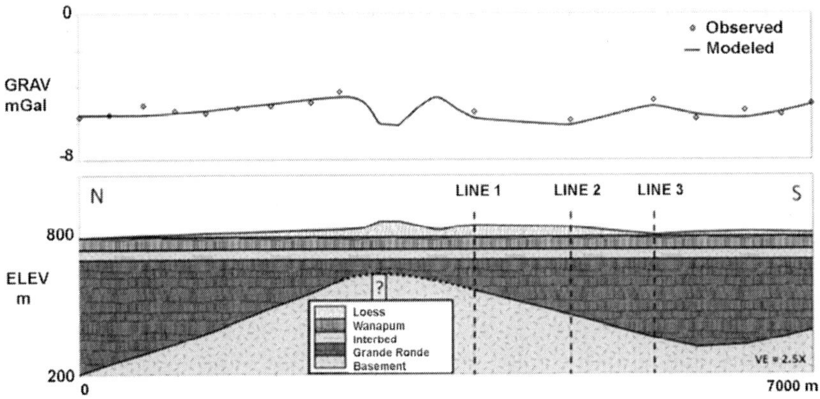

Figure 9. Gravity model through the Kamiak Gap from north to south. Modeled thickness of Grande Ronde at intersection of Line 1 (Figure 8) is 125 m, at intersection of Line 2 is 224 m, and at intersection of Line 3 is 316 m. The dashed line between basement and Grande Ronde north of Line 1 below the topographic divide indicates uncertainty in the model as there is approximately 1.35 km between survey point locations.

2.3. Thickness of the Aquifer System

Comparing the areas of the two normal curves of the histogram of borehole gravity data from CRBG strata at the Hanford site (Figure 7), we find that 29% of the basalt section there is aquiferous intraflow units; the remainder is massive basalt with negligible effective porosity. Based on the logs from deep wells in the Moscow area (Figure 6), a previous study [42] estimates that 25% of the Grande Ronde section consists of intraflow zones. A well construction report [34] for the DOE well (Figure 6) logged where water was seen in the drill core. About 28% of the core length contained water. Producing zones from the Grande Ronde in the Moscow-Pullman basin lie between elevations of 350 m and 716 m above sea level (Figure 6), a net thickness of 365 m. Thus, we estimate the effective thickness of the primary aquifer to be 25% to 29% of 365 m or 100 ± 2 m.

2.4. Specific Storage from Barometric Efficiency

Specific storage can be estimated from barometric efficiency if the porosity of the aquifer is known.

$$S_S = \frac{\eta \gamma_w}{E_w BE} \tag{1}$$

where η is porosity; γ_w is the unit weight of water in N/m^3; E_w is the elastic modulus of water (2.2×10^9 Pa); BE is barometric efficiency; and S_S is aquifer specific storage in m^{-1}. The barometric efficiency of aquifers in the CRBG is high because of the strength of the massive nearly impermeable flow interiors that separate the confined intraflow aquiferous zones. In the Grande Ronde aquifer system of the Moscow-Pullman basin, previous studies [43,44] have shown that BE ranges from 0.9 to 1.0. As described above, we estimate the mean porosity of the intraflow zones to be 0.355 ± 0.04. Using this value in Equation (1) above, along with the known BE of 0.95 ± 0.05, we find that the specific storage S_s of the aquiferous intraflow zones is $1.67 \times 10^{-6} \pm 0.18 \times 10^{-6}$ m^{-1}.

2.5. Specific Storage from Earthquake Seismology

Seismological theory [45,46,47] predicts that S_s for a confined aquifer can be found by comparing, as the Rayleigh waves from distant large earthquakes roll by, earthquake seismograms to water level oscillations as illustrated in Figure 10. To determine specific storage, the borehole amplification factor has to be known for the well. This factor depends on oscillation frequency, the borehole radius, initial height of the water column, screened aquifer thickness, transmissivity of the aquifer, and, to a minor extent, S_s [46]. Because our quantity to be determined, S_s is involved in the calculation of the borehole amplification factor, the method requires an iterative procedure for its solution. An initial guess of S_s is used to generate successively better approximations. However, because S_s has only a minor effect on the borehole response, convergence is quickly obtained.

The seismic Rayleigh wave response of municipal well Moscow 9 (see Figure 6 for location) was evaluated. This important supply well was shut down temporarily for pump repair for several months in 2012, giving an opportunity for the installation of a data logger. The well is cased except for 27 m of screen adjacent to several interconnected highly permeable flow top units within the Grande Ronde aquifer. The top of the aquifer is at a depth of 198 m below land surface. The static level of the water rises to a height of 104 m above the top of this confined artesian aquifer. The borehole diameter above the screened intervals is 0.22 m. A transmissivity estimate from previous well tests was used [48]. Rayleigh waves from three moderately large 2012 earthquakes (magnitude >7.7) with epicenters in Haida Gwaii, Okhotsk, and the Philippines were observed. During the same time intervals as the Rayleigh wave were passing, the data logger in well Moscow 9 was collecting measurements at one minute intervals (Figure 10). Twenty water level measurements immediately after the Rayleigh wave first arrival for each of three separate earthquakes were used to find S_S.

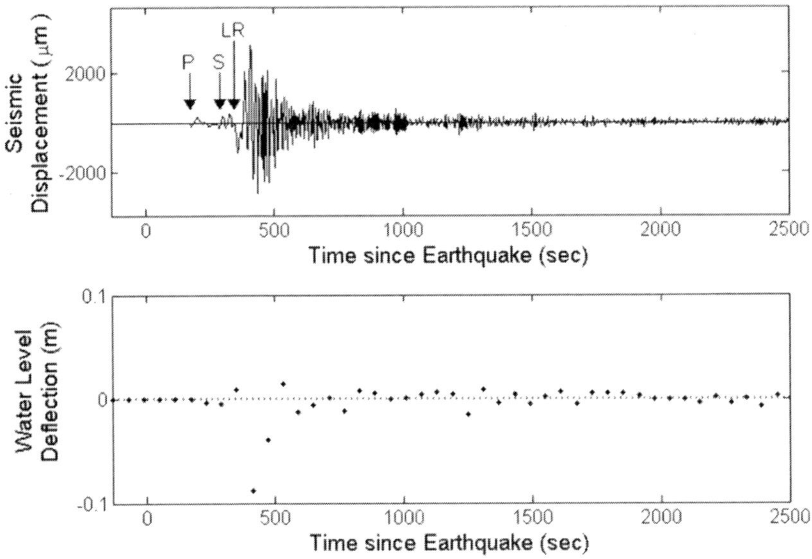

Figure 10. Water level oscillations observed in a water well near Moscow, Idaho while seismic waves from the 2012 M7.8 Haida Gwaii Earthquake passed through the aquifer.

Combining the data from the earthquakes, S_S was found to be $1.5 \times 10^{-6} \pm 0.2 \times 10^{-6}$ m^{-1}. However, this optimistic figure does not take into account sources of error that are difficult to quantify. The method assumes the aquifer is confined, uniformly porous, and free of heterogeneities.

2.6. Specific Storage from Earth Tides

Gravity theory predicts that earth tides (Figure 11) will result in harmonic subsurface dilatations within a few hundred meters of the earth's surface [46,49]. At inland sites, the lunar diurnal (O1) and lunar semi-diurnal (M2) are the most affected by the local hydraulic characteristics of the formations open to the wells [46]. On the other hand, the solar harmonics, K1 and S2, at periods of 1.0 and 0.5 days, contain the influence of atmospheric-pressure oscillations. At latitude θ, the resulting water level deflections for the O1 and M2 earth tides in an open water well in a confined aquifer are given by

$$A_{O1} = 1.56 \times 10^{-8} \sin(\theta) \cos(\theta)/S_s \text{ and } A_{M2} = 1.89 \times 10^{-8} \cos(\theta) \cos(\theta)/S_s \quad (2)$$

where A_{O1} and A_{M2} are the amplitudes at periods of 1.0758 and 0.5376 days [46,49].

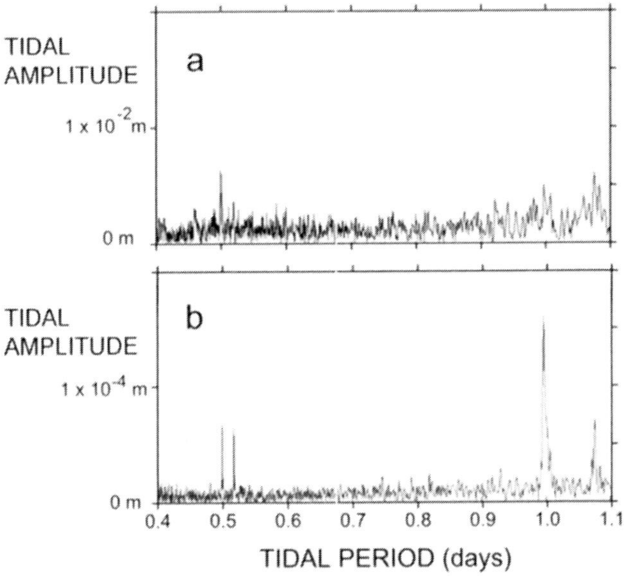

Figure 11. Water level deflections caused by earth tides in (**a**) DOE well and (**b**) WSU5 well. See Figure 6 for locations. The O1 and M2 harmonics have periods of 1.0758 and 0.5376 days. Note that the DOE tidal amplitudes are 50 times higher than those in Pullman well WSU5, suggesting that the primary aquifer system near Pullman is partially unconfined.

The DOE well, located midway between Moscow and Pullman (Figure 3), was the site of an experiment to measure earth tides in the study area. Groundwater levels were sampled at 10 min intervals for an entire year to produce the spectrum shown in Figure 11. The spectrum is noisy because of interference from nearby pumping wells. To agree with the S_s from barometric efficiency, the equations above predict that A_{M2} would be less than 0.001 m, an amplitude lost in the background noise apparent in Figure 11a On the other hand, we found A_{O1} at 0.0054 m to be five times above the background noise, resulting in a S_s of $1.44 \times 10^{-6} \pm 0.14$ m^{-1}, a value in agreement with our estimate from barometric efficiency.

2.7. Pressure State of the Aquifer System

Figure 11 also shows the earth tide response of well WSU5 (see Figure 6 for location) in Pullman. The amplitude of A_{O1}in this well was 7.0×10^{-5} m, 50 times smaller than the O1 amplitude in the DOE well. The amplitude A_{M2} was 6.4×10^{-5} m. The calculated S_s values from Equation (2) are 1.1×10^{-4} m^{-1} and 1.4×10^{-4} m^{-1} for the O1 and M2 harmonics respectively, far too large if the aquifer at this location is confined. Earth tides can occur in partially confined aquifers, and the strength of the response depends on the distance to the exposed phreatic water table [50]. The low tidal response of WSU5 compared to DOE is consistent with the idea that the primary aquifer system is unconfined at places near Pullman.

2.8. Storativity of the Aquifer System

Estimates of the storativity of the primary aquifer system based on previous well tests and ground water models in the Moscow-Pullman basin have ranged over six orders of magnitude (Figure 12). To better constrain this parameter, we made use of our estimates stated above of specific storage and effective thickness. In Moscow, where the aquifer system is clearly confined, these yield consistent values of specific storage of $1.5 \times 10^{-6} \pm 0.2 \times 10^{-6}$ m^{-1}. Multiplying the specific storage by the effective aquifer thickness of 100 m, we obtain a storativity of $1.5 \times 10^{-4} \pm 0.2 \times 10^{-4}$, a value that falls in the range of previously determined storativity values from well tests (Figure 12).

In Pullman, where the aquifer system is apparently unconfined at places, we suggest that the annual decrease in water levels is the result of extraction of water from storage in the aquiferous intraflow units with a storativity equal to the specific yield, which would be about 0.35 based on the results above. As shown in Figure 12, this storativity for the unconfined areas is much higher than values used in previous groundwater models of the primary aquifer system.

In summary, where the aquifer is confined, the storativity was found to be near 1.5×10^{-4}; where unconfined, 3.5×10^{-1}. Thus, for the same drop in hydraulic head, very little water is produced by water expansion from the confined units compared to the amount of water extracted from the unconfined area. For each m^3 of water produced from the confined units by expansion, more than 350 m^3 are extracted from storage in the uppermost unconfined aquiferous unit.

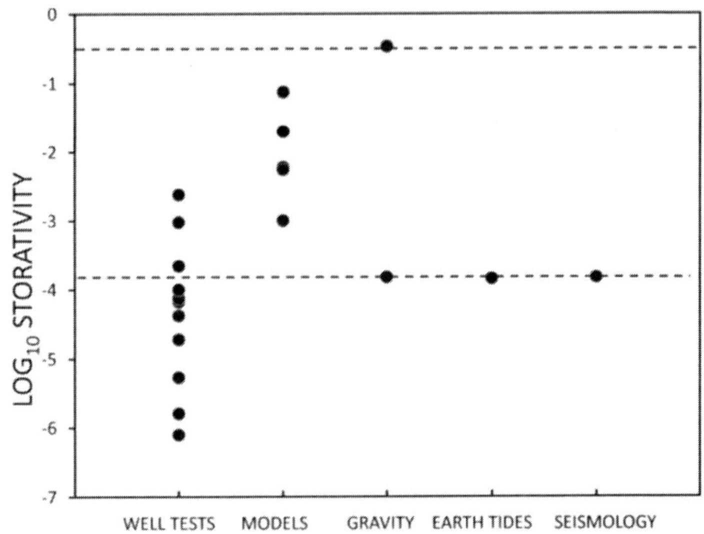

Figure 12. Values of storativity found by previous well tests [43] and groundwater models [7,8,31,43] compared to the values found using three geophysical methods described in the text. The dashed lines show the preferred values for storativity found in this study. The upper line is for the unconfined areas of the aquifer system and the lower value is for the confined units.

3. SYNTHESIS WITH PUMPAGE AND RECORDS OF WATER LEVEL DECLINE

3.1. Water Budget Model

The goal of this study was to employ geophysical methods to provide better estimates of certain parameters needed for future comprehensive groundwater models of the Moscow-Pullman basin. For the present study, however, we do employ a simple single cell model to test the sensitivity of the basin's recharge budget to our parameters. Single cell models employ infinite transmissivity, but finite storativity, and are commonly used in economic hydrology [51]. On an annual basis, we treat the Moscow-Pullman basin as a single cell with boundary conditions such that:

$$V_p = V_r + S\,A\,(\Delta h) \tag{3}$$

where V_p in m^3/year is the annual volume of water pumped; V_r in m^3/year is the annual volume of water replenished from outside the cell; S is the dimensionless storativity of the cell; A is the area of the cell in m^3; and Δh in m/year is the annual decline in water level. Two of the variables V_p and Δh in the above relation are well known from records kept by the Palouse Basin Aquifer Committee [6]. For the storativity and area of the cell, we use the estimates provided by the geophysical methods described above.

The annual volume of replenishment has two components: (1) that driven by Δh; and (2) that not directly associated with Δh [43]. In the first category are head-related changes in influx to the cell from the surrounding region. Also in this category are any changes in flux from overlying units due to the increased vertical hydraulic gradient. In the second category is decreased recharge due to depletion of the overlying hydraulically connected source (overlying leaky aquifers or surface streams). To better understand the effect a sustained drought may have on replenishment, it is useful to separate these two components. In our simple model, we employ a linear approximation such that:

$$V_r = S_c\,\Delta h + V_0 \tag{4}$$

where S_c in m^2/year is an annual volume replenished per unit annual drop in head for head-related sources and V_0 in m^3/year is the volume of annual recharge not directly head-related. Combining the above two equations and taking into account the possibility that the aquifer is unconfined in places:

$$Vp = (S_c + S\,A_c + S_y\,A_u)\,\Delta h + V_0 \tag{5}$$

where S and S_y are the storativities and A_c and A_u are the areas of the confined and unconfined portions, respectively. Thus, our simple model predicts a linear relationship between annual volume pumped and annual drop in water level, with an intercept equal to V_0 and a slope associated with storativity, area, and head-related replenishment.

Figure 13 shows a cross-plot of annual water volume pumped as a function of annual drop in water level for the primary aquifer in the study area as provided by the Palouse basin Aquifer Committee [6]. The data points (modified from [32]) are for years 1968–2010. The solid line in the cross-plot is the best-fit line to the data. The correlation coefficient (r^2) is 0.52, showing that the data can be reasonably fit by a straight line with slope of $30.6 \times 10^6 \pm 4.4 \times 10^6$ m and intercept of $58.34 \times 10^6 \pm 1.55 \times 10^6$ m^3/year.

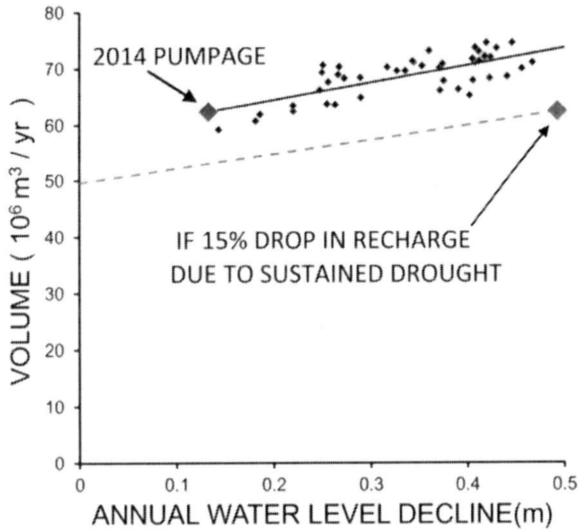

Figure 13. Annual water level decline *vs.* pumped volume for the Grande Ronde aquifer system. The data points show the historic data since 1968 along with a best-fit line. The large blue diamond is the current state. The red dashed line is the predicted state of the aquifer system in the event of a sustained drought causing a 15% drop in recharge. The red diamond indicates the expected annual water level decline if the current pumpage is maintained during the drought.

3.2. Model Results

We can make several inferences from the simple model. All the parameters in Equation (5) controlling the slope of the best-fit line are not known, so we consider a number of possibilities.

First of all, we find that the water budget is relatively insensitive to the parameters which we have derived by the geophysical methods described above. Regardless of values used for storativity, aquifer area, or head-related replenishment, there is a large amount of annual recharge unrelated to head (V_0 = 58.34 × 10^6 ± 1.55 × 10^6 m^3/year), accounting for 93.5% ± 2.6% of the current pumped volume, but unfortunately not enough to stop long term-water level declines. It is clearly impossible to model the aquifer without invoking a considerable amount of annual recharge.

Second, if the primary aquifer is in fact confined everywhere, then, from our geophysical results above, we can set S = 1.5 × 10^{-4} ± 0.2 × 10^{-4} and A_c = 620 km^2 in Equation (5), leading to S_c = 30.51 ± 0.01 m^2/year, a result very insensitive to our uncertainty in storativity. In this scenario, only 0.02% of the current pumped volume of 62.4 m^3/year, comes from storage. 6.5% comes from head-related replenishment, and the remainder from recharge. The relatively low percentage of head-related replenishment compared to recharge suggests that constant-head boundary conditions should be used with caution in future numerical models of the confined system. Also, the value of storativity in the confined system is unimportant and could be off by an order of magnitude without affecting the water balance by more than one per cent.

Third, if the primary aquifer is partly unconfined as the geophysical data suggest, we can estimate the maximum area of the phreatic water table. For this case, using our geophysical results, we can set S = 1.5 × 10^{-4} ± 0.2 × 10^{-4}, S_y = 0.355 ± 0.04, S_c = 0, A_c = 620 km^2 A_u, leading to an A_u of at most 86 ± 8 km^2 or 14% of the area of the Moscow-Pullman basin. Under this scenario, 6.5% of the current water being pumped comes from storage. Recharge accounts for 58.3 × 10^6 m^3/year, independent of head; the remaining 3.1 × 10^6 m^3/year of the water pumped is produced from storage almost entirely by gravity drainage of the uppermost aquiferous unit. In future models involving a phreatic water table, specific yield does become a relevant parameter, but, nonetheless, a very small contributor compared to recharge.

Fourth, the actual situation could well be a combination of the previous two scenarios, with a phreatic water surface over a smaller area releasing water from storage along with the drop in head inducing more recharge into the deeper aquifer systems from overlying units or decreasing the flux out of the study area to the west. However, these effects on a numerical model will all still be minor compared to head-unrelated recharge.

Finally, we can use our simple model to predict the response of the primary aquifer to a sustained drought (Figure 13). If we assume for example, that annual replenishment to the aquifer system is reduced by 15% as a result of an episode of continued low annual precipitation, the model shows that the annual water level decline would increase four-fold over the current value if current levels of pumpage were maintained.

4. CONCLUSIONS

This study clarifies a number of issues about the primary aquifer system within the Palouse basin and should lead to better numerical groundwater models, and more rational planning for the future not only for this aquifer system but for any CRBG aquifer should the 21st century prove to be as dry as predicted. The storativity of the deeper aquiferous basalt intraflow units is, in fact low, 1.5×10^{-4}, a value within the range of the many well tests that have taken place, but an order of magnitude or more below the values used for these units in previous groundwater models of the aquifer system. While the primary aquifer system is generally confined, in the vicinity of Pullman it is apparently unconfined with a phreatic water table and a storativity of 0.35, a value far higher than that used in previous groundwater models. Thus, the water that is derived from storage throughout the basin is at a cost of the lowering of the phreatic water table near Pullman and a general depressurization of the primary aquifer system throughout the Moscow-Pullman basin.

The derived values of aquifer thickness, area, and storativity derived from geophysical methods in this study will improve future groundwater models of the primary aquifer, but they will have only a second-order effect. The first order unknown is recharge. Although the mechanism remains elusive, currently more than 90% of the water pumped annually is replenished by recharge unrelated to head. Should that replenishment be halted by a prolonged drought, water level decline would accelerate. Furthermore, if the aquifer system is in fact, partially unconfined, water level drops may be sudden as the phreatic table encounters different units at depth. There is no guarantee that the primary aquifer system would continue to operate as a unified system as horizontal intraflow pathways get cut off as water levels decline.

Current studies of possible recharge galleries for the Moscow-Pullman basin are focused on the eastern side of the basin in hopes of recharging, not the primary aquifer system, but the overlying Wanapum system [27]. But this system provides only 10% of the net water supply, mostly in Moscow, and it is not clear that replenishing the Wanapum in Moscow will affect the primary aquifer. The geophysical results presented here suggest that the most logical location for future recharge galleries may be in the Pullman vicinity where the primary aquifer system appears to be unconfined and where its phreatic groundwater table is shallow. Two tributaries of the Palouse River have a confluence at Pullman. Redirecting the spring runoff in those tributaries into the aquifer system could restore historical water levels at Pullman and re-pressurize the confined units throughout the Moscow-Pullman basin.

ACKNOWLEDGMENTS

We thank Steve Robischon and the Palouse Basin Aquifer Committee for providing much of the hydrological data used in this paper; and John Bush, Dean Garwood, and the Idaho Geological survey for providing geological data. Comments by three anonymous reviewers have greatly improved this manuscript.

REFERENCES

1. U.S. Geological Survey. *Ground-Water Availability Assessment for the Columbia Plateau Regional Aquifer System, Washington, Oregon, and Idaho*; 2008. Available online: http://pubs.usgs.gov/fs/2008/3086/ (accessed on 22 July 2015).
2. Cook, B.I.; Ault, T.R.; Smerdon, J.E. Unprecedented 21st century drought risk in the American Southwest and Central Plains. *Sci. Adv.* **2015**, *1*.
3. Woodhouse, C.A.; Meko, D.M.; MacDonald, G.M.; Stahle, D.W.; Cook, E.R.A. 1200-year perspective of 21st century drought in southwestern North America. *Proc. Natl. Acad. Sci. USA* **2010**, *107*, 21283–21288.
4. Pederson, G.T.; Gray, S.T.; Woodhouse, C.A.; Betancourt, J.L.; Fagre, D.B.; Littell, J.S.; Graumlich, L.J. The unusual nature of recent snowpack declines in the North American Cordillera. *Science* **2011**, *333*, 332–335.
5. Vaccaro, J.J. Potential impacts of climate change on groundwater resources of the Columbia River basin. In Proceedings of the Pacific Northwest Science Conference, Portland, OR, USA, 16 June 2010; pp. 1–18.
6. Palouse Basin Aquifer Committee, 2013. Available online: http://www.webpages.uidaho.edu/pbac/ (accessed on 27 February 2015).
7. Barker, R.A. Computer simulation and geohydrology of a basalt aquifer system in the Pullman-Moscow basin, Washington and Idaho. Available online: https:// fortress.wa.gov/ecy/publications/documents/wsb48 (accessed on 22 July 2015).
8. Lum, W.E.; Smoot, J.L.; Ralston, D.R. Geohydrology and numerical model analysis of ground-water flow in the Pullman-Moscow area, Washington and Idaho. Available online: http://pubs.usgs.gov/wri/1989/4103/report.pdf (accessed on 22 July 2015).
9. Bauer, H.H.; Hansen, A.J. Hydrogeology of the Columbia Plateau regional aquifer system, Washington, Oregon, and Idaho. Available online: http://pubs.usgs.gov/wri/ 1996/ 4106/report.pdf (accessed on 22 July 2015).
10. Kahle, S.C.; Olsen, T.D.; Morgan, D.S. Geologic setting and hydrogeologic units of the Columbia Plateau Regional Aquifer System, Washington, Oregon, and Idaho. Available online: http://pubs.usgs.gov/sim/3088/pdf/sim3088.pdf (accessed on 22 July 2015).
11. Burns, E.R.; Snyder, D.T.; Haynes, J.V.; Waibel, M.S. Groundwater status and trends for the Columbia Plateau Regional Aquifer System, Washington, Oregon, and Idaho. Available online: http://pubs.er.usgs.gov/publication/sir20125261 (accessed on 22 July 2015).
12. U.S. Department of Energy. *Site characterization plan, Reference Repository Location, Hanford Site, Washington-consultation draft*; Office of Civilian Radioactive Waste Management: Washington, DC, USA, 1988.
13. Reidel, S.P.; Fecht, K.R.; Hagood, M.C.; Tolan, T.L. The geologic evolution of the central Columbia Plateau. In *Volcanism and Tectonism in the Columbia River flood-Basalt Province*; Reidel, S.P., Hooper, P.R., Eds.; Geological Society of America: Washington, DC, USA, 1989.
14. Drost, B.W.; Whiteman, K.J. Surficial geology, structure, and thickness of selected geohydrologic units in the Columbia Plateau, Washington. Available online: http://pubs.usgs.gov/wri/1984/4326/report.pdf (accessed on 22 July 2015).

15. Drost, B.W.; Whiteman, K.J.; Gonthier, J.B. Geologic Framework of the Columbia Plateau Aquifer System, Washington, Oregon, and Idaho. Available online: http://pubs.usgs.gov/wri/1987/4238/report.pdf (accessed on 22 July 2015).
16. Steinkampf, W.C.; Hearn, P.P., Jr. Ground-Water Geochemistry of the Columbia Plateau Aquifer System, Washington, Oregon, and Idaho. Available online: http://pubs.usgs.gov/of/1995/0467/report.pdf (accessed on 22 July 2015).
17. Vaccaro, J.J. Summary of the Columbia Plateau Regional Aquifer-System Analysis, Washington, Oregon, Idaho. Available online: http://pubs.usgs.gov/pp/1413a/report.pdf (accessed on 22 July 2015).
18. Tolan, T.; Lindsey, K.; Porcello, J. A Summary of Columbia River Basalt Group Physical Geology and its Influence on the Hydrogeology of the Columbia River Basalt Aquifer System: Columbia Basin Ground Water Management Area of Adams, Franklin, Grant, and Lincoln Counties, 2009. Available online: http://www.cbgwma.org/pdf/GWMA_Geology-Hydrogeology%20of%20CRBG_TEXT_June%202009.pdf (accessed on 27 February 2015).
19. McNab, W.H.; Avers, P.E. Ecological Subregions of the United States. 1996. Available online: http://www.fs.fed.us/land/pubs/ecoregions/ (accessed on 22 July 2015).
20. Leek, F. *Hydrogeological Characterization of the Palouse Basin Basalt Aquifer System, Washington and Idaho*; Washington State University: Pullman, WA, USA, 2006. [Google Scholar]
21. Hopster, D. *A Recession Analysis of Springs and Streams in the Moscow-Pullman Basin*; University of Idaho: Moscow, ID, USA, 2003. [Google Scholar]
22. Walters, K.L.; Glancy, P.A. Reconnaissance of Geology and of Ground-Water Occurrences and Development in Whitman County, Washington. Water-Supply Bulletin 26. 1969. Available online: https://fortress.wa.gov/ecy/publications/documents/wsb26.pdf (accessed on 22 July 2015).
23. Crosby, J.W., III; Chatters, R.M. *Water Dating Techniques as Applied to the Pullman-Moscow Ground-Water Basin*; Washington State University: Pullman, WA, USA, 1965.
24. Larson, K.R. *Stable Isotopes in the Pullman-Moscow Basin, Eastern Washington and North Idaho: Implications for the Timing, Magnitude, and Distribution of Groundwater Recharge*; Washington State University: Pullman, WA, USA, 1997.
25. Douglas, A.A. *Radiocarbon Dating as a Tool for Hydrogeological Investigations in the Palouse Basin*; University of Idaho: Moscow, ID, USA, 2004; Available online: http://www.webpages.uidaho.edu/pbac/Theses/Douglas_Thesis_Age_Dating_2004.pdf (accessed on 22 July 2015).
26. Carey, L.R. *Evaluation of Oxygen and Hydrogen Isotopes in Groundwater of the Palouse Basin and Moscow Sub-Basin*; University of Idaho: Moscow, ID, USA, 2011.
27. Fairley, J.P.; Solomon, M.D.; Hinds, J.J.; Grader, G.W.; Bush, J.H.; Rand, A.L. Latah County Hydrologic Characterization Report. Available online: http://www.researchgate.net/publication/262260160_Latah_County_Hydrogeological_Characterization (accessed on 22 July 2015).
28. MacQueen, J.D.; Mann, E. Borehole Gravity Meter Surveys at the Waste Treatment Plant, Hanford, Washington. Available online: http://www.pnl.gov/main/ publications/external/technical_reports/PNNL-16490.pdf (accessed on 22 July 2015).
29. Moisy, F. Discover Ezyfit: A Free Curve Fitting Toolbox for Matlab. Available online: http://www.mathworks.com/matlabcentral/fileexchange/10176-ezyfit-2-40/ content/ezyfit/demo/html/efdemo.html#1 (accessed on 27 February 2015).
30. Hyndman, R.D.; Drury, M.J. Physical properties of basalts, gabbros, and ultramafic rocks from DSDP Leg 37. *DSDP* **1977**, *37*, 395–401.
31. Smoot, J.L.; Ralston, D.R. *Hydrogeology and Mathematical Model of Ground-Water Flow in the Pullman-Moscow Region, Washington and Idaho*; Idaho Water Resources Research Institute, University of Idaho: Moscow, ID, USA, 1987; Available online: http://digital.lib.uidaho.edu/cdm/ref/collection/idahowater/id/310 (accessed on 22 July 2015).
32. Moran, K. Evaluation of the relationship between pumping and water level declines in the Grande Ronde Aquifer of the Palouse Basin. Available online: http://www. webpages.

uidaho. edu/pbac/ Presentations/ 2011/110915_ Moran_WL_Pumping_Relationships_report.pdf (accessed on 22 July 2015).
33. Foxworthy, B.L.; Washburn, R.L. *Ground Water in the Pullman Area, Whitman County, Washington*; Geological Survey Water-Supply Paper 1655. 1963. Available online: http://pubs.usgs.gov/wsp/1655/report.pdf (accessed on 22 July 2015).
34. Brown, J.C. *Well Construction and Stratigraphic Information: Pullman Test and Observation Well, Pullman, Washington*; Washington State University: Pullman, WA, USA, 1976; pp. 1–35.
35. Teasdale, E.W. *Hydrogeologic Basins in the Palouse Area of Idaho and Washington*; University of Idaho: Moscow, ID, USA, 2002.
36. Bush, J.H.; Garwood, D.L. Interpretation and use of well data in the Columbia River Basalt Group, Pullman, Washington. In Proceedings of 2003 Seattle Annual Meeting, Seattle, WA, USA, 2–5 November 2003.
37. Bush, J.H. *The Columbia River Basalt Group of the Palouse Basin with Hydrological Interpretations, western Latah County, Idaho and eastern Whitman County, Washington*; Report to Palouse Basin Aquifer Committee: Moscow, ID, USA, 2005.
38. Klein, D.P.; Sneddon, R.A.; Smoot, J.L. Magnetotelluric study of the thickness of volcanic and sedimentary rock in the Pullman-Moscow Basin of eastern Washington. Available online: http://pubs.usgs.gov/of/1987/0140/report.pdf (accessed on 22 July 2015).
39. Holom, D. *Ground Water Flow Conditions Related to the Pre-Basalt Basement Geometry Delineated by Gravity Measurements near Kamiak Butte, Eastern Washington*; University of Idaho: Moscow, ID, USA, 2006.
40. Conrey, R.; Beard, C.; Wolff, J. Columbia River Basalt flow stratigraphy in the Palouse Basin Department of Ecology test wells. Available online: http://www.webpages.uidaho.edu/pbac/pubs/2013_Conrey_Beard_Stratigraphy_in_Palouse_Basin_DOE_test_Wells.pdf (accessed on 22 July 2015).
41. Ryan, W.B.F.; Carbotte, S.M.; Coplan, J.O.; O'Hara, S.; Melkonian, A.; Arko, R.; Weissel, R.A.; Ferrini, V.; Goodwillie, A.; Nitsche, F.; *et al.* Global Multi-Resolution Topography synthesis. *Geochem. Geophys. Geosyst.* **2009**, *10*.
42. Ducar, S. *Properties of the Grande Ronde Aquifer in the Vicinity of Moscow, Idaho from the Synthesis of Aquifer Test Results with Seismic Groundwater Response*; Department of Geological Science, University of Idaho: Moscow, ID, USA, 2014.
43. Moran, K. *Interpretation of Long-Term Grande Ronde Aquifer Testing in the Palouse Basin of Idaho and Washington*; University of Idaho: Moscow, ID, USA, 2011.
44. Sokol, D. *Interpretation of short term water level fluctuations in the Moscow Basin, Latah County, Idaho*, Pamphlet No. 137. 1966. Available online: http://www.idahogeology.org/PDF/Pamphlets_(P)/P-137.pdf (accessed on 22 July 2015).
45. Cooper, H.H.; Bredehoeft, J.D.; Papadopulos, I.S.; Bennett, R.R. The response of well-aquifer systems to seismic waves. *J. Geophys. Res.* **1965**, *70*, 3915–3926.
46. Bredehoeft, J.D. Response of well-aquifer systems to earth tides. *J. Geophys. Res.* **1967**, *72*, 3075–3087.
47. Shih, D.C.F. Storage in confined aquifer: Spectral analysis of groundwater responses to seismic Rayleigh waves. *J. Hydrol.* **2009**, *374*, 83–91.
48. Folnagy, A.J.; Osiensky, J.L.; Kobayashi, D.; Sprenke, K.F. Specific storage from sparse records of groundwater response to seismic waves. *J. Hydrol.* **2013**, *503*, 22–28.
49. Merritt, M.L. Estimating Hydraulic Properties of the Floridan Aquifer System by Analysis of Earth-Tide, Ocean-Tide, and Barometric Effects, Collier and Hendry Counties, Florida. Available online: http://pubs.usgs.gov/wri/wri034267/wri03_4267.pdf (accessed on 22 July 2015).
50. Rojstaczer, S.; Riley, F.S. Response of the water level in a well to earth tides and atmospheric loading under unconfined conditions. *Water Resour. Res.* **1990**, *26*, 1803–1817.
51. Brown, G.; Deacon, R. Economic optimization of a single-cell aquifer. *Water Resour. Res.* **1972**, *8*, 557–564.

CHAPTER 13

Assessment of Groundwater Quality in the Gaza Strip, Palestine Using GIS Mapping

Basem Shomar1, Sami Abu Fakher2, Alfred Yahya1

1 Institute of Earth Sciences, University of Heidelberg, Heidelberg, Germany

2 Environmental Quality Authority, Department of GIS Mapping, Gaza, Palestine

ABSTRACT

A Geographical Information System (GIS) tool was used to construct thematic maps for groundwater quality in the Gaza Strip. Environmental data were integrated and an overall picture about the spatial variation in the groundwater quality of the Gaza Strip was defined. The integrated spatial maps helped to refine information on land use, soil types, depth to groundwater table, environmental "hot spots", and contaminant concentrations of the study area. The groundwater quality maps have been derived from the results of an eight-year monitoring program for major anions, cations and heavy metals. An environmental hot-spots map was derived from potential contaminating sources, showed direct and indirect influences on groundwater quality. The GIS maps showed not only contaminant distributions but also illustrated the need to improve the groundwater quality management methods. Several contaminants pose great problems in the water of Gaza. Integration of water data and GIS maps for all parameters revealed that there is probably no drinking water in Gaza according to the WHO standards. Moreover, the new maps of 2008 could be used as base-line for water planners and policy makers as well as guidelines for the Palestinian people to manage and protect their groundwater.

Increased water demand from population and economic growth, environmental needs, land use changes, urbanization, groundwater mining, deterioration of water quality, pollution from local and diffuse sources, environmental hot-spots and impacts on public heath and ecosystems are all factors that can create a severe water quality crisis as well as water shortage problems.

Keywords: Gaza, GIS, Groundwater quality

1. INTRODUCTION

A Geographic Information System (GIS) is an important tool for integrating spatial data with other information. It allows one to analyze the integrated data and to represent the information spatially facilitating planning of resource development, environmental protection and scientific research [1–5]. This capability makes GIS a powerful tool for groundwater assessments. GIS not only provides tools for interpolating measured values of water quality parameters from specific locations, but also enables one to link water quality with land use, soil characteristics, and other relevant information. In addition, GIS provides sophisticated map-generation capabilities, useful in communicating results of data analysis [6–8]. Although GIS has been used to describe associations between water quality and both natural and anthropogenic activities in the world, in Palestine it has been applied in a limited fashion, but it has not been well implemented for planning, resource management and environmental protection at a national or regional level.

The Gaza Strip region is a fragile ecosystem suffering from increasing environmental assaults due to escalating population growth and limited availability of natural resources to support development. Groundwater is, perhaps, the most precious natural resource in the Gaza Strip as it is the only natural source of fresh water. Therefore, groundwater contamination can pose serious health and economic threats to the population that relies on this water for drinking, agriculture, and industry uses. The aquifer of Gaza is extremely susceptible to surface-derived contamination because of the high permeability of sands and gravels that compose the soil profile of Gaza [9,10]. It has already deteriorated in terms of quantity and quality as a result of over-exploitation and direct and indirect contamination [11–15].

The main objective of the current study is to use GIS to compare water quality data and related information collected during an eight-year monitoring program for groundwater quality in the Gaza Strip. A secondary objective is to portray the contaminant distribution in the groundwater of the Gaza Strip in easily viewed maps for use by the public and decision makers.

2. STUDY AREA, MATERIALS AND METHODS

2.1. Study Area

The Gaza Strip is one of the most densely populated areas in the world (4138 people per km^2; [16]). For administrative purposes, the area has been divided into five regions: North, Gaza, Middle, Khan Younis and Rafah

(Figure 1). Approximately 85% of the population of the Gaza Strip drink from municipal groundwater wells and 15%, mostly in agricultural areas, use private wells to supply their drinking water [15].

The study area is part of the coastal zone in a transitional area between a temperate Mediterranean climate to the east and north and an arid climate of the Negev and Sinai deserts to the east and south. As a result, the Gaza Strip has a characteristic semi-arid climate. The hydrogeological features of the Gaza aquifer are well known. The coastal aquifer consists primarily of Pleistocene age Kurkar Group deposits, including calcareous and silty sandstones, silts, clays, unconsolidated sands, and conglomerates. Near the coast, coastal clays extend about 2-5 km in land, and divide the aquifer sequence into three or four sub-aquifers, depending upon the location. Towards the east, the clays pinch out and the aquifer is largely unconfined [17]. Within the Gaza Strip, the total thickness of the Kurkar Group is about 100 m at the shore in the south, and about 200 m near Gaza city. At the eastern Gaza border, the saturated thickness is about 60-70 m in the north, and only a few meters in the south near Rafah. Local perched water conditions exist throughout the Gaza Strip due to the presence of shallow clays [18].

2.2. GIS Mapping

The GIS application was conducted according to Asadi et al. [6] and Carrera-Hernández and Gaskin [19] with modifications and changes to suit the study area of the Gaza Strip.

2.2.1. Thematic Maps

The base map data used for the study included digitized data sets originally developed by the Ministry of Planning, the Water Authority, and the Environmental Quality Authority in Palestine. The data sets were originally digitized from 1:20,000 scale topographic maps. The original digitized maps are registered to the National Grids-Palestinian Grid 1923. The study area matches these maps which cover the entire Gaza Strip area. Thematic maps including the base map (Figure 1) and network maps (Figures 3-20) were prepared from 1:20,000 scale topographic paper maps using AutoCAD and Arc/Info GIS software. All the maps were scanned and digitized to generate a digital output.

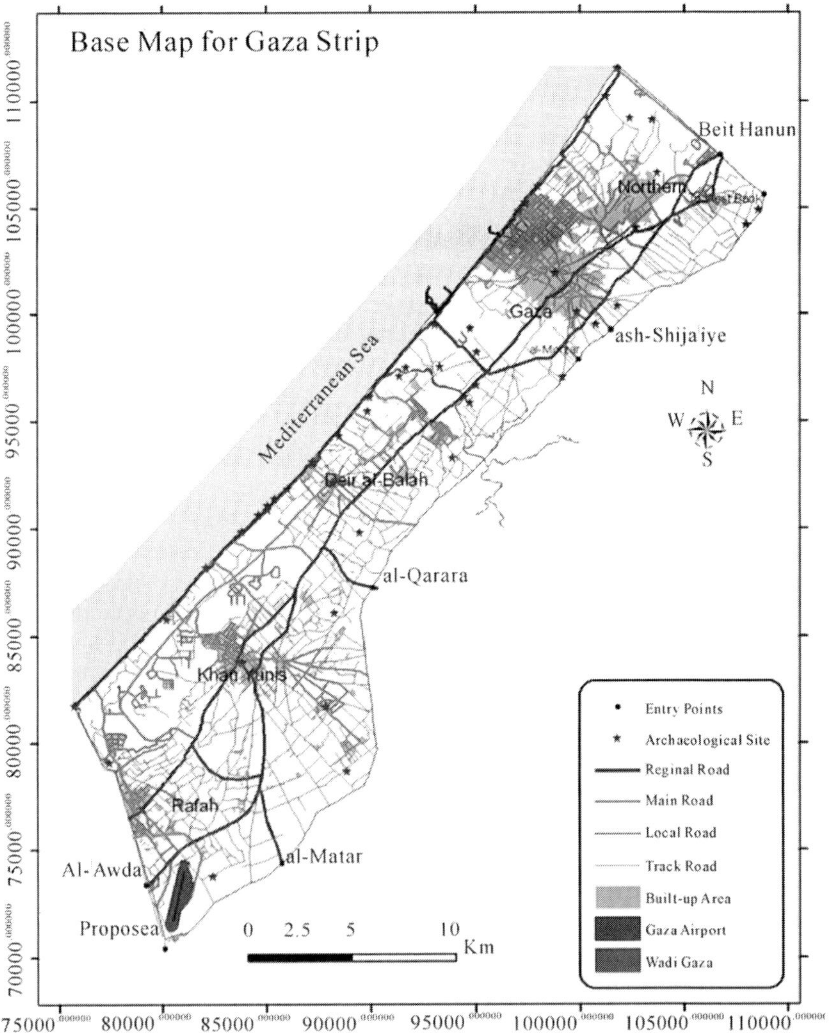

Figure 1. Gaza base map.

2.2.2. Water Quality Maps

The data used for the mapping water quality and other supporting maps were developed from the results of previously published studies [10,15,20–31]. Data for these studies were based on periodic fieldwork conducted by the Palestinian Water Authority for groundwater samples collected from predetermined locations of existing water wells in collaboration with Dr. Shomar who assisted with study design and was responsible for chemical analysis.

The sampling locations were integrated with the water data for the generation of spatial distribution maps of selected water quality parameters including electrical conductivity (EC), total dissolved solids (TDS), Cl^-, F^-, NO_3^-, SO_4^{2-}, total hardness, Ca^{2+}, Mg^{2+}, Na^+, K^+, total Fe, total Cr, and total Zn. The water data were linked to the sampling locations using the basic geodatabase creation function of ArcGIS 9.2 software.

The present study used the Inverse Distance Weighted (IDW) method for spatial interpolation of water pollutants or other parameters. This method uses a defined or selected set of sample points for estimating the output grid cell value. It determines the cell values using a linearly weighted combination of a set of sample points and controls the significance of known points upon the interpolated values based upon their distance from the output point thereby generating a surface grid as well as thematic isolines. Important water quality parameters and their distribution patterns were studied in Gaza metropolis also with the help of cartographic techniques [19]. Thus, GIS enables us to present the cause and affect relationship visually.

The depth to water table (Figure 6) is based on the monitoring results of 500 groundwater wells. Data were obtained from both field surveys and databases of the Palestinian Water Authority.

3. RESULTS

Owing to the large data set obtained from the analysis of 170 water samples for eight years, each having 27 parameters, this section focused on elements exceeding the World Health Organization (WHO) standards. Additionally, total Cr, Zn and Fe were presented to establish baseline values. Figures 7-20 showed the distribution of each parameter and the WHO standard. Maps showing land use, soil types, environmental hot-spots and socio-economic were included because these factors have direct and indirect impacts on groundwater quality.

3.1. Land Use

The area of the Gaza Strip is 363 km^2 of which about 25% is urbanized. Table 1 and Figure 2 show the distribution of land use in the Gaza Strip based on estimated figures for 2008 and the available literature. About 40% of the land is being used for agriculture, most of which is in the eastern half of Gaza where population densities are low (Figure 3). The land use data were obtained from the analysis of aerial photographs taken in 2008. The industrial sector was discussed in detail in Shomar [31]. With an average population density of 4,091 people/km^2, Gaza is one of the most densely populated areas in the world. About 80% live in the built-up areas shown on the map.

Table 1. Land use distribution in the Gaza Strip.

ID	Land use type	Area Km^2	Percent (%)
0	Airport	7.5	2.05%
1	Built-up	54	14.79%
2	Cultivated	157.5	43.15%
3	Existing Industrial Area	0.9	0.25%
4	Wastewater Treatment Site	0.45	0.12%
5	Fisheries Site	0.3	0.08%
6	Harbour	0.35	0.10%
7	Important Natural Resource 1	24	6.58%
8	Mawasi	14.5	3.97%
9	Natural Resource 2	62	16.99%
10	Nature Reserve	26.5	7.26%
11	Proposed Treatment Site	1.1	0.30%
12	Recreation	6.1	1.67%
13	Roads	9.8	2.68%
	Total Area	**365**	**100%**

3.2. Environmental Hot Spots

In addition to the major pollution point sources of overloaded wastewater treatment plants, unprotected solid waste dumping sites and Wadi Gaza

(Figure 4), hot spots appear sporadically in many locations due to Gaza's inability to maintain adequate infrastructure. For example, frequent electricity outage or blackout causes paralysis of wastewater pumping stations and results in untreated wastewater infiltrating homes and streets. The lack of gasoline and diesel causes solid waste to accumulate in the streets without transportation to the dumping sites [31].

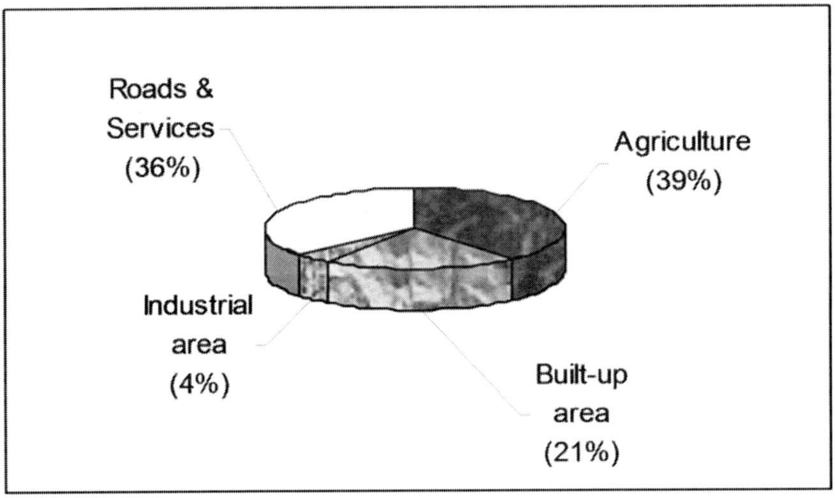

Figure 2. Major land use sectors in the Gaza Strip.

3.3. Soil Types

The Gaza Strip has several major soil types (Figure 5) including Arenosolic, Calcaric, Rhegosolic, and Calcaric Fluvisolic soils. Arenosolic (sandy) soils of dune accumulations are Regosols without a marked profile. The soils are moderately calcareous (5-8% $CaCO_3$), with low organic matter, and are physically suitable for intensive horticulture. Calcaric Arenosols (loessy sandy soils) can be found some 5 km inland in the central and southern part of the Strip, in a zone along Khan Younis toward Rafah, parallel to the coast. This belt forms a transitional zone between the Arenosolic soils and the Calcaric (loess) soils. Typical Calcaric soils are found in the area between the city of Gaza and the Wadi Gaza and contain 8-12% $CaCO_3$. Arenosolic Calcaric (sandy loess) soils are transitional soils, characterized by a lighter texture.

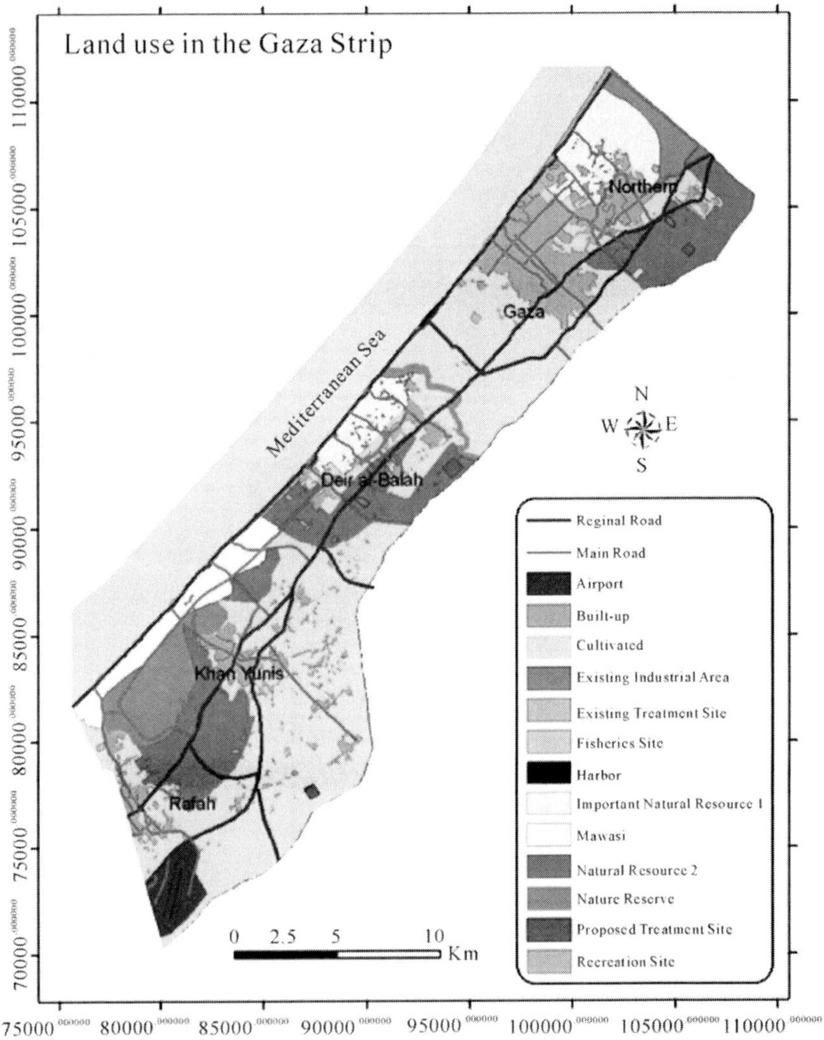

Figure 3. Land use in the Gaza Strip.

These soils can be found in the depression between the Calcareous (Kurkar) ridges of Deir El Balah. Apparently, windblown sands have been mixed with Calcareous deposits. Deposition of these two types of windblown materials originating from different sources has occurred over time and more or less simultaneously. These soils have a rather uniform texture. Another transitional form is the Arenosols over Calcaric soils. These are loess or loessial soils (sandy clay loam) that have been covered by a layer (0.20-0.50 m) of dune sand. These soils can be found east of Rafah and Khan Younis.

Fluvisols (alluvial) and Vertisols (grumosolic), which are dominated by loamy clay textures, are found on the slopes of the northern depressions between Beit Hanoun and Wadi Gaza. Drilling east of El Montar ridge revealed that alluvial deposits of about 25 m in thickness occur. At some depth, calcareous concentrations are present. The $CaCO_3$ content can be approximately 15-20%. Some of the soils have been strongly eroded, and the reddish-brown subsoils may be exposed on the tops of ridges and along slopes. The alluvial sediments are underlain by a calcareous layer.

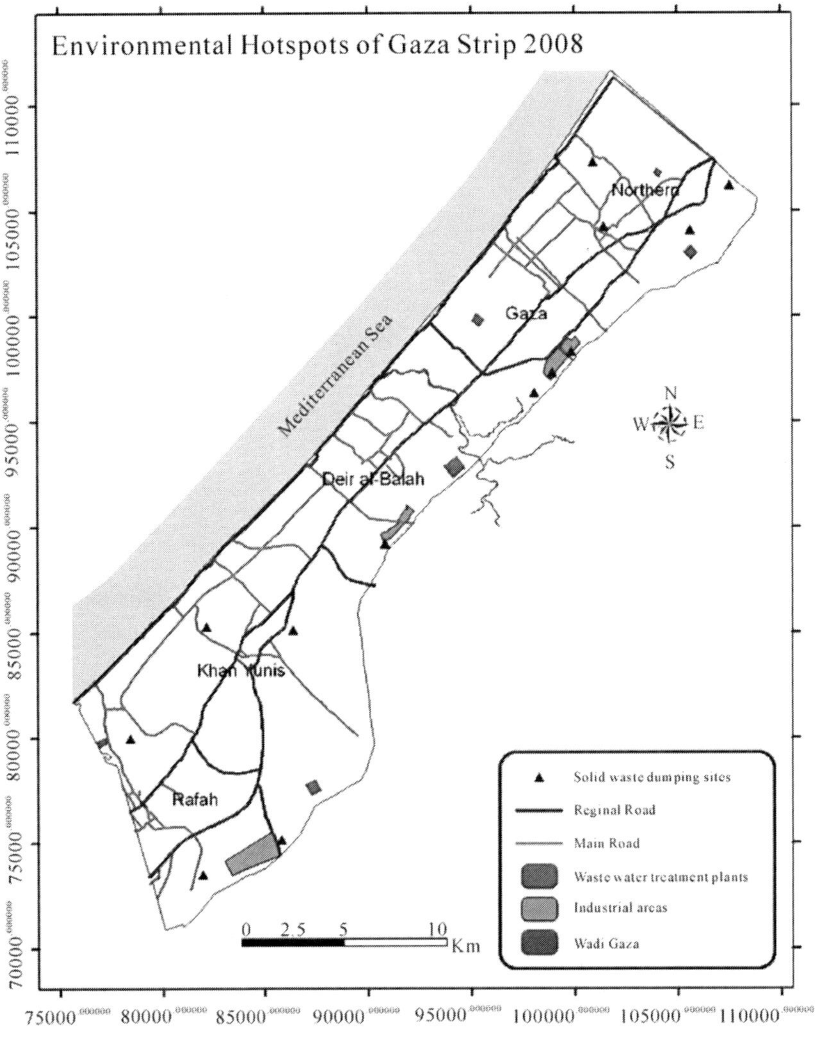

Figure 4. Environmental hot spots.

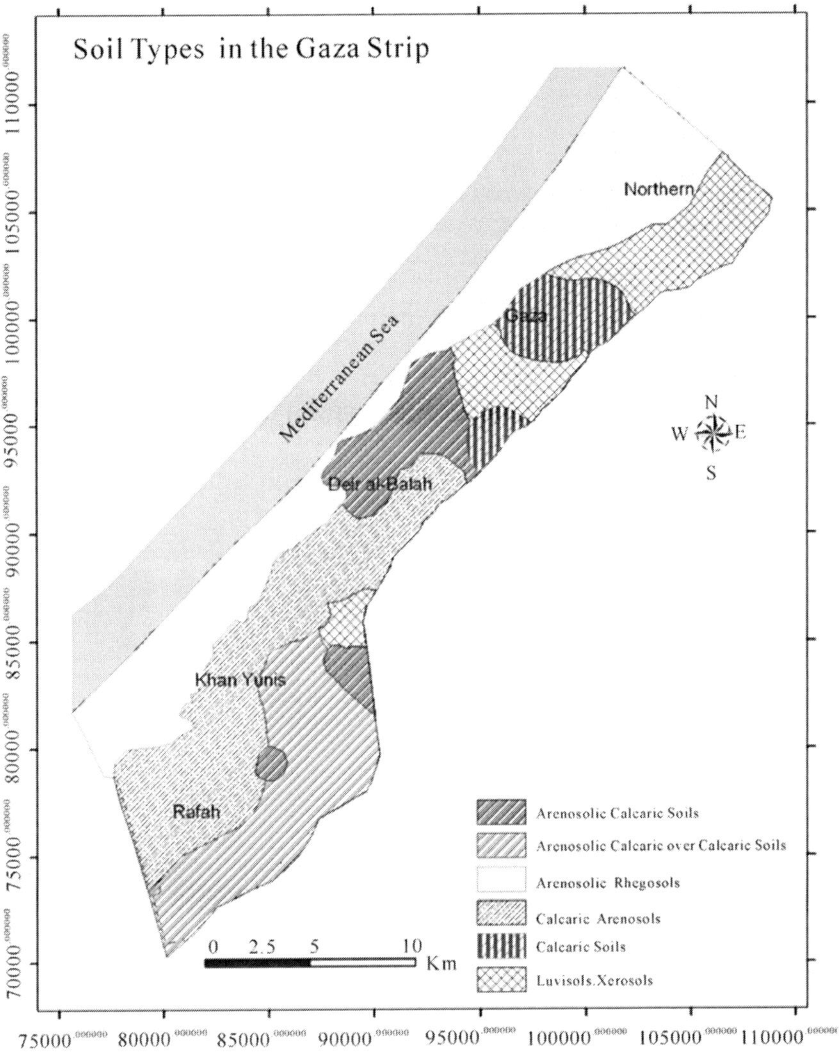

Figure 5. Soil types in the Gaza Strip.

3.4. Water Table

The depth to water table (Figure 6) varies between few meters in the west (very closed to the sea) to about 120m at some locations in the east.

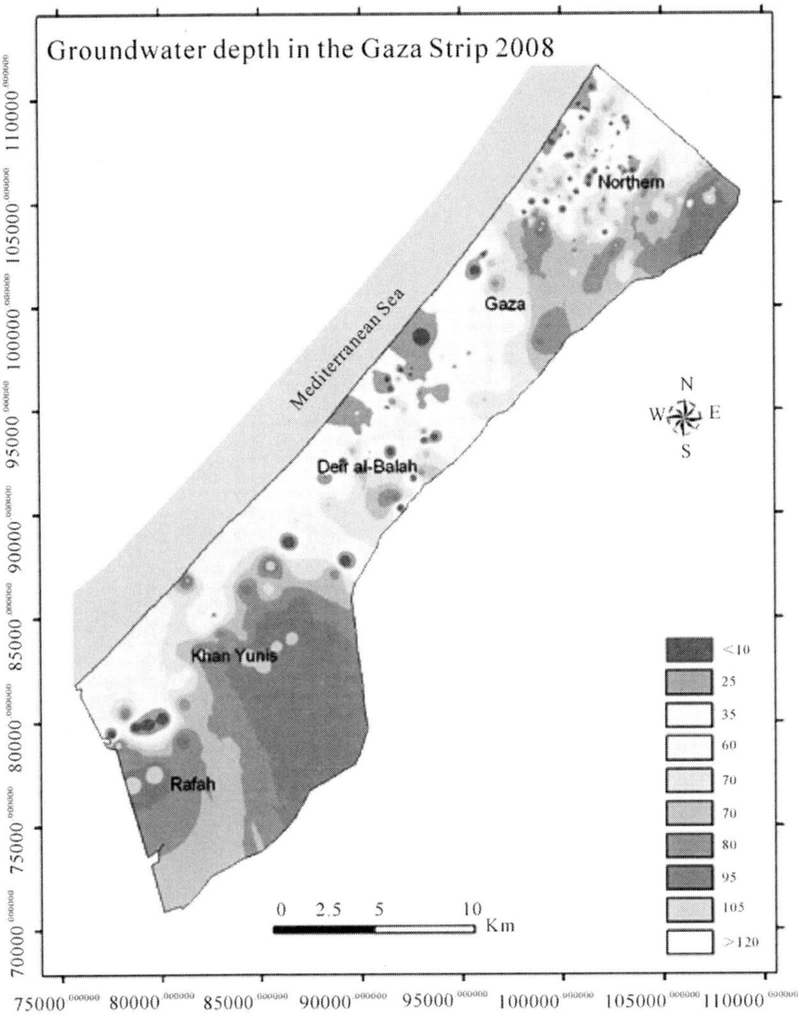

Figure 6. Groundwater depth in the Gaza Strip 2008.

3.5. Electric Conductivity

With relatively small, localized exceptions, the EC of municipal wells increases from north to south (Figure 7). The lowest EC value was 1,198 μS/cm and the highest was about 3,800 μS/cm. The most deteriorated and salty water was in the eastern regions of Khan Younis and Rafah with an average EC in the private groundwater wells of 5,000 μS/cm.

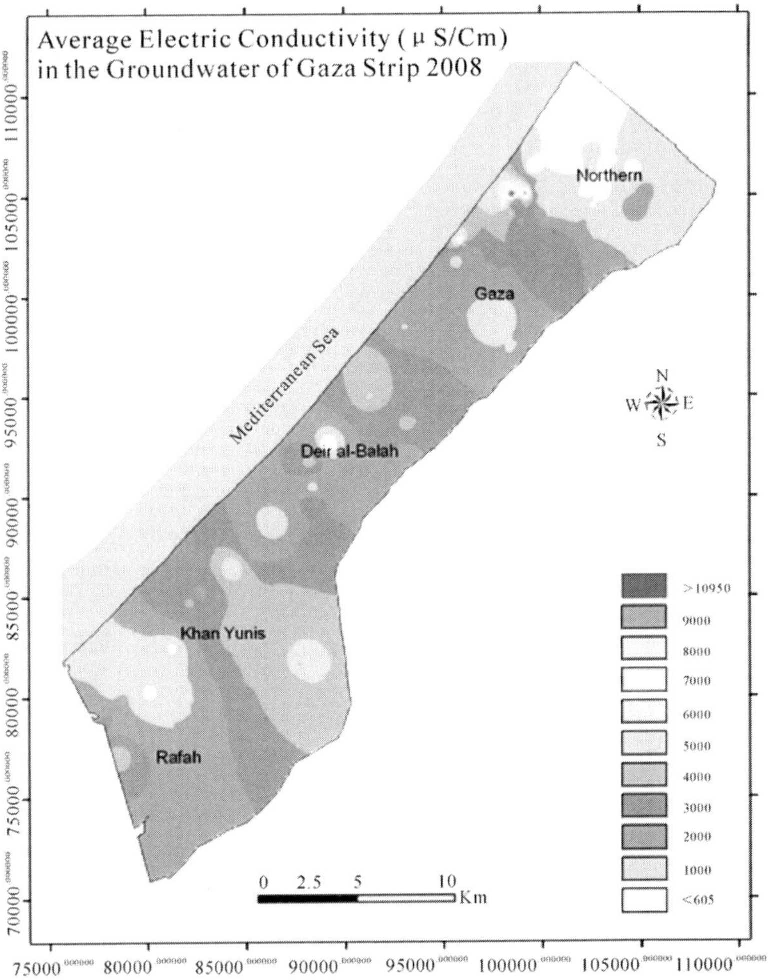

Figure 7. Electric conductivity in the groundwater of the Gaza Strip 2008.

3.6. Total Dissolved Solids

Figure 8 illustrates that groundwater in most of the Gaza Strip exceeds the WHO TDS standard, which is 1000 mg/L. The TDS and EC maps show similar patterns as both parameters indicate the concentration of dissolved solids in water. The high TDS value in the eastern parts of Khan Younis (3000-4000 mg/L) makes water in the area undrinkable. More than 50% of the sampled groundwater showed TDS of more than 2000 mg/L.

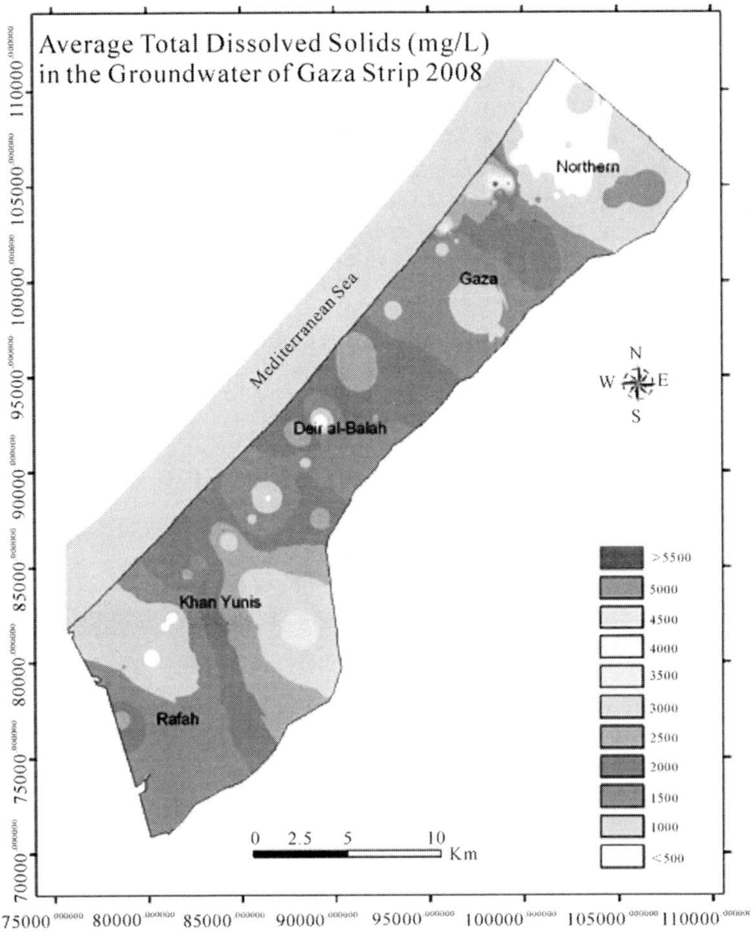

Figure 8. Total dissolved solids in the groundwater of the Gaza Strip 2008.

3.7. Anions (Chloride, Nitrate, Fluoride and Sulfate)

All wells in Gaza had at least one parameter of Cl^-, NO_3^-, F^- and SO_4^{2-} exceeding the WHO standards of 250, 50, 1.5 and 250 mg/L, respectively. Chloride concentrations (Figure 9) corresponded to EC. The lowest value of Cl^- at a municipal well was 35 mg/L whereas the highest value was 2652 mg/L for a well in Khan Younis. Water meeting the WHO Cl^- standard was found in less than 5% of sample wells, primarily in the northern parts and scattered in more isolated areas in the rest of Gaza.

The map of nitrate (Figure **10**) for the year 2008 confirms previous findings [29] that almost 90% of the groundwater wells of the Gaza Strip have NO_3^- concentrations two to eight times higher than the WHO standards.

Except for the north area, the average concentration of fluoride in the groundwater in the Gaza Strip (Figure **11**) is higher than the WHO standards (which is 1.5 mg/L). The most fluoride contaminated areas are Khan Younis (average 2.7 mg/L) and Rafah (average 2 mg/L) and this is consistent with the previous study of Shomar et al. [21]. The F concentration increases from north to south.

Most of the wells in Gaza have SO_4^{2-} concentrations exceeding the permissible WHO standard (Figure **12**). The highest levels of SO_4^{2-} were in Khan Younis and the southeast, where the average concentration is 380 mg/L.

3.8. Cations (Hardness, Calcium, Magnesium, Sodium and Potassium)

Most of the cations Ca^{2+}, Mg^{2+}, Na^+ and K^+ show concentrations higher than the WHO standards of 50, 30, 200 and 10 mg/L, respectively.

Dissolved calcium and magnesium in water are the two most common minerals that make water "hard". Based on the water hardness classification of 0 to 60 $mgCaCO_3/L$ as soft, 61 to 120 $mgCaCO_3/L$ as moderately hard, 121 to 180 $mgCaCO_3/L$ as hard, and more than 180 $mgCaCO_3/L$ as very hard, most groundwater in Gaza is hard to very hard (Figure **13**).

As water hardness is determined primarily by Ca^{2+} and Mg^{2+}, not surprisingly, the areas with highest levels of Ca^{2+} and Mg^{2+} (Figures 14 and 15) also have the hardest water. The average concentration of Ca^{2+} was 93 mg/L while the average concentration of Mg^{2+} was 48 mg/L. Areas between Gaza and the northern region and middle region wells showed the highest levels of both Ca^{2+} and Mg^{2+} and the results were 262 and 128 mg/L, respectively.

The lowest Na^+ levels were found in the north, and the highest levels were in the areas of Khan Younis and Rafah (Figure **16**).

Most wells had average value of K^+ that was less than 5 mg/L; however, few wells showed levels of K^+ more than 15 mg/L (Figure **17**).

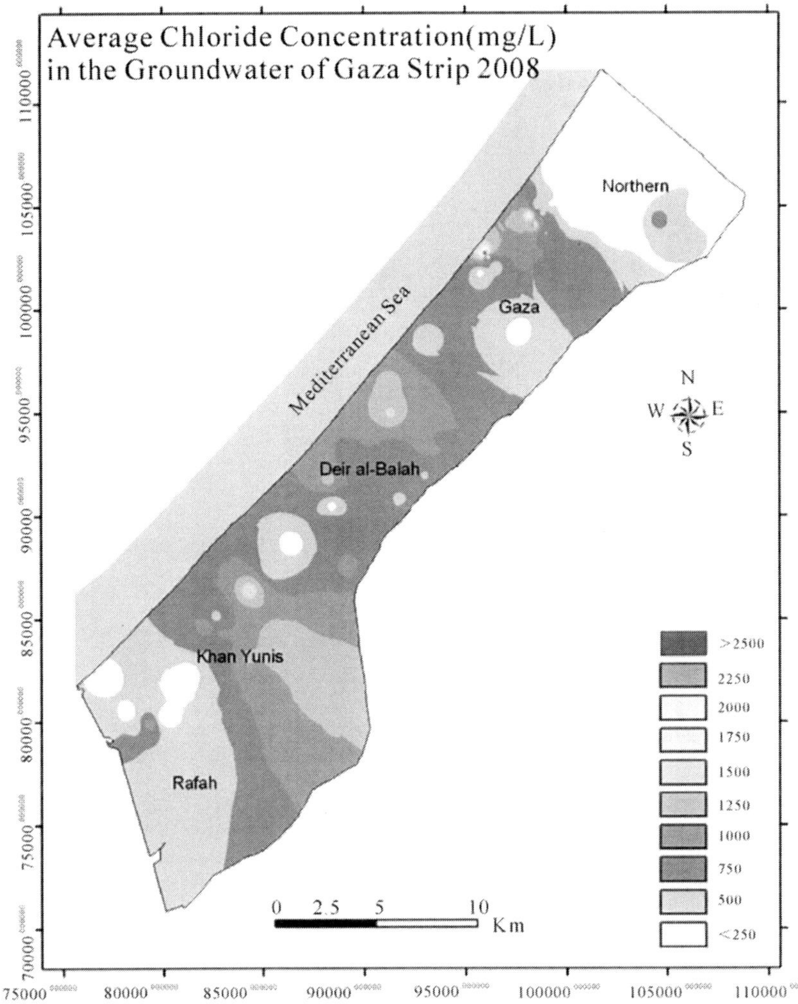

Figure 9. Chloride concentrations in the groundwater of the Gaza Strip 2008.

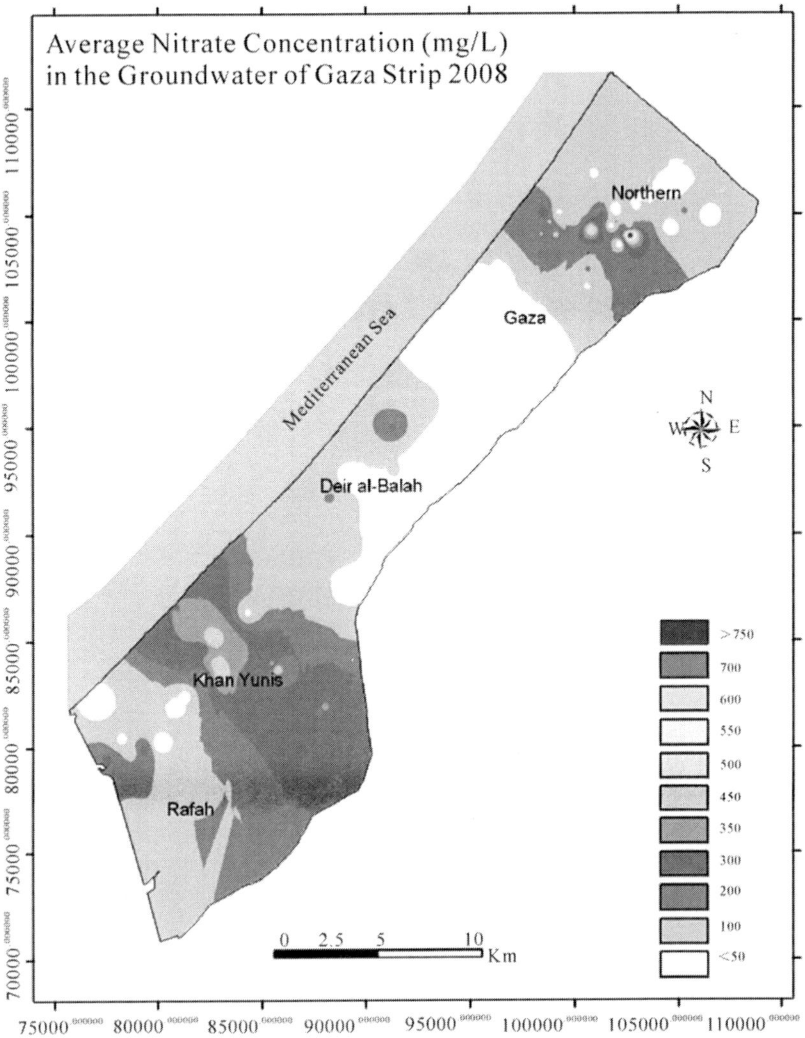

Figure 10. Nitrate concentrations in the groundwater of the Gaza Strip 2008.

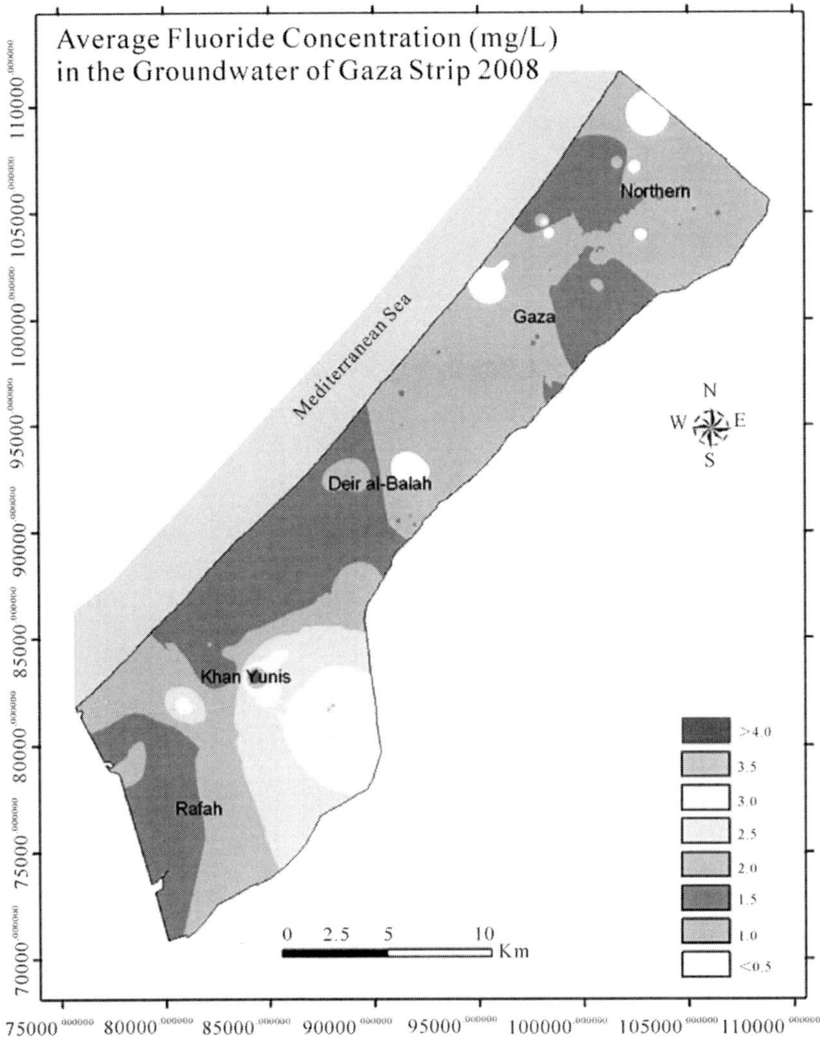

Figure 11. Fluoride concentrations in the groundwater of the Gaza Strip 2008.

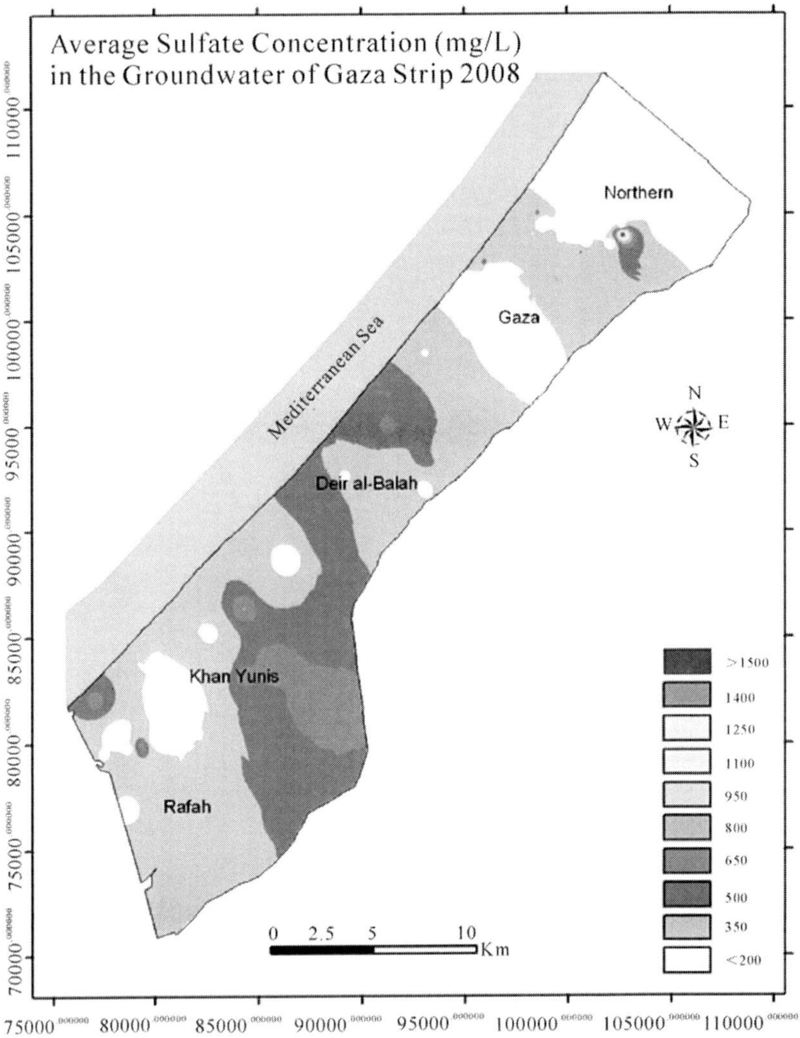

Figure 12. Sulfate concentrations in the groundwater of the Gaza Strip 2008.

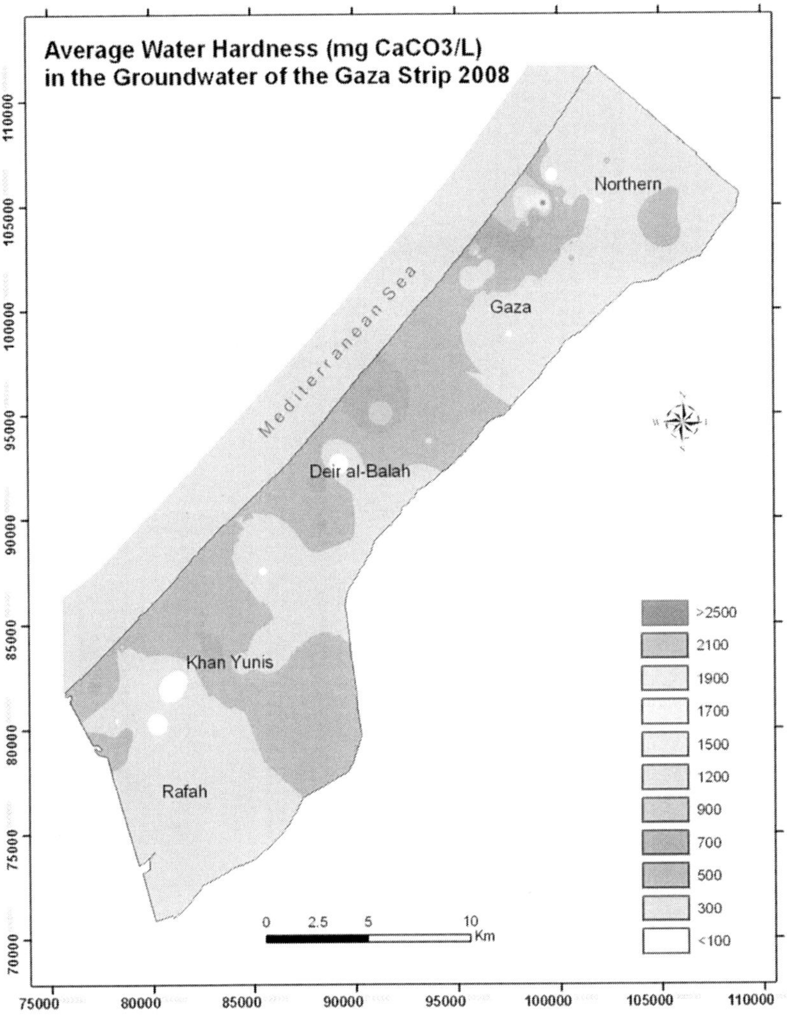

Figure 13. Groundwater hardness in the Gaza Strip 2008.

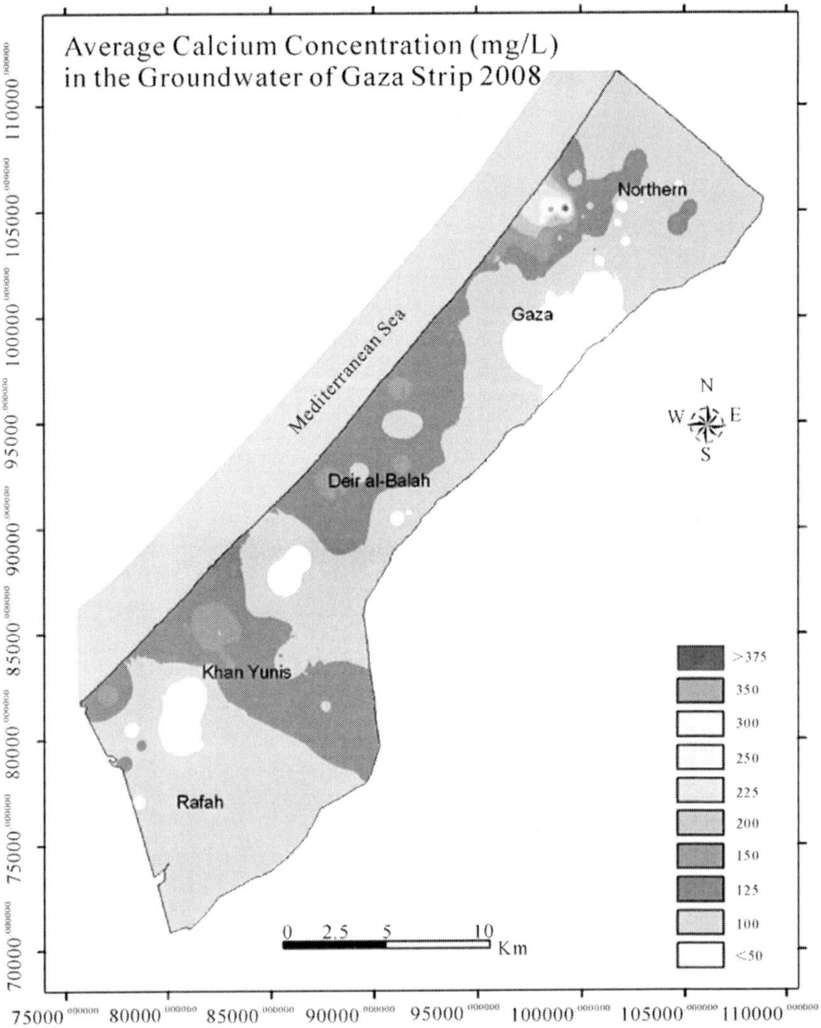

Figure 14. Calcium concentrations in the groundwater of the Gaza Strip 2008.

3.9. Trace Elements (Fe, Cr and Zn)

Total concentrations of Fe, Cr and Zn were detected in all wells of the Gaza Strip at concentrations lower than the WHO standards of 300, 50, and 3000 $\mu g/L$, respectively (Figures 18, 19 and 20). The average concentrations of

Fe, Cr and Zn in the groundwater of Gaza were 30, 75 and 15 $\mu g/L$, respectively.

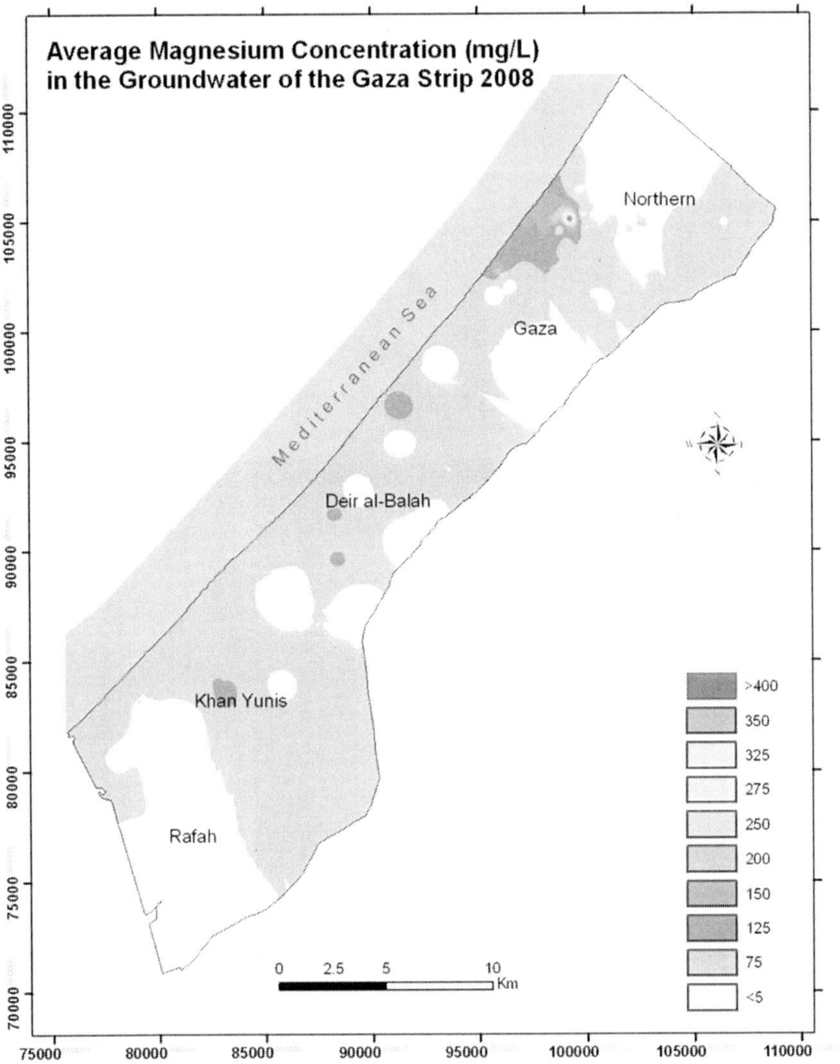

Figure 15. Magnesium concentrations in the groundwater of the Gaza Strip 2008.

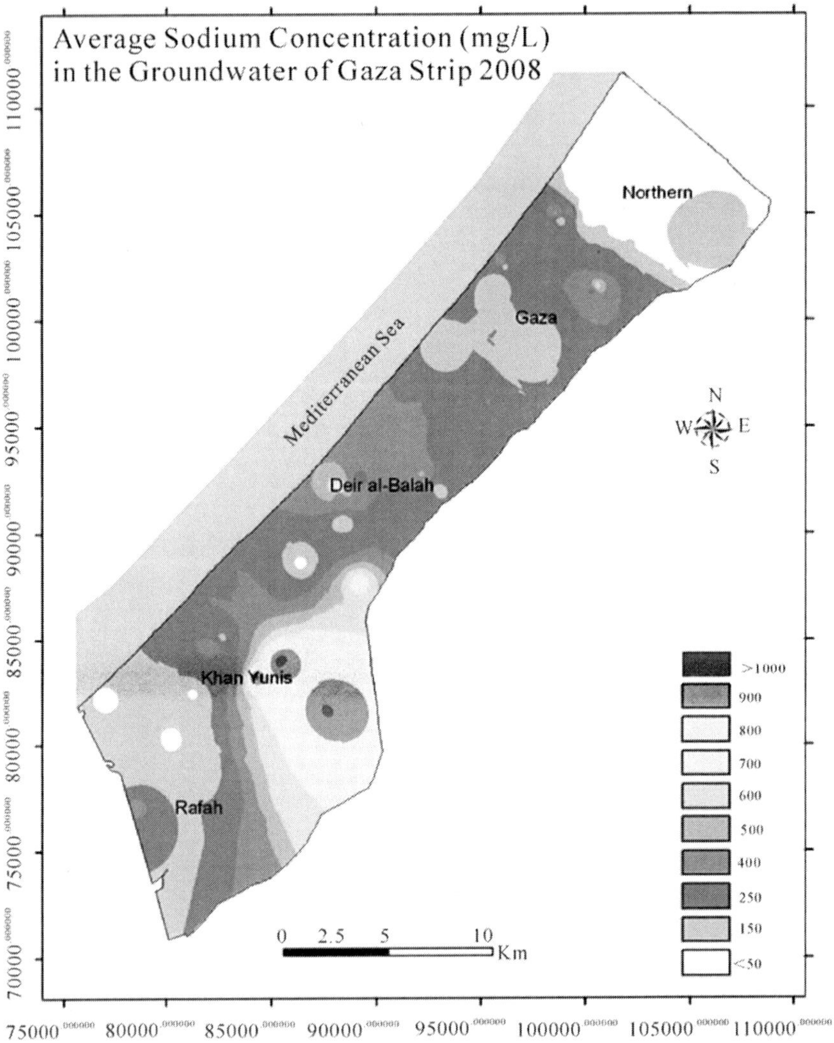

Figure 16. Sodium concentrations in the groundwater of the Gaza Strip 2008.

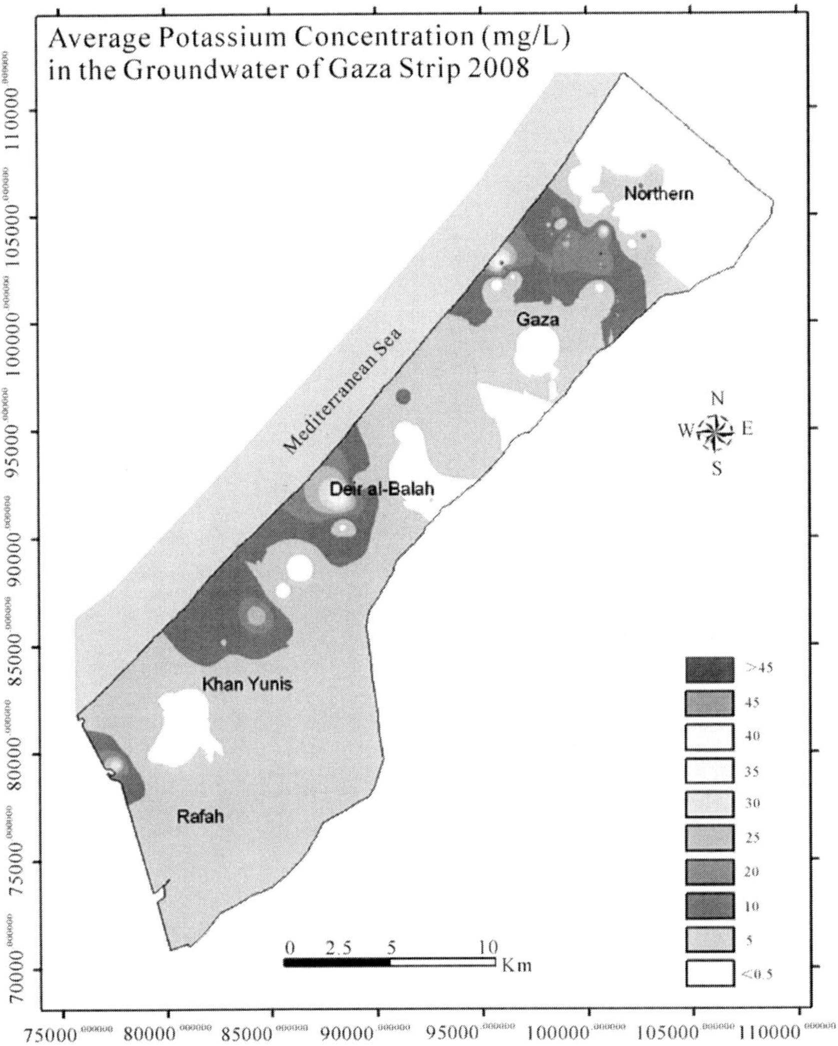

Figure 17. Potassium concentrations in the groundwater of the Gaza Strip 2008.

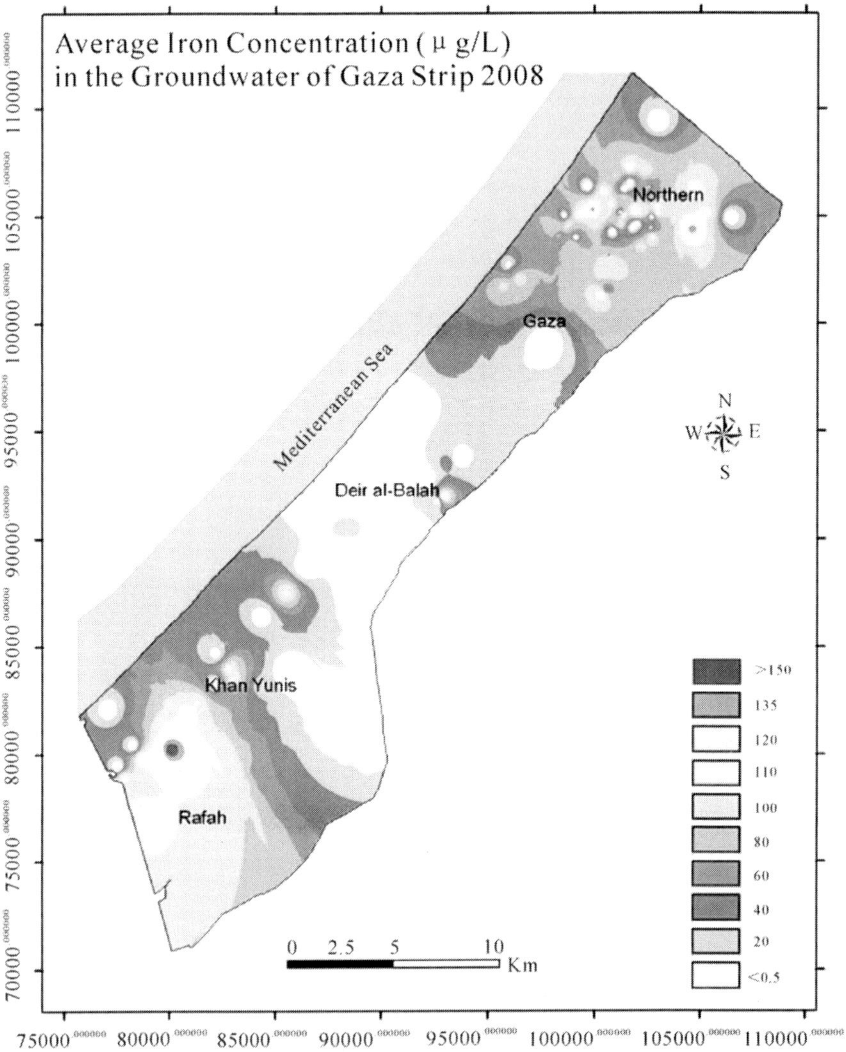

Figure 18. Iron concentrations in the groundwater of the Gaza Strip 2008.

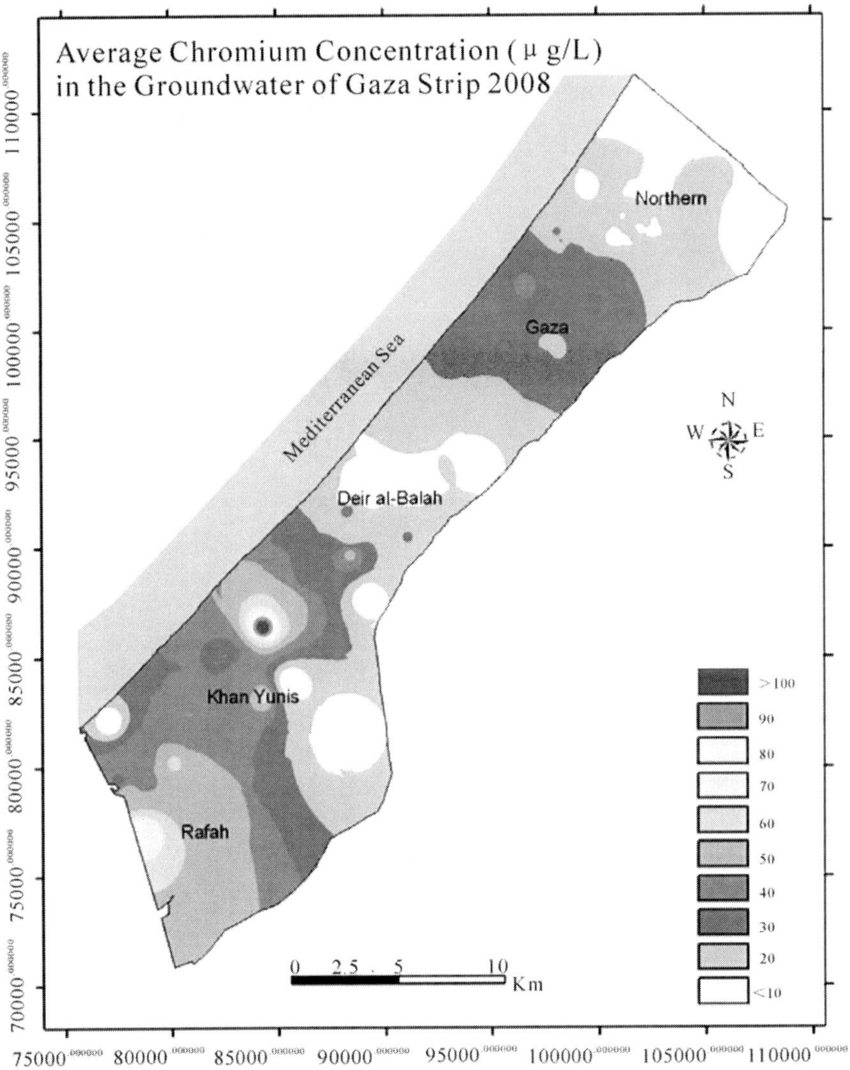

Figure 19. Chromium concentrations in the groundwater of the Gaza Strip 2008.

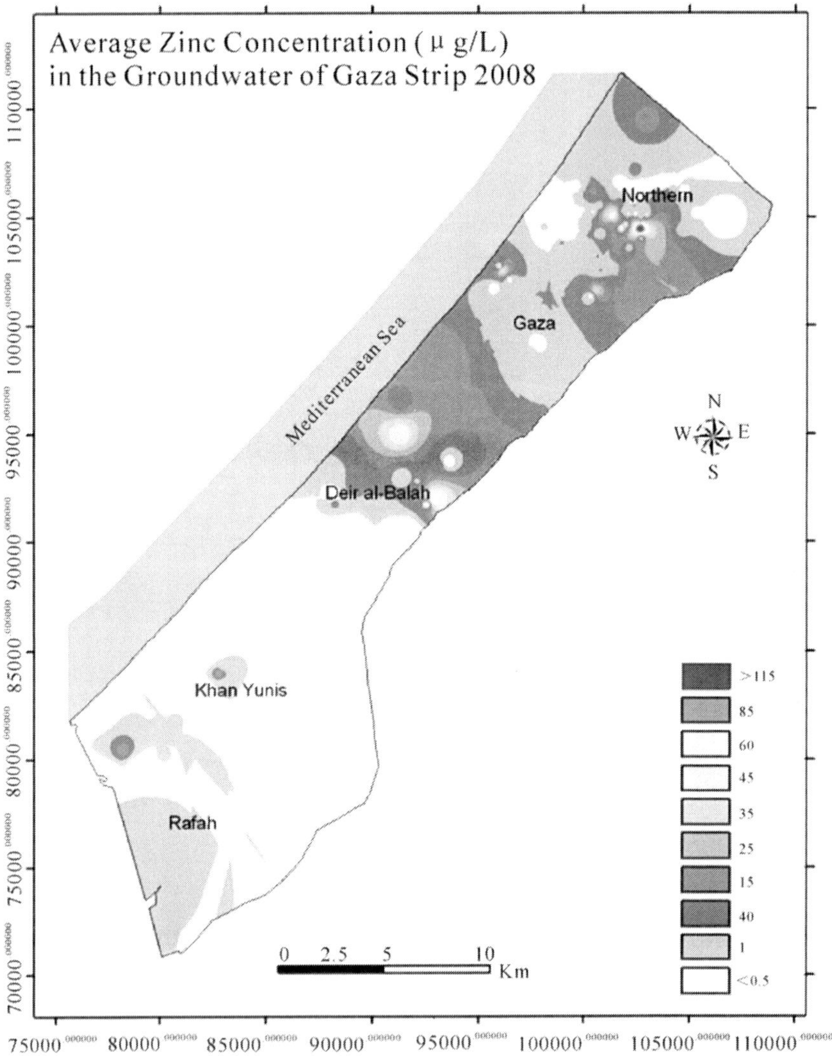

Figure 20. Zinc concentrations in the groundwater of the Gaza Strip 2008.

4. DISCUSSIONS

Deterioration of water quality and water scarcity is perennial problems in the region. As these maps indicate, the problem is particularly acute in Gaza where all groundwater wells have at least one parameter exceeding the WHO standards and about 90% of wells have salinity exceeding the WHO standard of 250 mgCl/L. The Gaza aquifer is impacted by contaminants from seawater intrusion, wastewater, manure and natural occurrence [29,32,33]. As the maps illustrate, as one goes from north to south in Gaza, the water quality deteriorates with city of Rafah having the poorest water quality. The U.S. Census Bureau estimates that Gaza's population is growing at about 4% per year making it among the fastest growing areas in the world [34]. The need for more water to meet the needs of the growing population, a dropping water table and significant challenges in maintaining and improving infrastructure for handling human wastes and managing agricultural and industrial pollutants sets the stage for continued deterioration.

4.1. Land Use and Environmental Hot-Spots

The sand dunes in Gaza protect the coastal areas against the sea and have a natural water cleaning capacity. This protection, however, is diminishing due to sand being removed without permission and extensive sand quarrying practices in the Gaza Strip [13]. Currently, about 60-80% of the domestic wastewater is discharged into the environment without treatment at the source, after collection from cesspits, through the effluent of the sewer system or at the overloaded treatment plants. Gaza's three wastewater treatment plants are outdated and overloaded with excess inflow of wastewater. For example, the largest, south of Gaza City, was designed to treat 42,000 cubic meters (CM) per day, the amount produced by 300,000 people, but now faces a daily inflow of more than 60,000 CM. As an emergency measure to prevent sewage from overflowing, barely treated wastewater is now piped to the coast, where the dark gray liquid flows along the beach. Additionally, 40% of Gazans are without access to a centralized sewage-disposal system contributing to the burgeoning cesspits. A 40-hectare lake of sewage that has formed in northern Gaza is a menace to people at the surface and to the aquifer beneath [35].

The treatment plants have been destroyed more than once as a result of the turbulent political situation. Beit Lahia wastewater treatment plant flooded in March 2007, killing several people, displacing thousands of people, and destroying homes and killing animals. Several groundwater wells in the areas surrounding the wastewater treatment plants have been closed completely due to the presence of faecal coliforms, detergents and elevated nitrate concentrations.

Solid wastes (including sludges) are disposed in dumping sites with no groundwater protection measures and elevated levels of several heavy metals were found in the agricultural wells in the eastern part of Gaza surrounding the central solid waste dumping site.

In the study of Shomar et al. [25], the water quality in the area of Wadi Gaza was seriously contaminated by several pollutants as the Wadi itself is currently polluted with wastewater and illegally dumped solid wastes.

4.2. Water Table and Groundwater Flow

Depth to water (Figure 6) indicates a regional groundwater flow direction from east to west. The most important source for recharge is the rainfall, which varies annually between 200 mm in the south and 400 mm in the north. Most of rainfall evaporates. The annual recharge varies between 20 and 40 MCM. Another 15 to 35 MCM laterally flows from the eastern boundaries of the Gaza Strip, while irrigation and leaky pipes are estimated to return 40 to 50 MCM, for a total annual recharge of 75 to 125 MCM [36]. The depth to water table generally shows a continuity of groundwater flow from east to west. Lateral inflow is important to the overall water balance in the Gaza Strip. The amounts of lateral inflow and outflow are subject to annual change due to varying hydrogeological parameters and human activities, such as rainfall and pumping [33]. However, average annual groundwater lateral inflow and outflow can be estimated based on different approaches. Since the groundwater level for the area of study is monitored monthly, a groundwater level contour map was created based on the data of the years 2006-2008.

4.3. Groundwater Quality

The groundwater aquifer's only natural output is the eight MCM per year that should flow into the Mediterranean, providing a crucial barrier against the intrusion of seawater. Thus, if no more than about 100 MCM were tapped from the aquifer per year, it could last indefinitely. But Gaza's 4000 wells remove as much as 160 MCM yearly. This estimated 60 MCM annual water deficit is why the water table is dropping rapidly and currently reaches 13 meters below sea level in some places. Saltwater from the Mediterranean, as well as deeper pockets of brine in the aquifer itself, flow in to fill the gap. As shown in Figure 9 (chloride concentrations), the saltwater intrusion is well under way in much of the region with "hot spots" in the coastal areas and to the south.

The occurrence of saline (brackish) water in both the south and east is most likely due to the fact that the annual rainfall in the south is lower than that in the north (200 and 400 mm, respectively). Also, the unconfined nature of parts of the aquifer in the Gaza Strip suggests an open system (unconfined) for the natural recharge, especially in the dunes area along the Mediterranean coast [32]. The other parts of the aquifer are of a confined to semi-confined nature [33]. Although the structural geology, which may play a significant role in this regard, is neither well documented nor well understood, hydrogeological barriers are assumed to be present, especially in the middle of the Gaza Strip. These barriers separate the two chemical facies in the north and south [37].

Since most of the wells do not meet all the WHO standards for drinking water, the water in Gaza is currently not suitable for drinking. The accelerating rate of saltwater intrusion [33] alone could make the Gaza aquifer unusable for agriculture, industrial and domestic nondrinking water uses within two or three decades. But there may be far less time available. The aquifer is also being contaminated with a cocktail of pollutants from Gaza's sewage and agriculture [26,35]. Given the large numbers of groundwater pollutants, an integrated approach to managing water resources is essential. Such an approach would include conservation, land use regulation, and control of human waste and agricultural and industrial pollutants.

5. CONCLUSIONS

1) No groundwater in Gaza meets all WHO drinking water standards and is, therefore, not safe for human consumption.

2) Areas of high nitrate concentrations are found in the vicinity of wastewater discharging areas, solid waste dumping sites and Wadi Gaza. Chloride is elevated in the coastal areas as a result of seawater intrusion and in the eastern areas as a result of upcoming and over pumping. Areas naturally contaminated with high concentrations of F^-, Ca^{2+}, Mg^{2+} and SO_4^{2-} occur as expected due to the underlying soil chemistry, geology and hydrogeology.

3) Integrating environmental and related data using GIS and using maps to illustrate areas of contamination can facilitate the development of an integrated approach toward groundwater protection in Gaza. Minimally, such an approach needs to include management of land use, wastewater and solid waste disposal, monitoring groundwater contamination, and regulating groundwater use.

4) Geography, politics, and war are conspiring to make the Gaza Strip a worst-case scenario for water-resource planners. Without immediate action, water that is currently unfit for human consumption will not be suitable for other uses likely.

ACKNOWLEDGMENTS

Basem Shomar, Sami Abu Fakher, Alfred Yahya This research was funded by Bundesministerium für Bildung und Forschung-BMBF, Germany through a project 02WT0534. The authors thank Prof. W. Shotyk, Prof. P. Outridge and Dr. Juliet VanEenwyk for their critical reading and comments on the manuscript. Special thanks to Mr. Naim Khatib for his support.

REFERENCES

1. H. Assaf and M. Saadeh, "Assessing water quality management options in the Upper Litani Basin, Lebanon, using an integrated GIS-based decision support system," Environmental Modeling & Software, Vol. 23, pp. 1327–1337, 2008.
2. V. Merwade, A. Cook, and J. Coonrod, "GIS techniques for creating river terrain models for hydrodynamic modeling and flood inundation mapping," Environmental Modeling & Software, Vol. 23, pp. 1300–1311, 2008.
3. U. Walz, "Monitoring of landscape change and functions in Saxony (Eastern Germany): Methods and indicators," Ecological Indicators, Vol. 8, pp. 807–817, 2008.
4. Yilmaz, "A case study for mapping of spatial distribution of free surface heave in alluvial soils (Yalova, Turkey) by using GIS software," Computers & Geosciences, Vol. 34, pp. 993–1004, 2008.
5. C. Zhang, L. Luo, W. Xu, and V. Ledwith, "Use of local Moran's I and GIS to identify pollution hot-spots of Pb in urban soils of Galway, Ireland," Science of the Total Environment, Vol. 398, pp. 212–221, 2008.
6. S. Asadi, P. Vuppala, and M. Anji Reddy, "Remote sensing and GIS techniques for evaluation of groundwater quality in municipal corporation of Hyderabad (Zone-V), India," International Journal of Environmental Research and Public Health, Vol. 4, pp. 45–52, 2007.
7. D. Ducci and G. Tranfaglia, "Effects of climate change on groundwater resources in Campania (Southern Italy)," Climate Change and Groundwater, Geological Society, Special Publications, London, Vol. 288, pp. 25–38, 2008.
8. D. Ducci, G. De Masi, and G. Delli Priscoli, "Contamination risk of the Alburni Karst System (Southern Italy)," Engineering Geology, Vol. 99, pp. 109–120, 2008.
9. Enshassi, "Environmental concerns for construction growth in Gaza Strip," Building and Environment, Vol. 35, pp. 273–279, 2000.
10. Shomar, A. Yahya, and G. Müller, "Occurrence of pesticides in the groundwater and the topsoil of the Gaza Strip," Water Air Soil Pollution, Vol. 171, pp. 237–251, 2006.
11. S. Abu Amr and M. Yassin, "Microbial contamination of the drinking water distribution system and its impact on human health in Khan Younis Governorate, Gaza Strip: Seven years of monitoring (2000–2006)," Public Health (in Press), 2008.
12. K. Alsharif, E. Feroz, A. Klemer, and R. Raab, "Governance of water supply systems in the Palestinian Territories: A data envelopment analysis approach to the management of water resources," Journal of Environmental Management, Vol. 87, pp. 80–94, 2008.
13. GCMEPMP, "Gaza coastal and marine environmental protection and management plan," Palestine: Ministry of Environmental Affairs, Jerusalem, pp. 1–112, 2001.
14. L. Hamdan, M. Zarei, R. Chianelli, and E. Gardner, "Sustainable water and energy in Gaza Strip," Renew Energy, Vol. 33, pp. 1137–1146, 2008.
15. B. Shomar, "Groundwater of the Gaza Strip: Is it drinkable?" Environmental Geology, Vol. 50, pp. 743–751, 2006.

16. PCBS, "Population, Housing and Establishment Census 2007," Palestinian National Authority, Palestinian Central Bureau of Statistics, Ramallah, Palestine, 2007.
17. Accessible at: http://www.pcbs.gov.ps/
18. PEPA, "Gaza Environmental Profile," Inventory of Resources: Part One, Gaza, Palestine: Palestinian Environmental Protection Authority, Gaza, Palestine, pp. 1–21, 1994.
19. MEnA, "Palestinian Environmental Strategy-Main Report," October. Gaza, Palestine: Ministry of Environmental Affairs, Gaza, Palestine, pp. 1–94, 2000.
20. J. J. Carrera-Hernández and S. J. Gaskin, "The Basin of Mexico Hydrogeological Database (BMHDB): Implementation, queries and interaction with open source software," Environmental Modeling & Software, Vol. 23, pp. 1271–1279, 2008.
21. B. Shomar, A. Yahya, and G. Müller, "Potential use of treated wastewater and sludge in the agricultural sector of the Gaza Strip," Journal of Clean Technologies and Environmental Policy, Vol. 6, pp. 128–137, 2004.
22. B. Shomar, A. Yahya, G. Müller, S. Askar, and R. Sansur, "Fluorides in groundwater, soil and infused-black tea and the occurrence of dental fluorosis among school children of the Gaza Strip," Journal of Water and Health, Vol. 2, pp. 23–36, 2004.
23. B. Shomar, "Inorganic and organic environmental geochemical issues of the Gaza Strip-Palestine," Shaker Verlag, ISBN 3-8322-3845-X, Aachen, Germany, 2005.
24. B. Shomar, G. Müller, and A. Yahya, "Geochemical features of topsoils in the Gaza Strip: Natural occurrence and anthropogenic inputs," Environmental Research, Vol. 98, pp. 372–382, 2005.
25. B. Shomar, A. Yahya, and G. Müller, "Chemical characterization of soil and water from a wastewater treatment plant in Gaza," Soil and Sediment Contamination, Vol. 14, pp. 309–327, 2005.
26. B. Shomar, A. Yahya, and G. Müller, "Seasonal variations of chemical composition of water and bottom sediments in the wetland of Wadi Gaza, Gaza Strip," Journal of Wetlands Ecology and Management, Vol. 13, pp. 419–431, 2005.
27. B. Shomar, "Trace elements in major solid-pesticides used in the Gaza Strip," Chemosphere, Vol. 65, pp. 898–905, 2006.
28. B. Shomar, "Sources of Adsorbable Organic Halogens (AOX) in Sludge of Gaza," Chemosphere, Vol. 69, pp. 1130–1135, 2007.
29. B. Shomar, "Water and wastewater in the Gaza Strip: environmentally as hot as politically," Zeitschrift für Geologische Wissenschaften, Vol. 35, pp. 165–176, 2007.
30. B. Shomar, K. Osenbrück, and A. Yahya, "Elevated nitrate levels in the groundwater of the Gaza Strip: Distribution and sources," Science of the Total Environment, Vol. 398, pp. 164–174, 2008.
31. B. Shomar, "Sources and build up of Zn, Cd, Cr and Pb in the sludge of Gaza," Environmental Monitoring and Assessment, Vol. 155, pp. 51–62, 2009.
32. B. Shomar, "Gaza Strip: Politics and environment," Water Policy (in Press), 2009.
33. M. Al-Agha, "Hydrogeochemistry and carbonate saturation model of groundwater, Khan Younis Governorate-Gaza Strip, Palestine," Environmental Geology, Vol. 47, pp. 898–906, 2005.
34. H. Baalousha, "Using CRD method for quantification of groundwater recharge in the Gaza Strip, Palestine," Environmental Geology, Vol. 48, pp. 889–900, 2005.
35. USCB, "U.S. Census Bureau, International Data Base, 2008," Accessible at: www.census.gov/ipc/www/idbnew. html
36. J. Bohannon, "Running out of water and time: Geography, politics, and war combine to make the Gaza Strip a worst-case scenario for water-resource planners," Science, Vol. 313, pp. 1085–1087, 2006.
37. CAMP, "Site, Geotechnical and hydrogeological investigation at the infiltration basins of Gaza wastewater treatment plant," Gaza Coastal Aquifer Management Program, Gaza, Palestine, 2002.
38. M. AL-Agha and H. EL-Nakhal, "Hydrochemical facies of groundwater in the Gaza Strip, Palestine," Hydrological Sciences, Vol. 49, pp. 359–371, 2004.

CHAPTER 14

Application of GIS and Remote Sensing to Groundwater Exploration in Al-Wala Basin in Jordan

Jawad T. Al-Bakri[1], Yahya Y. Al-Jahmany[2]

[1]*Department of Land, Water and Environment, Faculty of Agriculture, The University of Jordan, Amman, Jordan*

[2]*Department of Civil Engineering, College of Engineering, Majmaah University, Majmaah, Saudi Arabia*

ABSTRACT

Exploration of groundwater in countries with scarce water resources requires the implementation of effective tools that save time and money. In this study, geographic information systems (GIS) tools and remote sensing data were used to prepare and analyze digital layers of lithology, geological structure, drainage and topography to detect the most promising sites for groundwater exploration in an arid basin in Jordan. A separate map of existing wells was intersected with the generated maps to calculate the percentage of wells in each interval of density and count of lineaments and drainage. Different GIS functions of intersection and spatial query were then applied to produce the final map for the most promising sites for groundwater exploration. The possibility of using digital classification of remote sensing data for mapping the most promising sites for groundwater exploration was also investigated by applying unsupervised classification to a Landsat ETM+ image. Results showed that spatial distribution of the most promising sites for groundwater exploration was dependent on the interrelated factors of lithology, topography and geologic structure. The most promising sites were distributed within 4% of the study area. The highest percentage of groundwater wells was within the alluvial and wadi sediments, which were accurately detected by the digitally classified ETM+. The study showed that remote sensing and GIS provided efficient tools for mapping promising sites for groundwater exploration. However, the data of groundwater wells would contribute to refining the final locations of the most promising sites.

Keywords: GIS; Remote Sensing; Groundwater; Jordan

1. INTRODUCTION

Development of water resources represents a challenge for planners and decision-makers in countries with high population growth and water scarcity. In Jordan, the problem of water scarcity is being faced in all sectors and the current share of water is extremely low (145 m^3 per capita per year), as the country is dominated by an arid climate with low annual rainfall (**Figure 1**) [1]. Therefore, a considerable part of the agricultural production in the country depends on the available groundwater resources. The estimated safe annual yield from all renewable groundwater is 275 million cubic meters (MCM) while the annual withdrawal is 418 MCM [2]. The nonrenewable fossil aquifers are also providing 143 MCM per year for the coming 50 years [3].

At a population growth of 2.4%, an obvious increase in water demand is expected and the problem of water shortage will worsen with time [2,3], particularly under the adverse impacts of the expected climate change in the near future [3]. Without developing water resources, the expected annual water deficit will reach 488 and 769 MCM in years 2030 and 2050, respectively [3].

Currently, groundwater provides more than 45% of the total water resources in Jordan. The development of groundwater resources requires studies and surveys that identify areas with high potential for exploration to anticipate the inevitable geophysical surveys, which are expensive and time consuming. Groundwater exploration, however, has many problems such as the paucity of existing data and the high cost of data gathering [4]. Therefore, the integration of geologic and hydrogeological surveys with remote sensing data and geographic information systems (GIS) can contribute efficiently to these studies [4-6].

Remote sensing and GIS provide important data and tools for groundwater exploration [7-12]. Remotely sensed data of medium resolution are valuable sources for creating maps of geological structures [4,13,14]. Maps of lithology, slope and drainage patterns can be also extracted from remote sensing data of aerial photography and medium to high resolution satellite imagery, depending on the scale of the study and the available resources. Conversion of maps into digital layers will enable the analysis of these maps by GIS tools. Implementation of spatial functions, query and intersection will produce maps of potential locations for groundwater exploration, based on different criteria that integrate maps of lineaments, topography, and lithology.

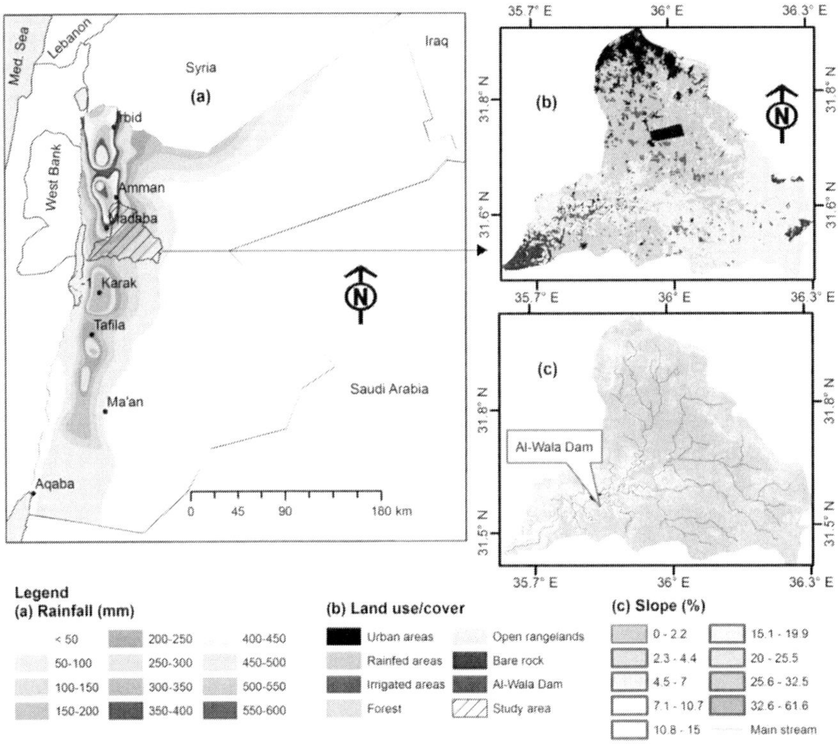

Figure 1. Location of the study area (a) and its land use/cover (b) and topography (c).

One approach for identifying potential sites for groundwater exploration is based on analyzing length density, frequency of lineament and drainage lines density to reflect different probabilities of groundwater potential in various parts of the area under consideration [7-9]. This approach was used in Jordan to identify promising sites with high potential for groundwater exploration [8,15]. Another approach for identifying these promising sites is based on analyzing hydro-geomorphological conditions to locate potential zones for groundwater recharge after applying a qualitative classification to the final map [11,14]. Reference [16] showed that this approach was useful in identifying areas for artificial recharge of groundwater, known as managed aquifer recharge (MAR).

In this study, remote sensing techniques and GIS tools are implemented to identify sites for groundwater exploration in Al-Wala basin in Jordan. The study aims to identify promising sites for groundwater exploration based on hydrogeological data. Also, it adds to the previous studies and efforts in Jordan by correlating locations with high potential with spectral classes derived from the digital classification of remotely sensed data of Landsat Enhanced Thematic Mapper Plus (ETM+). In terms of GIS approach, the study utilizes ground data to refine the selection of sites based on groundwater yield of the existing wells.

2. STUDY AREA

2.1. Location and Climate

The study covered Al-Wala basin in the middle parts of Jordan (**Figure 1**). The total area of the basin is 1900 km^2. The basin has a flat topography in the east and steep slopes in the west. Altitude of study area varies from 321 m below sea level at the eastern shoreline of the Dead Sea (lowest point on earth) to 997 m in the eastern parts of the study area.

The basin is dominated by an arid Mediterranean climate with relatively cold winters and hot summers. The mean monthly air temperature ranges from 6°C in January up to 22°C in August, with a mean annual air temperature of 15.0°C. With exception to northwestern parts, the mean annual rainfall is 100 - 200 mm, occurring mainly during winter season (October-April). The mean annual potential evaporation ranges from 1600 mm in the western parts to 2000 mm in the east and 2200 mm at the outlet near the Dead Sea. In terms of land use, about 47% of the basin is used as open rangelands. Agricultural activity is also taking place in the basin where rainfed cultivation of wheat and barley is practiced in 38% of the basin (**Figure 1**). Irrigated areas form 3% of the total area of the basin, while 7% of the basin is urban.

2.2. Geology

The basin is characterized by the strata of Belqa group and partly of Ajlun group (**Table 1**). The groups are mainly dominated by cretaceous carbonate rocks that outcrop in most of the study area [17]. The oldest rock that outcrops in the study area is the upper Wadi As Sir Limestone (WSL) formation, deposited in an agitated marine environment, from the Turonian age. The lower part is composed of marl, marly limestone, coquina, micrite, sand and dolomicrite with chert nodules that has an average thickness of 62 m. The upper part is composed of limestone, dolomitic limestone and fossiloferous limestone with an average thickness of about 4.5 m [18].

Table 1. Lithostratigraphic units and formations of the study area with percentage wells in each formation.

Group	Formation unit	Wells (%)
	Soil (S)	10.6
	Calcrete (Ct)	0.0
	Alluvial Gravels (Alg)	<0.1
	Alluvial Fans (Alf)	<0.1
	Alluvium and Wadi Sediments (Al)	18.9
	Travertine (Tr)	<0.1
	Fluviatile & Lacustrine Gravels (Pl)	25.6
Belqa	Basalt (B)	<0.1
	Wadi Shallalah Chalk (WSC)	<0.1
	Madaba Calcareous Breccia (MCB)	<0.1
	Umm Rijam Chert-Limestone (URC)	8.3
	Muwaqqar Chalk-Marl (MCM)	14.1
	Al Hisa Phosphorite (AHP)	11.1
	Amman Silicified Limestone (ASL)	1.8
	Wadi Umm Ghudran (WG)	6.7
	Wadi As Sir Limestone (WSL)	2.5
Ajlun	Fuhays/Hummar/Shuayb (F/H/S)	0.2
	Na'ur Limestone Formation (NL)	<0.1

An abrupt change in sedimentation resulting from an extensive Coniacian marine transgression marks the base of Belq group where a predominantly chalky unit of Wadi Umm Ghudran (WG) formation of ConiacianSantonian age overlies Ajlun group. This formation consists of Mujib Chalk, brecciated dolomitic limestone, and Dhiban Chalk with fish teeth outcropping all over the basin area. The formations of Belqa group include Al Hisa Phoshorite formation (AHP), (Campanian age), which consists of dark grayish chert, marly limestone and phosphate, and the Muwaqqar Chalk Marl formation (MCM) (Maastrichtian age), which consists of marl and chalk with chalky and micritic limestone concretions. Also, it includes Madaba Calcareous Breccia (MCB) (M. Eocene age), which consists of brecciated and partly conglomeratic clasts of phosphate, limestone, chert and chalk in a mudstone matrix located on hills, slopes and along some wadis in Madaba area.

The geology of the study area has some unique formations that include basaltic flows of Pleistocene, as volcanic tuffs restricted to the upper courses of Wadi Al Haidan. Quaternary Superficial Deposits (Pleistocene age) of Fluviatile and Lacustrine Gravels (PI) cover most of the wadis of the basin, particularly in the east. Aridic soils are dominating the low rainfall zones, while dark brown Mediterranean soils are dominating the high rainfall zones.

2.3. Hydrogeology

Groundwater of the basin flows from the eastern and northern highlands towards south and west till it reaches its downstream in the Dead Sea (**Figure 2**). Drainage pattern in the study area is modified by the direction of topographic displacement caused by the major faults so that it is directed towards faulted areas. The main wadis flow only in winter season and are directed ultimately westwards (**Figure 1**). The most important aquifers in the area are those within the upper cretaceous of AmmanWadi As Sir Limestone formations, with the intervening Wadi Umm Ghudran and overlaying Muwaqqar Chalk Marl formations, forming aquicludes [17,18].

The main source of groundwater is the aquifer system of Amman-Wadi As Sir formation known as B2/A7 aquifer system. Two recharge systems for this aquifer are recognized as follows:

1) Amman-Madaba, which is the main recharge system in the basin that includes the northwestern mountains in the basin. Groundwater of this recharge system flows southwards to Wadi Al-Wala and Al-Haidan (**Figure 2**).

2) Libb-Thiban, in which groundwater flows from the east and south towards Al-Haidan and through Al-Wala main stream.

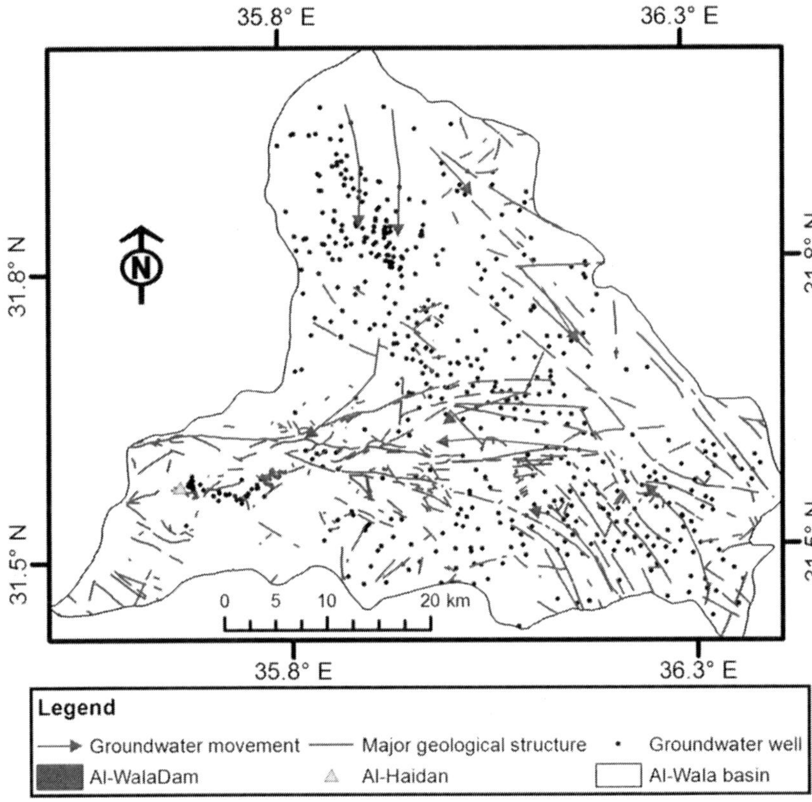

Figure 2. Major geological structures, groundwater wells and direction of groundwater movement in Al-Wala basin.

Ground water is the main source of water that is utilized for agriculture and municipal uses. The official records of Ministry of Water and Irrigation (MWI) for year 2010 show that the study area has 590 groundwater wells [19]. Fourteen springs are found in the downstream area in Al-Haidan. Water extracted from springs is used for both irrigation and domestic purposes.

Groundwater of the basin is highly important as water is pumped from Al-Haidan groundwater wells to supply drinking water to Madaba and southern parts of Amman. The MWI constructed Al-Wala dam, which became operational in the year 2003, to improve groundwater recharge. The current capacity of the dam is 9 MCM and the stored water is naturally and artificially recharging groundwater of Al-Haidan and other wells that are utilized for drinking and irrigation.

The future plan of MWI is to raise the height of the dam to store 26 MCM. The water will mainly be utilized for groundwater recharge in Al-Haidan and in the upstream area in the middle and the southern parts of the basin. Also, about 6 MCM will be utilized for irrigating fodder crops in the Jordan Valley [20]. Therefore, identifying the areas with high potential for groundwater exploration is of high importance for supporting agricultural activities in the area and for developing drinking water resources for the cities of Madaba and Amman.

3. METHODOLOGY

The methodology of this study was based on two approaches. The first was based on GIS, where layers of lithology, drainage patterns and lineaments were used to derive the most promising sites for groundwater exploration, as described by References [5,8]. Additionally, data on water quality and quantity were used to verify sites for groundwater exploration. The second part of the study was based on digital image processing (DIP) of remote sensing data to identify the spectral class that was highly correlated with the sites identified by the GIS tools. The first approach is based on the hypothesis that lineaments and drainage patterns would be the most important factors for evaluating the potential concentration of water in the fracture zone aquifer, while the second approach was proposed as a rapid method with minimum inputs for groundwater exploration. The following sections describe the methodology in more details.

3.1. Data Collection and Preparation

Hard copy maps of geology, topography and hydrology, covering the study area at a scale of 1:50,000, were acquired, scanned and geo-referenced in GIS. A satellite image of Landsat 7 ETM+ covering study area, except a small part in the east, was downloaded from the USGS website. The ETM+ imagery data which were used in this study included the visible (1, 2 and 3), the shortwave infrared (SWIR), the middle infrared (MIR) and the panchromatic bands. The spatial resolutions for these bands were 30 m for the visible to infrared and 15 m for the panchromatic. Geometric correction was carried out for the image using ground control points (GCP's) collected by a global positioning system (GPS). In total, fifteen points were used to correct the image using a second order transformation and the nearest neighbor resampling method [21].

A digital elevation model (DEM) at 30 m resolution, available from the data of the Advanced Spaceborne Thermal Emission and Reflection Radiometer (ASTER), was downloaded from the website of the provider (http://

asterweb.jpl.nasa.gov/gdem.asp). The DEM was used to generate a layer of streams network of the basin. The layer was overlaid over the geo-referenced topographic maps in GIS to correct the positional shifts. Line smoothing functions were also applied and manual editing was carried out for some of the streams which were either incorrectly delineated or had positional shifts. Data editing and improvement were also carried out for the layer of lineaments features (faults) using the panchromatic band of the ETM+, which was utilized to increase the details of this GIS layer.

Data on groundwater wells were obtained from the MWI. The data were transferred into spreadsheets and included the coordinates of wells, salinity of water and yield of well in cubic meters per hour (m^3/h). The yield was classified into three intervals following the criteria of the MWI for groundwater wells used in irrigation. The intervals were 35 m^3/h or less for low, 35 - 70 m^3/h for the medium and above 70 m^3/h for the high yield. The spreadsheet was transferred into a point layer in GIS and included the records for 560 wells.

3.2. Application of GIS

The output digital maps included layers of lineaments, stream (drainage) network and lithological formations with all attributes associated with each layer. **Figure 2** shows the lineament map, produced from this stage, and included all types of faults and strikes in the study area. The digital map of lineament was analyzed within GIS to identify types of structures in relation to ground water existence. Following this stage, all maps were rasterized to derive count (per·km^2) and length density (m/km^2) for both lineaments and stream network. Each of the output maps was reclassified into a new map with ten equal intervals of count and density.

The map of wells was intersected with the density maps of lineaments and drainage lines. The intersection results were then summarized in spreadsheets to calculate percentages of wells within each density interval of structure lines and drainage patterns. The same procedure of intersection was applied to determine the number and percentage of wells in each lithological unit. The intervals of count and density with maximum number of wells and the units of lithology were intersected to identify possible sites for groundwater exploration based on data of both medium and high yield wells. The output sites were based on the existence of 50% of wells or more within an interval or two for the medium and high yield groundwater wells. Wells with high levels of water salinity were excluded from the analysis. The final map was refined after considering altitudes, derived from the DEM, with maximum occurrence of medium and high yield of groundwater wells.

3.3. Application of Remote Sensing

The ETM+ imagery was transformed into three Principal Components (PC's), known as "Eigen Channels", which accounted for all variations of the selected bands of the ETM+. This was mainly performed to reduce data autocorrelation and dimensionality into fewer interpretable bands [21,22]. The unsupervised image classification was then applied to the output PC's using the Iterative Self-Organizing Data Analysis Technique (ISODATA) clustering algorithm to group the ETM+ grey levels into fewer classes, which were then identified in relation to lithological units using the existing geological maps and field visits.

The overall accuracy of the classification was assessed through a confusion matrix [21] by comparing a number of random points of classified pixels with existing geological maps and ground data. The output layer from the digital classification was intersected with the layer of the promising sites of groundwater wells to identify the class where most of these sites were located. The unsupervised classification was used as a rapid method that would automatically identify all possible spectral classes that corresponded to lithological formations in the study area. Therefore, results from DIP were intersected with results from GIS analysis of layers of lineaments and streams to assess the degree of agreement between the two methods.

4. RESULTS AND DISCUSSION

4.1. Lithology and Geological Structures

Results of intersecting the maps of wells and lithological formations are summarized in **Table 1**. Results showed that less than 26% of the wells were distributed in the fluviatile and lacustrine gravels, while about 19% of the wells in the study area existed in Alluvium and surrounding Wadi sediments (**Table 1**).

The intersection results also showed that about 76% of wells in the above mentioned formations were characterized by medium and high water yields. These results agreed with References [7,23] who found that most potentially excellent aquifers were found in valleys with thick sand or gravels covers. In Jordan, results of this study also confirmed the findings of References [8,15] who concluded that wells with high yield were found in thick coarse grained alluvial materials. These formations were expected to act as conduits for groundwater from deep to shallow aquifers. Therefore, most of the wells with good yield were found in these formations.

Table 2. Major fault groups in the study area.

Fault group name	Description
Wadi Zarqa Main Fault	Linear strike slip fault changes from a reverse to normal while moving from east to west.
Rumeil Fault	Normal fault extending for about 9 km with a downthrown of about 250 m near Wadi Rumeil area.
Wadi Al-Wala	Normal and reverse sets of faults with a downthrown to the south.
Az-Zafaran	A strike slip fault with dextral movement, starts with a reverse displacement and changes to a normal while moving from east to west. Displacement is more than 150 m near the desert highway where Muaqqar chalk marl is faulted against Amman Silicified Limestone formation.
Wadi Al Falij	Numerous small faults trending northeast-southwest in marble deposits.

Analysis of lithological formation showed that structure features were observed above ground with considerable displacements found in Paleozoic formations. These were totally or partially absent in Mesozoic or Cretaceous lithological formations. This could be attributed to the stronger effect of stresses and faulting on older rock formations, as explained by Reference [5]. The stresses caused a deformation in the rock layers and produced geological structures that were identified and mapped as lineaments on the ETM+ imagery. Summary of the major fault in the study area is shown in **Table 2**.

In the study area, most sedimentary rocks had been deposited as horizontal strata. In various parts of the study area, subsequent uplifting and tilting caused the strata to be inclined. Furthermore, more than 70% of the main geologic structures were faults with tics showing downthrows side, faults inferred beneath superficial deposits and lineaments that were visible on the ETM+ imagery. All these structures were expected to form zones of high permeability layers which would recharge groundwater aquifers.

4.2. Distribution of Wells within Count and Density Intervals

Results showed that wells with high yield were mainly located in Al-Haidan in the downstream area, while wells with low and medium yields were distributed all over the study area (**Figure 3**). Results of intersecting wells with counts and densities of lineaments and streams are presented in **Table 3**. In terms of lineaments count, wells were mainly found within the intervals of low count. Although this was unexpected, however it could be attributed to the fact that the lineaments count was already very low (less than 3.4 lineaments per km^2). On the other hand, the relationship between lineaments density and groundwater wells showed that 50% of groundwater wells were located within the lowest density interval. This could be explained by factor of topography, as many of the low and medium yield wells were located in flat areas in the northern part of the study area (**Figure 3**), where geological structures were not detected. In this part of the study area slope was less than 2% which would promote water infiltration and groundwater recharge [8,24]. In addition to topography, the high rainfall and the lithological formations (Holocene-recent alluvial and wadi sediments) enhanced groundwater recharge.

Although a considerable number of wells were found in the northern parts of the study area, however, most of the wells were characterized by low or medium yield. The high yield wells, on the other hand, were mainly within the moderate to high density of lineaments in the middle parts of the study area or in the low lineament density in the downstream area. In the middle parts of the study area, the high densities of lineaments were acting as points of groundwater recharge. In both locations, 69% of the wells were located in areas with an altitude of 650 - 750 m above sea level. None of the wells with high yield, on the other hand, occurred in altitudes of more than 750 m. This could be attributed to the direction of groundwater movement which was towards the southwestern parts of the study area (**Figure 2**). Similar findings of the impacts of topography on groundwater recharge were also indicated by References [11,15].

The combined impacts of topography, the high density of lineament and the alluvial deposits resulted in the presence of high yield wells in the middle parts of the study area around the main stream of Al-Wala. These results agreed with References [8,15] which indicated strong relationships between lineament density and locations of groundwater wells. In this study, none of the medium to high yield wells occurred in intervals that succeeded 5725 m/km^2 (Interval 6) (**Table 3**).

Querying groundwater wells with the map of drainage density showed that most of the groundwater wells were within the intervals of moderate drainage count and density (**Table 3**). The medium yield wells were mainly located within interval 5, while the high yield wells were mainly located within interval 4. Generally, drainage pattern is a reflection of the infiltration/precipitation ratio.

This ratio is also controlled largely by permeability, which is in turn a function of the rock type and fracturing of the underlying rock or surface bedrock [7]. When comparing two terrain types, the one that contains the greatest drainage density is usually less permeable [7]. Therefore, this fact would imply that in groundwater exploration it would be important to combine the lineament map with drainage and lithological formations maps to identify the target sites of groundwater exploration, as explained in the next section. By this, variations in lithology, geologic history and response of bedrocks to stresses from folding to fracturing would be considered in groundwater exploration.

4.3. Intersection of Maps for Identifying Promising Sites

Results of intersecting the intervals of lineament and drainage count and density with the lithology and altitude are presented in **Figure 4**. The output of intersection presents the locations of promising sites for groundwater exploration.

The map includes two classes that present medium and high probability sites based on the data of groundwater wells.

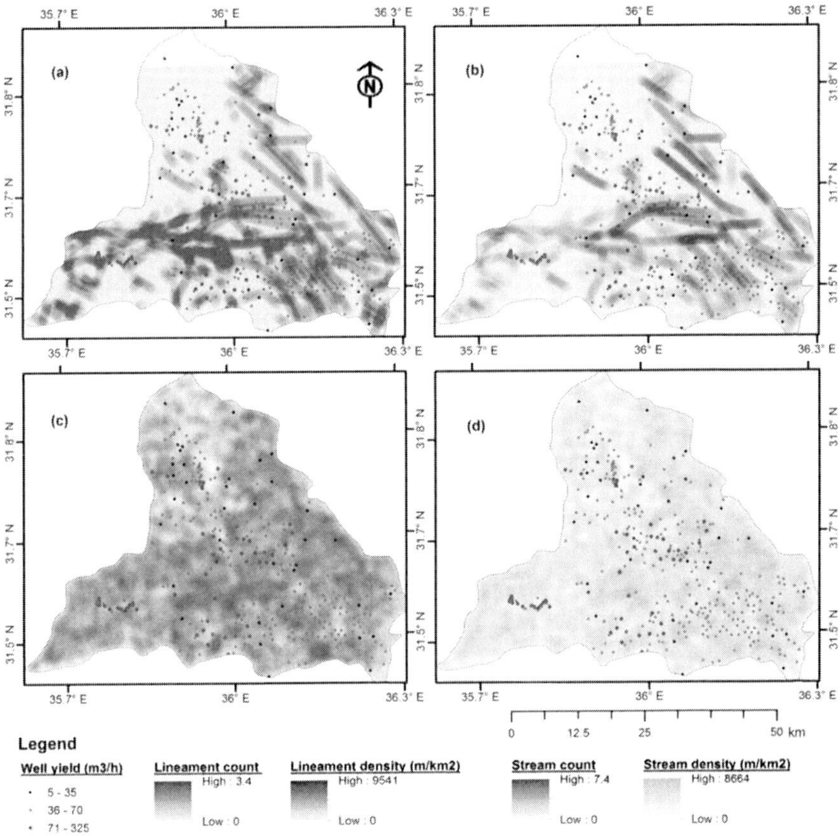

Figure 3. Distribution of groundwater wells within the layers of lineament count (a); lineament density (b); stream count (c) and stream density (d).

Table 3. Percent of wells within the different intervals of lineament and stream counts (per·km^2) and densities (m/km^2).

Interval	Lineament count	Wells (%)	Lineament density	Wells (%)	Stream count	Wells (%)	Stream density	Wells (%)
1	0.00 - 0.34	36.7	0 - 954	50.3	0.00 - 0.75	1.3	0 - 866	1.9
2	0.34 - 0.67	12.2	954 - 1908	20.2	0.76 - 1.49	4.3	867 - 1733	5.6
3	0.68 - 1.02	20.5	1909 - 2862	12.0	1.50 - 2.24	8.0	1734 - 2599	15.7
4	1.02 - 1.35	9.6	2863 - 3816	7.2	2.25 - 2.98	14.7	2600 - 3466	33.1
5	1.36 - 1.68	12.2	3817 - 4770	5.3	2.99 - 3.73	24.5	3467 - 4332	23.2
6	1.69 - 2.02	5.3	4771 - 5725	3.7	3.74 - 4.47	28.0	4333 - 5199	13.1
7	2.03 - 2.36	2.4	5726 - 6679	0.8	4.48 - 5.22	14.9	5200 - 6065	5.6
8	2.37 - 2.70	0.5	6680 - 7633	0.3	5.23 - 5.96	3.2	6066 - 6931	1.6
9	2.71 - 3.03	0.5	7634 - 8587	0.0	5.97 - 6.71	0.8	6932 - 7798	0.5
10	3.04 - 3.37	0.0	8588 - 9541	0.3	6.72 - 7.45	0.3	7799 - 8664	0.0

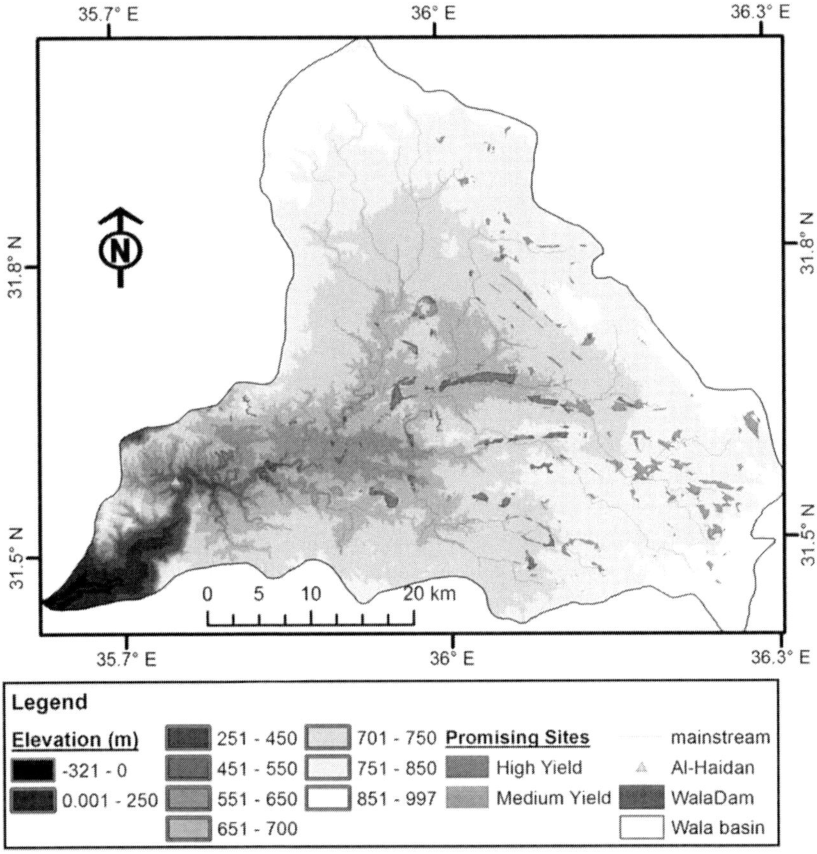

Figure 4. Locations of promising sites for groundwater exploration in Al-Wala based on the intersection of lineaments, lithology, drainage and altitude.

Results from this GIS approach showed that promising sites for groundwater exploration were distributed in different parts of the study area. However, most of the high yield sites were in the middle part of the study area on the main stream of Al-Wala. This part of the study area, mainly located within the formations of the fluviatile and lacustrine gravels (PI), was characterized by medium density of lineaments and streams, with altitudes ranging between 680 to 710 m. A small area of high yield was predicted in the southwestern part of the study area near Al-Wala dam. This area was classified as a high yield site due to its lower altitude and its occurrence on alluvial deposits (PI) with relatively moderate lineament densities (2863 - 3816 m/km^2).

The medium yield sites, on the other hand, were mainly distributed within the alluvium and wadi sediments (Al) and at altitudes range of 750 - 850 m. In terms of areas, the high yield sites had a total area of 26 km^2 while the medium yield sites had a total area of 50 km^2. These figures represent 1.4% and 2.6% of the study area for both high and medium yield sites, respectively.

4.4. Digital Classification of ETM+

Initial results of the unsupervised classification of the first three PC's, which accounted for all variations in the ETM+ image, produced a layer with ten spectral classes. The output layer (**Figure 5**) was compared with the map of lithological formations in the study area, and subsequently the final number of classes in the output layer was merged to 5, after considering class separation using the method of transformed divergence [21]. **Table 4** summarizes the final classes and the major lithological units represented by each class. Classification accuracy varied from one class to another, with an overall accuracy of 76%. The variation in class accuracy was mainly attributed to the spectral mixing between classes and the accuracy of the original geologic maps.

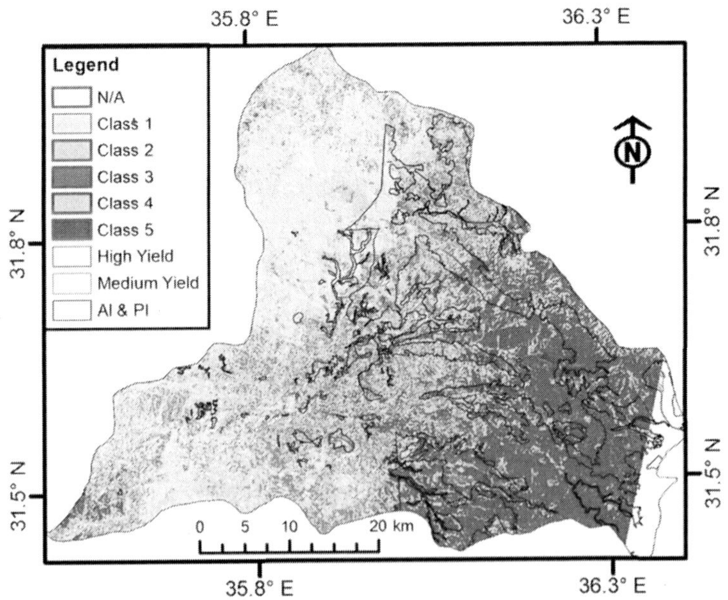

Figure 5. Distribution of alluvial formations and the most promising sites for groundwater exploration within the spectral classes of the unsupervised classification of the ETM+ image.

Distribution of groundwater wells within the classified PC image (**Table 4**) showed that about 28% of groundwater wells were mainly located within class 5, representing alluvial formations and mapped with high accuracy. This figure was very close to results obtained from map intersection within GIS, which showed that 30% of groundwater wells were located within the alluvial formations.

Results showed that 71% and 52% of the promising sites for medium and high yield, respectively, were located within class 5. Also, 65% of the alluvial formations were located within this class. Therefore, outputs from digital classification would provide important thematic layers that could serve the purpose of identifying promising sites for groundwater exploration in basins similar to Al-Wala, particularly in locations where digital layers of geology might not be available.

Other factors and data, however, should complement the DIP techniques. This was evident by the results of this study as the outputs from the unsupervised classification were not related to promising locations in the downstream area, where topography and lineament density were the main factors in identifying promising sites for groundwater exploration.

5. CONCLUSIONS AND RECOMMENDATIONS

Results from this study showed the significant roles of GIS tools and remote sensing data in identifying promising sites for groundwater exploration. The study provided an example that integrated real data on groundwater quantity and quality to refine the selection of the target sites. These data showed that sites with possible high yield were related to moderate intervals of lineament and stream densities. The other important data that affected the spatial distribution of the promising sites were lithology and topography.

The study showed that GIS tools were very useful in refining the intervals of lineament and stream densities and in identifying the most promising sites based on data of existing groundwater well. Remote sensing data, on the other hand, could provide important information on geology and topography. Therefore, these tools and data sources would be recommended for groundwater exploration, as they would provide timely and cost effective tools for identifying and narrowing the target areas for groundwater exploration before carrying out further investigations.

Table 4. Lithological description of the major classes produced from the unsupervised classification of principal components and the percentage of wells in each class.

Class	Description	Lithological Formations	Mapping Accuracy (%)	Wells** (%)	Wells* (%)
1	Lithologic formations of late Cretaceous.	WG, ASL, AHP, MCM	73.8	29.8	33.7
2	Ajlun Group formations from Cenomanian to Turonian.	NL, F/H/S, WSL	68.9	7.5	2.7
3	Lithologic formations of the Pleistocene sediments of the geologic age, middle to upper Belqa Group.	PL, Tr	66.7	20.7	25.6
4	Lithologic formations of Tertiary (middle part of Belqa Group).	URC, MCB, WSC, B	70.8	14.1	8.3
5	Holocene-Recent alluvial and wadi sediments.	Al, Alf, Alg, S	95.0	27.9	29.5

Based on intersection with thematic classes of unsupervised classification; *Based on intersection with lithological map (Table 1**).

ACKNOWLEDGEMENTS

Jawad T. Al-Bakri, Yahya Y. Al-Jahmany (2009-2012) "Assessment and monitoring of desertification in Jordan using remote sensing and bioindicators".

REFERENCES

1. "Annual Report of Jordan Meteorological Department," JMD (Jordan Meteorological Department), Amman, 2010.
2. "Water for Life: Jordan's Water Strategy 2008-2022," MWI (Ministry of Water and Irrigation), Amman, 2009.
3. J. T. Al-Bakri, M. Salahat, A. Suleiman, M. Suifan, M. R. Hamdan, S. Khresat and T. Kandakji, "Impact of Climate and Land Use Changes on Water and Food Security in Jordan: Implications for Transcending 'The Tragedy of the Commons'," Sustainability, Vol. 5, No. 2, 2013, pp. 724-748. http://dx.doi.org/10.3390/su5020724
4. Shaban, "Support of Space Techniques for Groundwater Exploration in Lebanon," Journal of Water Resource and Protection, Vol. 2 No. 5, 2010, pp. 469-477.http://dx.doi.org/10.4236/jwarp.2010.25054
5. R. M. Teeuw, "Groundwater Exploration Using Remote Sensing and a Low-Cost Geographical Information System," Hydrogeology Journal, Vol. 3, No. 3, 1995, pp. 21- 30. http://dx.doi.org/10.1007/s100400050057
6. Travaglia and O. Ammar, "Groundwater Exploration by Satellite Remote Sensing in the Syrian Arab Republic," FAO Publications, Rome, 1998.
7. Edet, C. Okereke, S. Teme and E. Esu, "Application of Remote Sensing Data to Groundwater Exploration: A Case Study of the Cross River State, Southeastern Nigeria," Hydrogeology Journal, Vol. 6, No. 3, 1998, pp. 394-404. http://dx.doi.org/10.1007/s100400050162
8. El-Naqa, N. Hammouri, K. Ibrahim and M. El-Taj, "Integrated Approach for Groundwater Exploration in WadiAraba Using Remote Sensing and GIS," Jordan Journal of Civil Engineering, Vol. 3, No. 3, 2009, pp. 229-243.
9. A. Dar, K. Sankar and M. A. Dar, "Remote Sensing Technology and Geographic Information System Modeling: An Integrated Approach Towards the Mapping of Groundwater Potential Zones in Hardrock Terrain, Mamundiyar Basin," Journal of Hydrology, Vol. 394, No. 3-4, 2010, pp. 285-295.http://dx.doi.org/10.1016/j.jhydrol.2010.08.022

10. M. Gupta and P. K. Srivastava, "Integrating GIS and Remote Sensing for Identification of Groundwater Potential Zones in the Hilly Terrain of Pavagarh, Gujarat, India," Water International, Vol. 35, No. 2, 2010, pp. 233-245. http://dx.doi.org/10.1080/02508061003664419
11. N. Hammouri, A. El-Naqa and M. Barakat, "An Integrated Approach to Groundwater Exploration Using Remote Sensing and Geographic Information System," Journal of Water Resource and Protection, Vol. 4 No. 9, 2012, pp. 717-724.http://dx.doi.org/10.4236/jwarp.2012.49081
12. S. Mukherjee, "Role of Satellite Sensors in Groundwater Exploration," Sensors, Vol. 8, No. 3, 2008, pp. 2006- 2016. http://dx.doi.org/10.3390/s8032006
13. M. Yazdi, N. Sadati, A. A. Matkan and D. Ashoorloo, "Application of Remote Sensing in Monitoring of Faults," International Journal of Environmental Research, Vol. 5, No. 2, 2011, pp. 457-468.
14. S. Ganapuram, G. T. Vijaya Kumar, M. Krishna, E. Kahya and M. Cüneyd Demirel, "Mapping of Groundwater Potential Zones in the Musi Basin Using Remote Sensing Data and GIS," Advances in Engineering Software, Vol. 40, No. 7, 2009, pp. 506-518.http://dx.doi.org/10.1016/j.advengsoft.2008.10.001
15. Al-Dababseh "An Integrated Approach to Groundwater Exploration of Northern Wadi Araba Basin Using Remote Sensing and Geographic Information System," MSc Thesis, The University of Jordan, Amman, 2003.
16. M. Alraggad and H. Jasem, "Managed Aquifer Recharge (MAR) through Surface Infiltration in the Azraq Basin/ Jordan," Journal of Water Resource and Protection, Vol. 2 No. 12, 2010, pp. 1057-1070. http://dx.doi.org/10.4236/jwarp.2010.212125
17. NRA (Natural Resources Authority, Jordan), "Geological Map of Madaba, Sheet 3153 II, Scale 1:50000," National Geological Mapping Project 1988-1990, NRA, Amman, 1992.
18. N. Al-Hunjul, "The Geology of Madaba Area," Natural Resources Authority Press, Jordan, 1995.
19. MWI (Ministry of Water and Irrigation, Jordan), 2010. http://www.mwi.gov.jo/mwi/homepage.aspx
20. MWI (Ministry of Water and Irrigation, Jordan), "Sustainable Fodder Crop Production," MWI, Jordan Valley Authority (JVA), Community Action Plan of the Badia Restoration Program (BRP), Ministry of Environment, Jordan, 2013.
21. J. R. Jensen, "Introductory Digital Image Processing: A Remote Sensing Perspective," 3rd Edition, Prentice Hall, Upper Saddle River, 2005.
22. R. A. Schowengerdt, "Remote Sensing: Models and Methods for Image Processing," 3rd Edition, Academic Press, London, 2006.
23. F. El-Baz, "Groundwater Concentration beneath Sand Fields in the Western Desert of Egypt: Indications by Radar Images from Space," Egyptian Journal of Remote Sensing and Space Sciences, Vol. 1, 1999, pp. 1-24.
24. N. S. Rao, "Groundwater Potential Index in a Crystalline Terrain Using Remote Sensing Data," Environmental Geology, Vol. 5, No. 7, 2006, pp. 1067-1076. http://dx.doi.org/10.1007/s00254-006-0280-7

CHAPTER 15

Groundwater Abstraction for Irrigation and Its Impacts on Low Flows in a Watershed in Northwest Germany

Hartmut Wittenberg

Institute of Ecology, Leuphana Universität, C13.015, Scharnhorststr. 1, Lueneburg D21335, Germany

ABSTRACT

Low flows of the Ilmenau River (1434 km^2) in northwest Germany have decreased by about 25% over the last 50 years. In the same period, moderate climate changes have taken place and annual groundwater abstractions for sprinkler irrigation have increased by up to 50 hm^3 (million m^3), with a strong variation due to the respective prevailing weather conditions. Time-series analyses with multiple regression analysis allow detecting and quantifying different influences on low flows. It is also shown that farmers allocate irrigation water volumes carefully according to seasonal precipitation and temperatures. Decline of groundwater levels in summer and the low flow situation are aggravated by the cumulative effect of higher irrigation in drier years. Groundwater recharge and recovery of the water table have been observed subsequently during the winter season.

Keywords: low flows; groundwater; irrigation; multiple regressions; Germany

1. INTRODUCTION

In spite of their high impact on environment, forestry, agriculture, waste water disposal, power plants, waterways, *etc.*, drought periods in Central Europe were seldom the focus of public attention. The hot and dry 1990s, and in particular the drought year 2003, changed this situation as it became clear that in spite of its temperate climate, Germany can be hit severely by low flows and drought [1], stimulating hydrological research, e.g.,[2]. Like other hydrological processes and water balances of river basins, low flows are not only subject to climate variability and catchment characteristics [3], but also anthropogenic impacts. Due to the superposition of various influences in hydrological data, the effect of single factors is not easily quantified. The present study aims at establishing relationships between the variation and decrease of low flows in a river in northern Germany, meteorological data, and groundwater abstractions for irrigation.

2. THE STUDY AREA

The Ilmenau River in the north German lowlands (Figure 1) is the largest river of the Lueneburg Heath and a tributary of the Elbe River, which drains into the North Sea. Daily average flows have been recorded since 1956 at the Bienenbüttel gauging station. The catchment area of 1434 km^2 corresponds roughly to the county of Uelzen. The town of Uelzen is situated in its center at approximately 53° N, 10.5° E, and 35 m a.s.l. The landscape was shaped by the Saale glaciation, with sandy and loamy soils and a slightly hilly topography with altitudes between 20 and 136 m a.s.l. With annual temperature means of around 9 °C and precipitation of about 700 mm (see also in paragraph 5), the climate is temperate and humid. Snow in winter melts at the latest in the early spring and has no direct impact on low flows in summer. Mean flow is 9.2 m^3/s (runoff depth 202 mm) and is mainly fed by the shallow groundwater aquifer. The region is characterized by agriculture with wheat, potatoes, and sugar beets as the main crops. A large sugar factory is located in Uelzen.

To ensure high crop yields in dry years, sprinkler irrigation was introduced in the 1950s [4]. Because of the low water retention capacity of the soils and sometimes scarce summer rainfalls, irrigation is necessary for agricultural production [5]. With an average annual low flow in the Ilmenau River of more than half of the mean flow, there is no indication of a serious water shortage affecting the biosphere. Nevertheless, the public is very attentive and irrigation farmers find themselves generally suspected of wasting water. The quantitative and temporal analysis of irrigation as it is actually carried out by farmers in the area is therefore a further aim of this study.

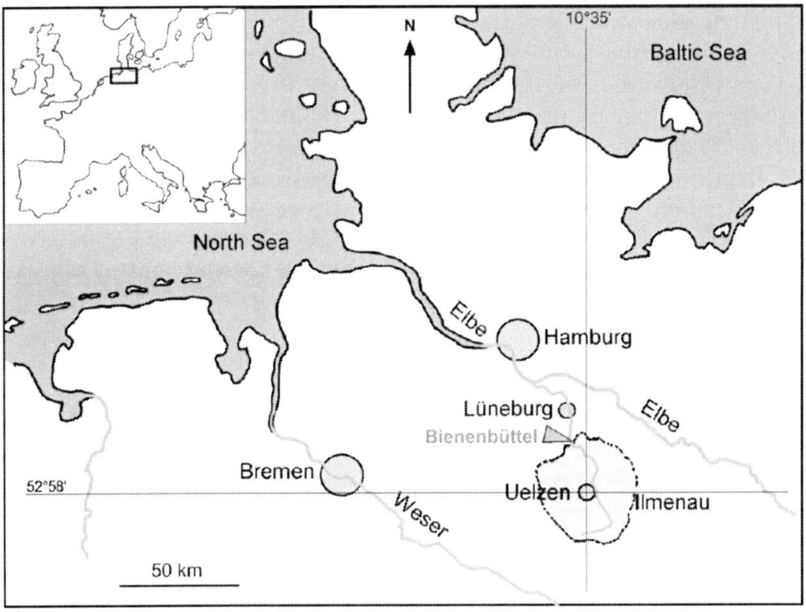

Figure 1. Map of northwest Germany with the Ilmenau River basin and Bienenbüttel gauging station.

3. IRRIGATION IN THE ILMENAU BASIN

Sprinkler irrigation introduced in the 1950s is still considered by farmers to be the most suitable and economical method for growing crops under regional conditions [6]. It is typically performed by turntable machines with rain-gun trolleys. The trolley is pulled along the irrigation track by its PE (polyethylene) hose, which is wound up by the drum. Slow rotation is provided by a turbine driven by the irrigation water. Field strips of several hundred meters in length and about 30 m width can be irrigated in one application with a typical depth of 30 mm. Figure 2 shows a unit in operation.

Water is taken from the Quaternary aquifer at a depth of about 100 m by pumping stations owned and operated by the farmers. All wells, groundwater withdrawals, and irrigated areas need permission from the local environmental and water authorities [7]. Applications must be backed by hydro-geological expertise. The irrigation water depth for the respective areas is limited to 70 mm per summer over a seven-year average, *i.e.*, higher abstractions in dry years can be compensated by lower ones in wetter summers. Nearly 700 km^2 (50%) of the county is farmland. Since the 1950s, when irrigation was introduced, irrigated areas were enlarged to about 550 km^2. The irrigation depth of 70 mm would then need a volume of 38.5 hm^3 per year. Depending on the summer weather, the actual volumes varied in the observed years between less than 10 and about 50 hm^3. For the catchment area of 1434 km^2 the maximum abstraction corresponds to a 35 mm depth, which is about 5% of mean annual basin precipitation (1955–2012) or more than a quarter of the precipitation in a very dry summer (May–October 1959: 131 mm). Effects on low flows and groundwater storage must be expected. Figure 3 shows the development of registered areas and of the groundwater abstractions for irrigation 1977–2012.

Figure 2. Sprinkler irrigation with pipe drum and rain-gun trolley.

With the objective of ensuring a sustainable management of water resources, the water and environmental authorities observe this development. The chamber of agriculture initiated an interdisciplinary study of irrigation possibilities for different climate change scenarios and adapted technologies [8].

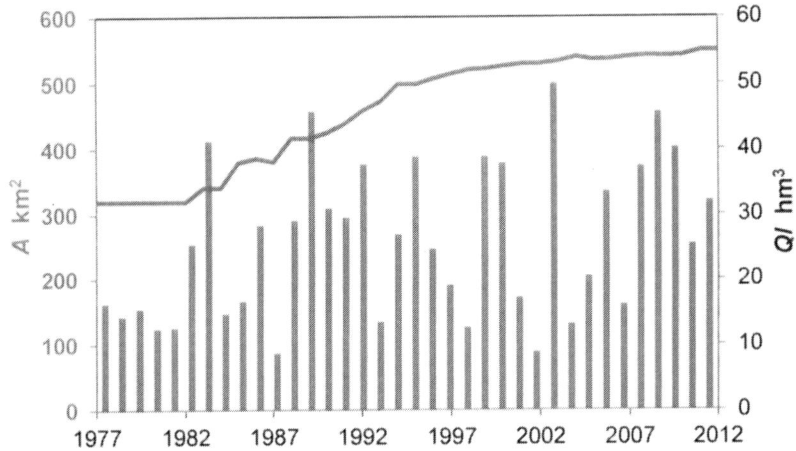

Figure 3. Development of irrigation areas and groundwater abstraction in the catchment; registration of areas and groundwater abstractions began in 1977, data: County of Uelzen.

4. LOW FLOWS, CLIMATE, AND GROUNDWATER ABSTRACTIONS

At the Bienenbüttel gauging station, daily flows have been recorded since 1956. In assessing low flows and water levels, the absolute minima of each year are of little significance. Since the damage potential of low flow events also depends on their duration, the lowest arithmetic mean LQ_{xd} of x consecutive daily flow values of every year is used as the assessment variable. The annual values LQ_{xd} are determined for durations of typically $x = 3, 7, 20$ days for each water budget year. In northern Germany, the relevant low flow events are to be expected in the summer months (May–October).

The time series 1956–2012 of annual low flows LQ_{20} at the Bienenbuettel/Ilmenau gauging station in Figure 4 reveals a significant (t-test) linear downward trend. The strong fluctuations can be interpreted as a sequence of different negative and positive trends. It seems that after a decline until the end of the 1970s, there is no further clear trend, but instead there is a high variability in annual low flows. The physical causes of these variations and inconsistencies of the low flow hydrograph can be due to both weather processes and anthropogenic influences. Mean annual basin precipitation is also depicted in Figure 4, derived by the Thiessen polygon method using the data of 13 stations in and around the catchment.

It increased statistically by about 100 mm during the observation period while evapotranspiration was stimulated by the warming of mean summer temperatures of about 1 K. In Germany, the hydrological year is divided into the winter half-year (November to April) and the summer half-year (May to October). The latter represents roughly the vegetation season. Mean values of catchment precipitation and air temperatures in the two seasons in Lueneburg are given in Table 2. Essential impacts are to be expected from groundwater abstractions for irrigation, which are recorded by water meters at all irrigation wells. Annual total values for the catchment expressed in m³/s are also shown in Figure 4.

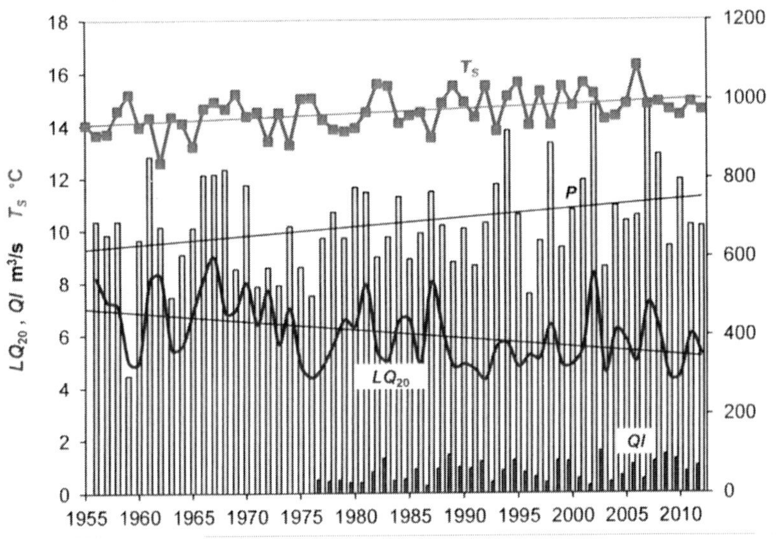

Figure 4. Annual basin precipitation *P*, summer temperature *Ts*, 20-day low flows LQ_{20}, and groundwater abstractions *QI*, Ilmenau Basin at Bienenbüttel, 1977–2012.

5. MULTIPLE REGRESSION ANALYSIS—LOW FLOWS

Multiple regression is a classic method of data analysis [9] but is rarely used in contemporary hydrology. Even state-of-the-art studies on low flows [2] mostly use regression only in its simple version and for trend analysis. Unlike the simple trend analysis of low flows which relates all changes to time, multiple regression enables the assessment of statistical associations among the response variable and primarily independent multiple covariates. The relationship established in Equation (1) assumes annual low flows:

$$LQx_t = LQ_0 + a \cdot PW_t + b \cdot PS_t + c \cdot PJ_{t-1} + d \cdot TW_t + e \cdot TS_t + f \cdot QI_t \quad (1)$$

where,

- LQx_t:→ x days low flow in the year t in m³/s;
- LQ_0:→ base value of low flow in m³/s;
- PW_t:→ winter precipitation in the year t in mm;
- PS_t:→ summer precipitation in the year t in mm;
- PJ_{t-1}:→ precipitation of preceding water year in mm;
- TW_t:→ mean air temperature, winter of year t in °C;
- TS_t:→ mean air temperature, summer of year t in °C;
- QI_t:→ groundwater abstraction for irrigation in year t in mm.

With the data set of 1977–2012, there is an over-determined system of 36 equations. Reducing the system by the Gauss-Jordan algorithm determines the seven unknowns, *i.e.*, the base value and the six coefficients a to f, according to the least squares criterion. For the 20-day period, low flows are obtained:

$$LQM_{20t} = 4.48 + 0.0030\,PW_t + 0.00094\,PS_t + 0.0033\,PJ_{t-1} + 0.022\,TW_t - 0.049\,TS_t - 0.102\,QI \quad (1a)$$

The coefficients, factors, or weights represent the sensitivity of low flow values to unit changes of the variables. Table 1 shows the correlation coefficients of the variables to one another and to the dependent variable low flow. Correlations with $|R| > 0.33$ are significant at the 95% level. In addition, the coefficients concerning the registered irrigation areas AI are listed. These values are expected to have an influence on the assessment of applied water volumes in Section 7.

The overall correlation coefficient between the $n = 36$ calculated values and the data is $R = 0.865$ ($R^2 = 0.75$). As expected, it is considerably higher than all partial coefficients and thus demonstrates the advantage of multiple regression over simple regression analysis. The coefficient of variation, *i.e.*, the average deviation between data and values by Equation (1a), is 9%.

Table 1. Partial correlation coefficients of variables, time series 1977–2012.

variable	*PW*	*PS*	*PJt-1*	*TW*	*TS*	*QI*	*AI*
PW	1						
PS	0.127	1					
Pt-1	0.194	−0.092	1				
TW	0.490	0.168	−0.020	1			
TS	0.186	−0.194	0.181	0.336	1		
QI	0.033	**−0.618**	0.320	0.199	**0.472**	1	
Ai	0.096	0.259	0.312	0.222	0.310	0.324	1
LQ_{20}	0.265	**0.578**	0.116	−0.059	−0.317	**−0.735**	−0.220

The groundwater abstraction for irrigation *QI* has a significant negative correlation with summer precipitation *PS* and a positive one with summer temperature *TS*, an indication for disciplined and frugal water allocation by the farmers. Higher rainfall reduces the need for irrigation, while higher temperatures increase evaporation and irrigation requirements.

As indicated by the correlation coefficients, the information of the values *PS* and *TS* with relevance for low flows seems to be largely included in the time series *QI*. The influence of winter temperature *TW* on low flows is negligible (see Equation (1a)). Therefore, a further multiple regression for LQ_{20} with variables as in Equations (1) and (1a), but without *PS, TW,* and *TS* was performed:

$$LQ_{20t} = 4.22 + 0.0032\, PW_t + 0.0033\, PJ_{t-1} - 0.1092\, QI \tag{1b}$$

The correlation coefficient between the observed and calculated low flows is $R = 0.863$ and the coefficient of variation is $CV = 9.2\%$. Thus, the result is practically the same as for Equation (1a) and the assumption is confirmed.

Table 2 contains the mean values from 1977 to 2012 of the input variables in the first line. By multiplying with their coefficients or weights according to Equations (1a) and (1b), respectively, their contribution to the total value of the low flow is obtained (following lines).

First of all, the coefficients are statistical weights that do not necessarily represent partial physical flows. However, since the coefficients of the three parameters used in both Equations (1a) and (1b) have similar to equal values, it is concluded that the computed partial flows are not purely mathematical regressive values but do stand for real water flows. It is worth noting the high contributions of winter and the previous year's precipitation to low flows because of their importance for groundwater recharge.

Table 2. Coefficients and partial flows of LQ_{20} according to Equations (1a) and (1b).

variable	LQ_0 m³/s	PW mm	PS mm	$Pj\text{-}1$ mm	TW °C	TS °C	QI mm	ΣLQ_{calc} m³/s
mean values		336	376	708	3.79	14.7	18.1	
coefficients 1a		0.0030	0.00094	0.0033	0.022	−0.0494	−0.1023	
flow	4.48	1.01	0.35	2.34	0.08	−0.73	−1.85	5.68
coefficients 1b		0.0032	-	0.0033	-	-	−0.1092	
flow	4.22	1.08	-	2.34	-	-	−1.98	5.68

In this study, one focus is on the influence of groundwater abstraction. As shown in Table 2, the respective average annual reduction of low flow LQ_{20} is about 1.9 m³/s or 25%. Observed low flows from 1977 to 2012 and values computed by Equation (1b) are depicted in Figure 5. Annual amounts of groundwater abstraction are shown for comparison. By omitting the last term of Equation (1b), the influence of groundwater abstraction QI is eliminated and a time series of hypothetical uninfluenced low flows obtained. The resulting hydrograph is also shown in Figure 5.

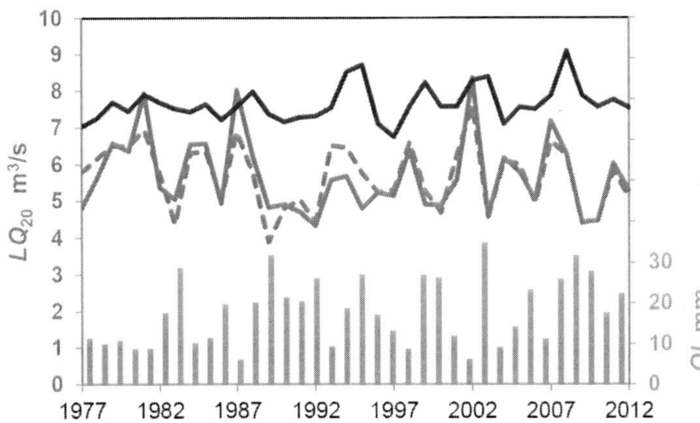

Figure 5. Observed and computed (Equation (1b), dashed line) annual low flow values, computed low flows uninfluenced by abstractions (above, dark blue), and groundwater abstractions *QI*.

As reflected by the above equations, the reduction of annual low flows is higher in drier years with more groundwater withdrawal than in moderate years. This is also shown by the empirical probability curves in Figure 6. These curves were determined according to the method of "plotting positions".

Data and the computed uninfluenced low flows are respectively sorted in decreasing order, and statistical return intervals T are determined for every value according to Equation (2):

$$T = (n+1)/(n+1-m) \tag{2}$$

with

- T:→ return interval in years;
- n:→ number of data, length of data set;
- m:→ ranking from 1 (greatest value) to n (lowest value).

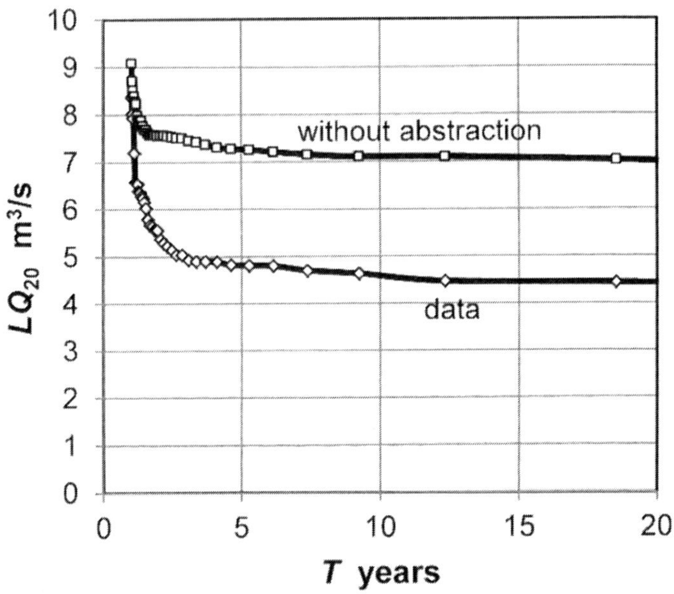

Figure 6. Frequency analysis of 20-day low flows.

The difference between probable low flows with and without groundwater abstraction increases with the recurrence interval T. The regressions and calculations described above were also carried out for time series of low flows of three- and seven-day durations. Comparable results and relationships following the same pattern have been obtained.

6. GROUNDWATER IMPACTS

Low flow occurs when direct runoff has ceased and the river only carries baseflow, *i.e.*, groundwater outflow from the adjacent aquifer. Therefore, the lowest annual low flow coincides with the lowest groundwater level. Groundwater withdrawal lowers the level but also reduces the outflow, preserving, in a sense, the aquifer. At the end of the irrigation period, with lower evapotranspiration in the autumn, rainfall leads to a recharge and recovery of the aquifer. The respective hydrographs of monthly precipitation *P* and abstraction *QI* and daily groundwater levels *GWL* for a sub-catchment of the Ilmenau Basin in the summer season of 1994 are shown in Figure 7. Daily values of mobile groundwater storage *GWS* obtained in another study [10] are also depicted. These values were computed from baseflow separated from total flow by using a nonlinear reservoir algorithm.

In the same study, it was found that, under regional conditions, only about 70% of irrigation water is lost for the groundwater, since 30% infiltrates back or prepares the soil for infiltration of the next rainfalls. These findings match the results obtained in a different approach [11].

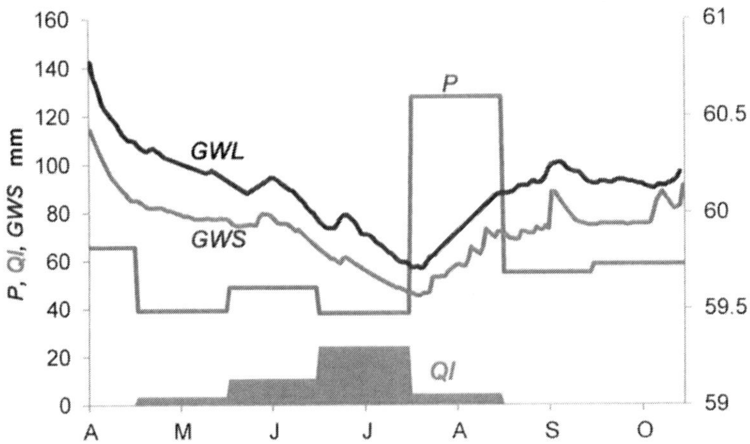

Figure 7. Monthly rainfall *P*, groundwater abstraction *QI*, daily values of groundwater level *GWL*, and computed mobile groundwater storage *GWS*, Eisenbach catchment, summer 1994.

While low flows are lowered by groundwater abstractions, the long-term volumes and water levels of the aquifer are little affected. Depressions are only observed at some places at higher parts of the watershed. Water quantity does not pose substantial problems. However, the preservation of long-term groundwater quality by a careful use of fertilizers and agrochemicals remains an important issue [12].

7. MULTIPLE REGRESSION—GROUNDWATER ABSTRACTION

Except for some smaller test areas, groundwater abstraction data are available per irrigation season (April–October) of every year, while precipitation and temperature data are available monthly and per season. As given in Table 1, groundwater abstraction *QI* is significantly correlated with summer rainfall *PS*, summer temperature *TS*, and irrigated areas *AI*. The corresponding regression had the following result:

$$QI_t = -11.6 - 0.0743\,PS_t + 2.56\,TS_t + 0.000434\,AI_t \qquad R = 0.82 \tag{3a}$$

While the correlation coefficient is high, the coefficient of variation, *i.e.*, the mean deviation of calculated values from abstraction data, is 26% and thus not quite satisfying. This is mainly due to a fuzziness of the term summer rainfall (April–October). For irrigation requirements and management, it makes a big difference whether rains fall early in the growing period rather than during or even after harvest. A first step to improving the relationship was the substitution of the values of total summer precipitation (April–October) with those of the main irrigation period (May–August). A correlation coefficient of R = 0.86 and a mean deviation of 23% were obtained.

However, rainfall depth has more or less importance depending on the month of the irrigation period. Monthly data of groundwater uptake in some smaller test areas in the watershed [13] from 1990 to 2011 showed the following temporal distribution of volumes: April 3%, May 17%, June 30%, July 29%, August 15%, September 4%, and October 2%. Monthly rainfalls were multiplied by the percentages. The sum of these products for every year represents a weighted mean value for replacing the total summer rainfall. Indeed, the partial correlation of groundwater abstraction with the new variable is closer (R = −0.73) than with summer rainfall (R = −0.62; Table 1). The improved relationship is:

$$QI_t = -9.02 - 0.419\,Pwei_t + 2.48\,TS_t + 0.0004\,AI_t \qquad R = 0.89 \tag{3b}$$

The result is shown in Figure 8. The annual weighted means of seasonal rainfall P_{wei} are plotted for comparison. The mean deviation between data and computed values is 21%. This seems high, but it must be seen in the context that the variation coefficient (standard deviation/mean) of the annual abstractions is more than double (47%). It is also noticeable that this variation is essentially preserved in the computed series of *QI* (42%), though they are based on variables with much lower variation coefficients, P_{wei} (23%), *TS* (4%), and *AI* (19%). Other than with single regression, where variation is determined (and damped) by the independent variable, this is possible by the superposition effects of multiple regression.

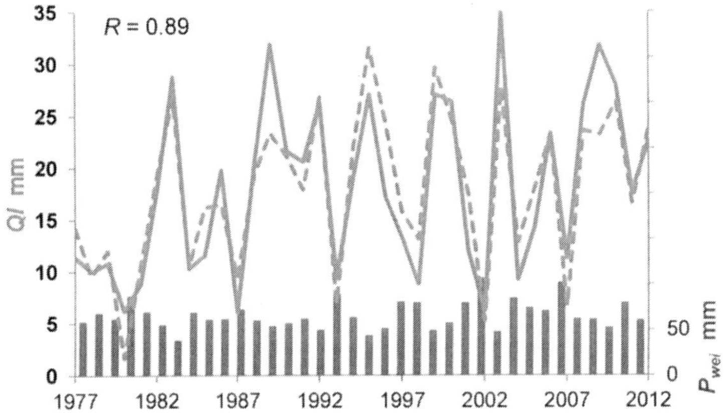

Figure 8. Recorded and computed (Equation (3b), dashed line) annual groundwater abstractions *QI*, weighted means of seasonal rainfall P_{wei}.

8. CONCLUSIONS

Multiple regression analysis allows distinguishing and quantifying the influence of different climatic and man-made variables such as precipitation, temperature, and groundwater withdrawal on annual low flows. It was found that groundwater abstraction for irrigation in the Ilmenau Basin accounts for an average decrease of low flows of about 25%. The groundwater levels and long-term volumes recover during the winter season. The annual abstraction volume is highly correlated with rainfall depth in the vegetation season, summer air temperature, and irrigation area. It is thus evident that groundwater abstractions in the Ilmenau Basin are based closely on actual irrigation requirements.

ACKNOWLEDGMENTS

Thanks go to the Water Authority Landkreis Uelzen (Claudia Boik and Carolin Kuschel) for groundwater abstraction data and advice, to the water autority NLWKN Lüneburg (Daniel Gauglitz) for flow data and to Ulrich Ostermann, District Association of Water and Soil Associations Uelzen, for advice and communication about the temporal water distribution.

REFERENCES

1. Droughts Germany. Available online: http://www.climateadaptation. eu/ germany/ droughts/ (accessed on 20 June 2015).
2. Hisdal, H.; Stahl, K.; Tallaksen, L.; Demuth, S. Have streamflow droughts in Europe become more severe or frequent? *Int. J. Climatol.* 2001, *21*, 317–333.
3. Van Loon, A.F.; Laaha, G. Hydrological drought severity explained by climate and catchment characteristics. *J. Hydrol.* 2015, *526*, 3–14.
4. Brühmann, G. Untersuchung des Wasserhaushaltes im Uelzener Becken (Water balance investigation in the Uelzen Basin). *Wasser Boden* 1982, *2*, 51–57.
5. Riediger, J.; Breckling, B.; Nuske, R.; Schröder, W. Will climate change increase irrigation requirements in agriculture of Central Europe? A simulation study for Northern Germany. *Environ. Sci. Eur.* 2014, *26*.
6. FAO Sprinkler Irrigation. Available online www.fao.org/docrep/s8684e/s8684e06.htm (accessed on 20 June 2015).
7. Ostermann, U. Grundwasserentnahme für landwirtschaftliche Beregnung aus Sicht einer Unteren Wasserbehörde (Groundwater abstraction for agricultural irrigation from the perspective of a water authority). *Mitt. NNA* 1998, *3*, 47–50.
8. LWK, Landwirtschaftskammer (Chamber of Agriculture) Niedersachsen. *No Regret—Genug Wasser für die Landwirtschaft?! (Enough Water for Agriculture?!)*; Schulz, E., Ed.; LWK Niedersachsen: Uelzen, Germany, 2008.
9. Gauss, C.F. *Theoria Combinationis Observationum Erroribus Minimis Obnoxiae (Theory of Combination of the Observations Subject to Smallest Errors)*; H. Dieterich: Göttingen, Germany, 1823; pp. 1–58.
10. Wittenberg, H. Effects of season and man-made changes on baseflow and flow recession: Case studies. *Hydrol. Process.* 2003, *17*, 2113–2123.
11. Wessolek, G.; Renger, M. Einfluss der Beregnung auf den regionalen Wasserhaushalt (Influence of irrigation on the regional water balance). *Wasser Boden* 1987, *3*, 112–114.
12. Lewandowski, J.; Meinikmann, K.; Nützmann, G.; Rosenberry, D. Groundwater—The disregarded component in lake water and nutrient budgets. Part 2: effects of groundwater on nutrients. *Hydrol. Process.* 2015, *29*, 2922–2955.
13. Ostermann, U.; Water and Soil Association, Uelzen, Germany. Personal Communication, 2014.

CHAPTER 16

Impact of agrochemicals on ground water

Prashant Kumar Shrivastava

Department of Geology, Govt. V.Y.T. P.G. Autonomous College, Durg 491 001, C.G. India

ABSTRACT

Pesticide use has grown because not only must our exploding population be supplied with food, but crops and food are grown for export to other countries. India has become one of the largest producers of food products in the world, partly owing to our use of modern chemicals (pesticides) to control the insects, weeds, and other organisms that attack food crops. But, as with many things in life, there's a hidden cost to the benefit we get from pesticides. We've learned that pesticides can potentially harm the environment and our own health. Water plays an important role here because it is one of the main ways that pesticides are transported from the areas where they are applied to other locations, where they may cause health problems. Pesticide contamination of groundwater is a subject of National importance because groundwater is used for drinking purpose by about 50 percent of the Nation's population. This especially concerns people living in the agricultural areas where pesticides are most often used, as about 95 percent of that population relies upon groundwater for drinking water. Before the mid-1970s, it was thought that soil acted as a protective filter that stopped pesticides from reaching groundwater. Studies have now shown that this is not the case. Pesticides can reach water-bearing aquifers below ground from applications onto crop fields, seepage of contaminated surface water, accidental spills and leaks, improper disposal, and even through injection waste material into wells.

Keywords: Pesticide, Insects, Contamination, Ground water, Aquifer

1. INTRODUCTION

The term "pesticide" is a composite term that includes all chemicals that are used to kill or control pests. In agriculture, this includes herbicides (weeds), insecticides (insects), fungicides (fungi), nematocides (nematodes), and rodenticides (vertebrate poisons). In areas where intensive monoculture is practiced, pesticides were used as a standard method for pest control. Unfortunately, with the benefits of chemistry have also come disbenefits, some so serious that they now threaten the long-term survival of major ecosystems by disruption of predator-prey relationships and loss of biodiversity. Also, pesticides can have significant human health consequences. Agricultural use of pesticides is a subset of the larger spectrum of industrial chemicals used in modern society. The American Chemical Society database indicates that there were some 13 million chemicals identified in 1993 with some 500 000 new compounds being added annually. The amount of pesticide that migrates from the intended application area is influenced by the par ticular chemical's properties: its propensity for binding to soil, its vapor pressure, its water solubility, and its resistance to being broken down over time. Factors in the soil, such as its texture, its ability to retain water, and the amount of organic matter contained in it, also affect the amount of pesticide that will leave the area. Some pesticides contribute to global warming and the depletion of the ozone layer. Protecting water quality is a top environmental issue (Paningbatan et al. 1993).

2. PESTICIDES AND WATER QUALITY

Using pesticides effectively while maintaining water quality presents an important challenge. As citizens, we must recognize the significant role of pesticides in maintaining a high quality of life. We must acknowledge that the effective production of food and fiber relies on pesticides to control weeds, insects, and plant diseases that interfere with the growth, harvest and market ability of crops. As pest control operators and homeowners - rural as well as urban - we must acknowledge the importance of pesticides in controlling pests in our homes, restaurants, hospitals, parks, ornamental plantings, golf courses, etc. But at the same time we must be aware that pesticide applications can affect water quality. Human and environmental health may be threatened when excessive concentrations of pesticides enter surface or ground water.'Pesticides and water quality' is a complex subject,both technically and politically; but if current and future expectations for community life, agriculture, industry, wildlife and natural habitats are to be met, input from an educated public is essential. A basic knowledge of the subject is important to allow informed participation in the ongoing debate (Hodgson et al. 1996)

Ground water is a widely distributed natural resource found beneath the earth's surface. Many people have the mistaken impression that ground water occurs as underground rivers and reservoirs. However, most ground water occurs in tiny voids (spaces) between grains of sand and gravel, between silt and clay, or in cracks and fractures in bedrock.

3. GEOLOGY OF GROUND WATER

The geology of a particular location dictates the depth and volume of ground water. Usable ground water available to supply wells and springs comes from geologic formations called aquifers, which may be shallow (near the earth's surface) or very deep (hundreds of feet below the surface). As a general rule, fresh water aquifers tend to lie 60-300 feet below ground. Aquifers are composed of various materials such as rock, sand, and gravel that reflect local geology. Some consist of unconsolidated (loose) deposits of sand, clay, silt, or gravel containing water in the voids between particles and rock fragments. Other aquifers occur as cracks in bedrock or consolidated (solid) materials such as igneous rock (granite, basalt), sedimentary rock (limestone, siltstone, sandstone), or metamorphic rock (slate). Aquifers are characterized as either confined or unconfined. Confined aquifers lie below a layer of less permeable clay or rock-a confining layer-which greatly slows the vertical movement of water. The water in confined aquifers can be recharged from water that moves into the waterbearing zone from distant areas where there are no confining layers. Unconfined aquifers do not have a confining layer and are 'open' to water moving down from surfaces directly above. The water surface of unconfined aquifers-the water table-fluctuates with changes in atmospheric pressure, rainfall and other factors. Unconfined, unconsolidated aquifers are particularly vulnerable to contamination because, typically, they are quite shallow and surface water can infiltrate quickly down to the water table (ground water) in certain soils. (Damalas et al. 2011).

4. DOWNWARD MOVEMENT OF WATER

Between top soil and water-saturated soils, voids of unconsolidated materials fill with water and air, forming the vadose (unsaturated) zone. The portion of the vadose zone near the soil surface is where plants root, vegetation decays, and animals burrow; it is in this area that most terrestrial plants and soil organisms reside. The lower portion of the vadose zone hosts less biological activity. Precipitation either runs off sloping land or infiltrates only the upper few inches of soil, then percolates downward and permeates the upper vadose zone. As water enters soil voids, a variety of physical processes pull it into the vadose zone, replacing air. The water table is defined as the area that separates the vadose and saturated zones. Water below the water table is ground water. All soils can store water in voids. A soil's ability to store and transfer water downward in saturated or unsaturated conditions is a function of numerous interrelated processes and features. For example, the nature of soil particulates and the way they aggregate influence features such as porosity and how water is attracted to soils. Soils with small voids can store more water than those with larger voids. Under saturated flow, porosity and the pull of gravity greatly influence water movement. Under unsaturated conditions, attraction of water to soil surfaces (matrix potential), movement along a maze-like flow path, and very small pores (capillaries) influence water movement. Natural ground water movement is often (but not always) in the direction defined by local topography. Horizontal flow of ground water generally is slow and is measured in inches per day or feet per year, depending on the porosity and the permeability of the materials making up the aquifer. (Medina 1996)

In most cases, a period of a few months to several years is required for a pesticide to leach considerable distances through soil to reach ground water. Therefore, a pesticide generally needs to be both persistent and mobile to reach most aquifers. The extent of pesticide movement through soil depends on the degree of interaction between the pesticide and soil particles, soil microorganisms, and weather. The amount of pesticide that will reach low soil depths varies dramatically with slight environmental fluctuations, making estimation difficult. A judgment can be made on the overall likelihood of ground water contamination by comparing the mobility and persistence of a chemical to those of similar pesticides previously detected in ground water at multiple use sites. Many factors affecting the environmental fate of a pesticide are not well understood. Site-specific behavior of pesticides in soils cannot be predicted unless actual field data are available for comparison. A Health Advisory Level (HAL) is considered the maximum level of a drinking water contaminant, in milligrams per liter (parts per million, or ppm) or micro- grams per liter (parts per billion, or ppb), that would not be expected to cause noncarcinogenic health risks over a given duration of exposure.

5. RISK ASSESSMENT OF PESTICIDES FOR WATER QUALITY CONCERNS

Understanding the toxicological properties of a drinking water contaminant is necessary when calculating a health advisory. Toxicological profiles for pesticides generally are derived from animal tests because human testing is not possible. Data from human epidemiological studies can be used, but such data generally are unavailable. Use of pesticides in developing countries is extremely variable, from nil in large parts of Africa, to extremely heavy dosage in intensive agricultural areas of Brazil and plantations of Central America. In their review of the limited research literature on pesticide use and impacts in Africa, Calamari and Naeve (1994) conclude that, "The conconcentrations found in various aquatic compartments, with few exceptions are lower than in other parts of the world, in particular in developed countries which have a longer history of high pesticide consumption and intense use. Generally, the coastal waters, sediments and biota are less contaminated than inland water environmental compartments, with the exception of a few hot spots." The effects of past and present land-use practices may take decades to become apparent in groundwater. When weighing management decisions for protection of groundwater quality, it is important to consider the time lag between application of pesticides and fertilizers to the land and arrival of the chemicals at a well. This time lag generally decreases with increasing aquifer permeability and with decreasing depth to water. In response to reductions in chemical applications to the land, the quality of shallow groundwater will improve before the quality of deep groundwater, which could take decades. (Castaneda et al. 1996)

Pesticides are mostly modern chemicals. There are many hundreds of these compounds, and extensive tests and studies of their effect on humans have not been completed. That leads us to ask just how concerned we should be about their presence in our drinking water. Certainly it would be wise to treat pesticides as potentially dangerous and, thus, to handle them with care. We can say they pose a potential danger if they are consumed in large quantities, but, as any experienced scientist knows, you cannot draw factual conclusions unless scientific tests have been done. Some pesticides have had a designated Maximum Contaminant Limit (MCL) in drinking water set by the U.S. Environmental Protection Agency (EPA), but many have not. Also, the effect of combining more than one pesticide in drinking water might be different than the effects of each individual pesticide alone. It is another situation where we don't have sufficient scientific data to draw reliable conclusions (Helling et al. 1996).

Protection of ground and surface water quality is critical to economic viability, as well as human health and environmental quality. Although pesticides are essential in the production of an adequate, economical food supply, rural (agricultural) as well as urban uses loom as possible sources of water contamination. Detection of pesticides in water aroused public interest in the environmental impact of agricultural chemicals; and the resulting heightened concern is reflected in strict legislation which impacts the pesticide industry significantly.

6. CONCLUSIONS

In the interest of minimizing risks associated with pesticides, significant public resources have been allocated for the development and implementation of rational, pesticide use policies based on solid scientific evidence. Compliance involves extensive, detailed, expensive laboratory research and field studies to determine the behavior and environmental fate of pesticides-that is, in deriving the solid scientific evidence and it follows that manufacturers must commit significant financial resources to product development en route to the marketplace. A pesticide's route and rate of entry into the environment, as well as its degradation characteristics, are key to understanding and predicting its potential impact on surface and ground water. Preregistration research data also play a significant role in determining use pattern and hazard statement language for the pesticide label. Research findings also influence the stringency of post-registration monitoring program.

REFERENCES

1. Castaneda, A.R. and Bhuiyan, S.I. 1996. Groundwater contamination by ricefield pesticides and some influencing factors. Jour. Environ. Sci. Health. A 31(1) : 83- 99.
2. Damalas, Christos A. and Ilias G. Eleftherohorinos, 2011. Pesticide exposure, safety issues and risk assessment indicators. International Journal of Environmental Research and Public Health. Web of Science.
3. Helling, C.S. and Gish, T.J. 1986. Soil characteristics affecting pesticide movement in groundwater. In: Evaluation of Pesticides in Groundwater, W.J. Garmer, R.C. Honeycutt and H.N. Niggs (Eds.). ACS Symposium Series, American Chemical Society, Washington, D.C., United States.
4. Hodgson, Ernest, Levi and Patricia, E. 1996. Pesticides: An important but underused model for the environmental health sciences. Environmental Health Perspectives. Supplement 1, Vol. 104. Academic Search Premier.

5. Medina, J.R., Calumpang, S.M.F. and Medina, M.J.B. 1991. Insecticide residues in selected well waters in Calamba and Calauan, Laguna. The Philippine Agriculturist. 74 (2) : 195-206.
6. Paningbatan, E.P. and Medina, J.R., Calumpang, S.M.F. and Medina, M.J.B. 1993. Movement of monocrotophos and endosulfan in saturated soil. The Philippine Agriculturist. 76 (3) : 259-268.

CHAPTER 17

Multifactorial Fuzzy Approach for the Assessment of Groundwater Quality

Natarajan Venkat Kumar, Samson Mathew, Ganapathiram Swaminathan

Civil Engineering Department, National Institute of Technology, Tiruchirappalli, S. India

ABSTRACT

This paper describes the utility of fuzzy Simulink model to assess the groundwater quality levels in Tiruchirappalli city, S. India. Water quality management is an important issue in the modern times. The study aimed at examining the influence of multiple parameters of ground water on potable quality. The data collected for Tiruchirappalli city have been utilized to develop fuzzy Simulink approach. This is illustrated with seventy nine ground water samples collected from Tiruchirappalli Corporation, S. India. The chara cteristics of the groundwater groundwater for this plain were monitored during the years 2006 and 2008. The quality of groundwater at several established stations within the plain were assessed using Fuzzy simulation. The results of the calculated fuzzy logic Simulink model and the monitoring study have yielded good agreement. Groundwater quality for potability indicated high to moderate water pollution levels at Srirangam, Ariyamangalam, Golden Rock and K. Abisekapuram depending on factors such as depth to groundwater, constituents of groundwater and vulnerability of groundwater to pollution. Fuzzy logic simulation approach was a practical, simple and useful tool to assess groundwater quality. This approach was capable of showing the water quality assessment for drinking on fuzzy Simulink model

Keywords: Groundwater, Potability, Fuzzy Simulation, GIS, Tiruchirappalli

1. INTRODUCTION

The quality, quantity and availability of drinking water are one of the most important environmental, social and political issues at global level. There are quality variation specifications of water for its intended use. Groundwater chemistry is affected from many factors such as movement through the rocks, recycling by irrigation practices, natural or artificial recharge and discharge.

Therefore, the determination of water quality is of primary prerequisite for public, agricultural and industrial supplies. Monitoring of water quality and qualitative decision-making on the basis of a data is challenge for environmental engineers and hydrologists as every step from sampling to analysis contains uncertainties. The regulatory limits for various pollutants/contaminants in drinking water proposed by various regulatory bodies like Bureau of Indian Standards (BIS) and World Health Organization (WHO) [1,2] are having limitations due to variation in intake to water by individuals during various seasons throughout the year. A prescribed limit from any regulatory body contains uncertainties as these are the extrapolated values from the data either from animal experiments or very trivial epidemiological studies [3-5].

The applications of artificial intelligence techniques have been used to convert human experience into a form of understandable by computers. Fuzzy logic provides a powerful and convenient formalism for classifying environmental conditions and for describing both natural and anthropogenic changes. Whereas traditional indices are based either on crisp sets with discontinuous boundaries between them, or on continuous variables whose values are only meaningful to experts, fuzzy sets make it possible to combine these approaches. From fuzzy logic it can be used to classify and quantify environmental effects of a subjective nature and it even provides formalism for dealing with missing data. The fuzzy memberships developed in this paper were used as water quality indices, but it is also possible to "defuzzify" them and to obtain a more traditional type of index [6].

A new groundwater modeling technique was developed in which fuzzy set theory is combined with finite difference modeling methods. Additionally, a rule based fuzzy-set approach to risk analysis of nitrate contaminated groundwater was presented in the study of Dahab [3]. Recently, neuro fuzzy techniques have been applied to predict groundwater vulnerability using GIS [4,7,8].

In this present paper, a fuzzy Simulink system was developed to assess the groundwater quality of Tiruchirappalli city corporation at previously selected seventy nine sampling stations. The applied water quality assessment system involves the selection of water quality parameters and index values to from BIS and WHO standards. Ranges of values were assigned to these parameters to form the input Membership Functions (MBF). In this application, twelve input variables, namely as pH, Ec, Ca^{++}, Mg^{++}, Na^{+}, Cl^{-}, Alkalinity, Total Hardness, SO_4, F, NO_3, and Coli forms were used to describe Water Quality Index (WQI) [9]. The input membership functions of the selected parameters have three linguistic terms defined as Acceptable, Non Acceptable and Rejected levels of groundwater.

The groundwater hydrochemistry records of the study area were used for the preparation of water quality maps. These maps were obtained by geostatistical (Kriging) methodology and the results are presented in the form of equal ion concentration lines [10]. The MATLAB V. 2008 (a) software was also used to analyse the data.

2. STUDY AREA, MATERIALS AND METHODS

2.1. Study Area

The Base map of Tiruchirappalli city was drawn from Survey of India Topo sheets Nos. 58 J/9, 10, 13 and 14 and satellite imagery (IRS-1C and LISS III) is lies between 10° 48' 18" North: 78° 41' 7" East. The general topology of Tiruchirappalli is flat and lies at an altitude of 78 m above sea level. Tiruchirappalli is fed by the rivers Cauvery and Kollidam. There are reserve forests along the river Cauvery. Golden Rock and the Rock Fort are the prominent hills. The southern/south-western part of the district is dotted by several hills which are thought to be an offset of the Western Ghats Mountain range and the soil is considered to be very fertile. For the sample collection, seventy nine bore well locations were identified. These locations were identified in such a way that the bore wells were evenly distributed over the study area.

The water samples were collected for periods between March 2006 and December 2008. The water from these bore wells were used for drinking, house hold utilities and bathing by the residents. The Laboratory tests were conducted on these samples for 16 different physicochemical potable water quality parameters as per the standard procedure [1,2,11] criteria are adopted for testing these samples.

2.2. Thematic Maps

The base map data used for the study included digitized data sets originally developed by Survey of India, the Tiruchirappalli city corporation. The data sets were originally digitized from 1:20,000 scale topographic maps. The work maps were prepared from 1 : 20,000 scale topographic paper maps using AutoCAD, Arc GIS 9.2 and Surfer V.7.

The groundwater hydrochemistry records of the study area were used for the preparation of maps. These maps are obtained by geostatistical (Kriging) methodology and the results were presented in the form of equal ion concentration lines [10,12,13].

The groundwater quality data were used as the hidden layer for the preparation of base maps. These features were the boundary lines between mapping units, other linear features (streets, rivers, roads, etc.) and point features (bore well points, etc.). The contours were developed for pH, EC, Cl^-, Na^+, Ca^{++}, Mg^{++}, Total Hardness, Alkalinity, F, SO_4^-, Coli form and NO^{-3} for the seasonal conditions of the study period between 2006 and 2008.

The monitoring and sampling program was initiated in 2006 and finalized the year 2008. A total of seventy nine monitoring stations were established of them represented groundwater conditions. The groundwater stations had different depths to groundwater. The sampling locations of all the stations are shown in **Figure 1**. A total of seventy nine separate ground water quality monitoring sessions were realized during the study period during the months of June, August, October and November of the year 2006-2008 and March, June and October of the year 2006-2008.

2.2.1. Potable Water Quality Maps

The data used for the mapping water quality assessment for potability were developed from the laboratory water quality analysis. Data for these studies were based on the sampling conducted by the first author for groundwater samples collected from predetermined locations of existing bore wells in Tiruchirappalli city. The data were linked to the sampling bore well locations using geodata base creation of Arc GIS 9.2 and surfer software.

The decision on the water quality assessment for potability gives that the water is desirable, acceptable and not acceptable as per the guidelines from BIS and WHO [1, 12] regulatory bodies. But, in the border line cases of water quality parameters, it becomes a Herculean task as different types of uncertainties are involved at various part of experimental and measurement process right from sampling, sample storage, processing and analysis. The sets of the monitored data and limits should not be as crisp set, but as fuzzy sets. One way of avoiding the difficulty in uncertainty handling in water quality assessment is to introduce a margin of safety or degree of precaution before applying a single value to drinking water quality standards as the same technique was also used by other workers in the field of environmental science [14-16]. These methodologies based on fuzzy set theory were utilized with real environmental water quality assessment to handle the uncertainties in imprecise environment in decision-making on the potability water quality can be handled. Keeping the importance of uncertainty handling in the drinking water quality assessment and versatility of the fuzzy set theory in decision-making. An attempt was made to classify the under ground water from Tiruchirappalli corporation, South India for the drinking purposes.

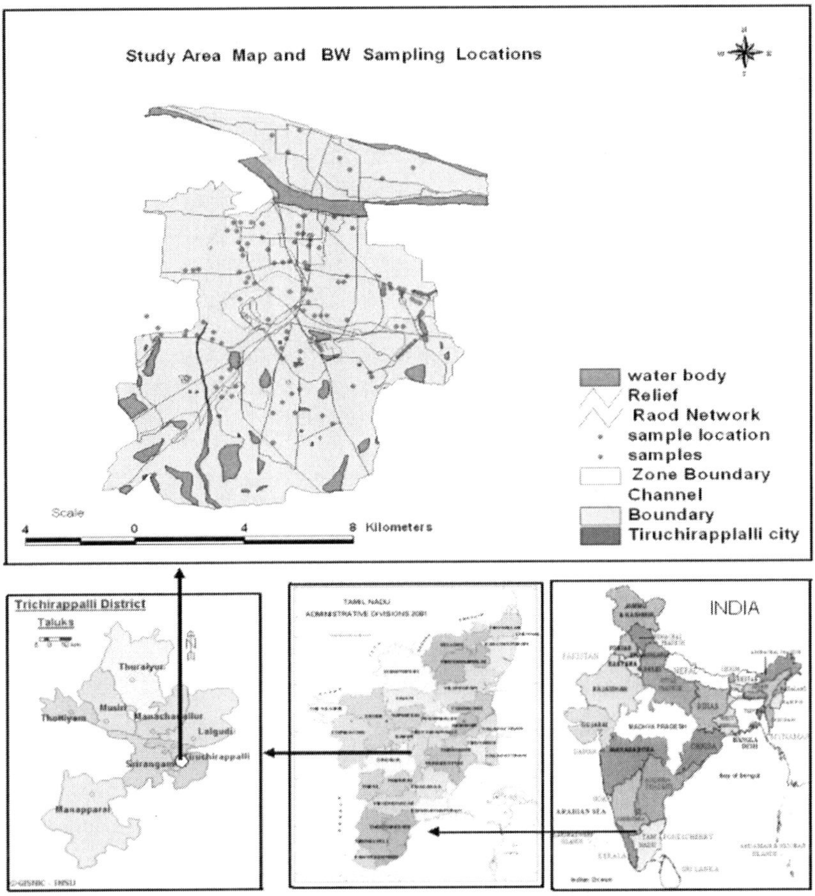

Figure 1. Study area map with sampling locations.

2.3. Fuzzy Sets

The theory of fuzzy sets was first introduced by [17] to model uncertainty in subjective information. Fuzzy sets are defined as sets whose members are vague objects. Data can generally be received in terms of linguistic judgments and beliefs (natural language), which can then be converted to the form of fuzzy sets in order to provide a base for logical and mathematical reasoning [18]. A fuzzy set is represented by a membership function defined on the universe of discourse. The universe of discourse is the space where the fuzzy variables are defined.

The membership function gives the grade, or degree, of membership (m) within the set, of any element of the universe of discourse. The membership function maps the elements of the universe on to numerical values in the interval (0, 1). A membership function value of zero implies that the corresponding element is definitely not an element of the fuzzy set, while a value of unit means that the element fully belongs to the set. A grade of membership in between corresponds to the fuzzy membership to set [13].

2.3.1. Fuzzy Rule Evaluation

Fuzzy rules appear no different to standard rules. They take the familiar form "IF x is a, THEN y is b", where x and y are linguistic variables, and where a and b are linguistic values. Under classical logic, the THEN implication is true of the IF condition, known as the antecedent, is evaluated as true. For fuzzy rules, the implication is set to be true to the same degree as the antecedent. Considering two linguistic variables length and width, and the rule "IF length is a, THEN width is b", If a is instantiated to a value then b is assigned to the value such that the membership of b is the same as that for a, each to their respective fuzzy sets.

This process is known as monotonic selection. Antecedents with more than one statement are not a problem as the earlier rules for fuzzy union (disjunction, OR) and intersection (conjunction, AND) are applied. Similarly for consequents with multiple parts, the resultant antecedent membership is applied to all parts of the consequent.

2.3.2. Matching between Fuzzy Values

The fuzzy number for field data (A) on parameters and the fuzzy numbers (A') characterizing linguistic terms are matched together to arrive at a measure called a Degree of Match (DM) defined by: Erosy et al. (1992), [13]. DM ff (AA') = ∫ μ A ∩ A' (x) dx / ∫ μA' (x) S dx, x ÎX. in which x denotes the universe, and μA ∩ A' , x is membership grade for A∩A'.

Furthermore, if A and A' are the discrete possibility distributions the measure is defined as:

$$DM\ ff\ (AA') = S\ \mu\ A \cap A'\ (x)\ /\ S\ \mu A'\ (x),\ x\ \hat{I}X.$$

2.4. Fuzzification

Fuzzification is the process of decomposing a system input and/or output into one or more fuzzy sets. Many types of curves can be used, but triangular or trapezoidal shaped membership functions are the most common. In this present study trapezoidal functions were used for the Input variables and triangular functions were utilized for output variables. Fuzzy sets span a region of input (or output) value graphed with the membership. Any particular input is interpreted from this fuzzy set and a degree of membership is interpreted.

The membership functions should overlap to allow smooth mapping of the system. The process of fuzzification allows the system inputs and outputs to be expressed in linguistic terms so that rules can be applied in a simple manner to express a complex system.

2.4.1. Defuzzification

After fuzzy reasoning a linguistic output variable which needs to be translated into a crisp value. The objective is to derive a single crisp numeric value that best represents the inferred fuzzy values of the linguistic output variable. Defuzzification is such inverse transformation which maps the output from the fuzzy domain back into the crisp domain. Some defuzzification methods tend to produce an integral output considering all the elements of the resulting fuzzy set with the corresponding weights. Other methods take into account just the elements corresponding to the maximum points of the resulting membership functions [13]. The Mean of Maximum (MoM) method is used only in some cases where the Center of Maximum (CoM) approach does not work. In the CoM method, only the peaks of the membership functions are used. The defuzzified crisp compromise value is determined by finding the place where the weights are balanced.

Thus the areas of the membership functions play no role and only the maxima are used. The crisp output is computed as a weighted mean of the term membership maxima, weighted by the inference results.

2.4.2. Fuzzy Logic Simulation System

In fuzzy logic system, the knowledge of the human is put in the form of a set of fuzzy linguistic rules. These rules would produce approximate decisions, just as a human would. The human expert observes quantities by observing the inputs, and leads to a decision or output using his judgment. The human expert can be replaced by a combination of a fuzzy rule-based system (FRBS) and a block called a defuzzifier.

The inputs are fed into the FRBS, where physical quantities are represented into linguistic variables with appropriate membership functions. These linguistic variables are then used in a set of fuzzy rules within an inference engine, resulting in a new set of fuzzy linguistic variables. In defuzzification stage, the variables are combined and changed to a crisp output which represents an approximation to actual output [19].

Simulink models and sub models representing the complex interaction between various parameters are framed and used for twelve selected parameters. The grouping of parameters and rules designed for Simulink. A typical block diagram showing Simulink of first group water quality parameters viz. pH, EC, Cl^-, Na^+ is presented in **Figure 2**. Data collected from the study area for various seasons were used as the inputs for Fuzzy simulation model.

The simulation was used for the collected data for seasonal variations. Based on Field experience, Delphi technique [20] and expert knowledge 66 rules are designed for physico-chemical water quality parameters in Group I as shown in **Figure 2**.

The parameters of the kind Total Alkalinity, Total Hardness, Calcium and Magnesium were the members of second group parameters and were selected with 73 design IFTHEN rules used in the Simulink process. The schematic block diagram is shown in **Figure 3**. The health concern parameters such as Sulphate, Fluoride [21], Nitrate and Coli forms were selected in group three and 42, IF – THEN rules were utilized based up on expert's knowledge and shown in **Figure 4**. Out puts from all the three groups that is the results obtained from first, second and third groups were combined and framed as final simulation model as shown in **Figure 5**.

The quality assessment of groundwater of Tiruchirappalli city, a total of 27 rules was fired for the final assessment of groundwater quality. The results from all the three groups are aggregated to assess the final classification of water as shown in block diagram **Figure 5**. The processes are applied to all the seasonal water samples and the results obtained are as shown in **Figure 6**.

The rule based decision on expert's perception was fired using Mamdani implification of maximum and minimum operator [22]. To assess the drinking water quality of the groundwater samples, 181 rules are fired.

2.5. Approach towards Groundwater Quality Assessment for Potability

A fuzzy rule based system is generated in which users classify the water according to given data in Desirable, Acceptable, Not acceptable, Rejected quality with respect to different parameters, all connected using AND operator [23]. The user's feedback is also taken with respect to overall quality for different parameters connected by AND operator. For example, one of the feedbacks taken may be like this, If TDS = good AND pH = medium and Sulphate = good then, overall water quality = High. After this, Delphi's technique [20] is applied to converge the feedback of various users to a single value; a degree of match is computed between the user's perception and field data for different parameters. For every type of water quality viz. good (Desirable) medium (Acceptable) or bad (Not Desirable). The water quality for which degree of match is the highest is considered to rerepresent the quality of the water sample.

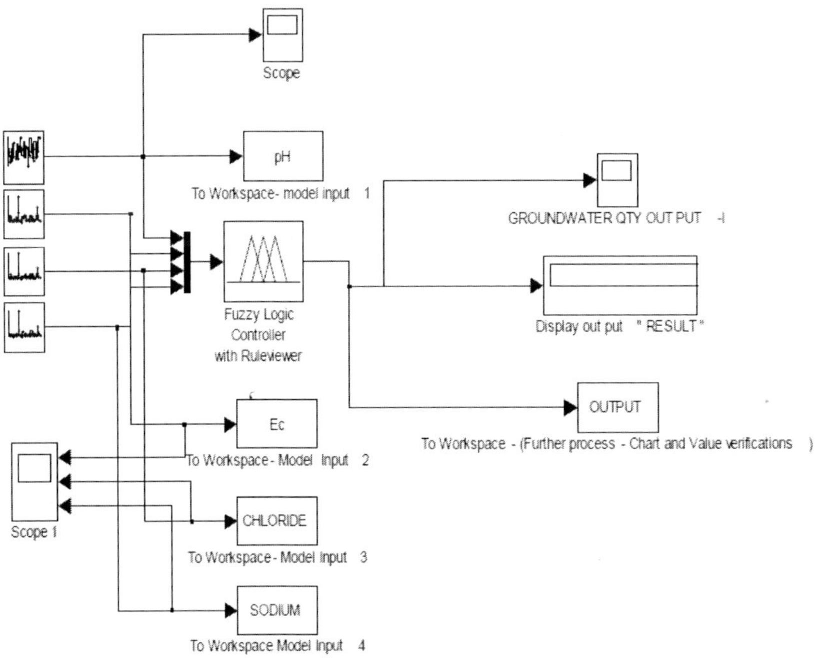

Figure 2. Block diagram for simulink process of FIP - First group water quality parameters.

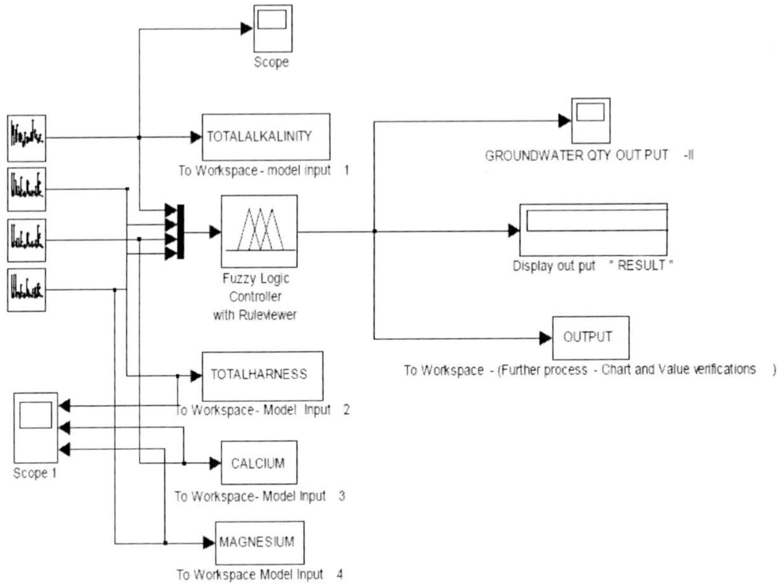

Figure 3. Block diagram for simulink process of FIP - Second group water quality parameters.

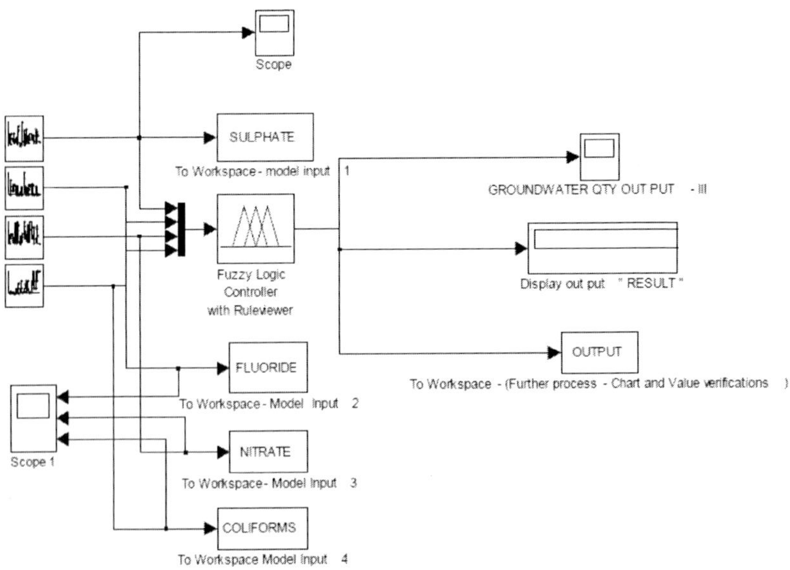

Figure 4. Block diagram showing simulink process of FIP - Third group water quality parameters.

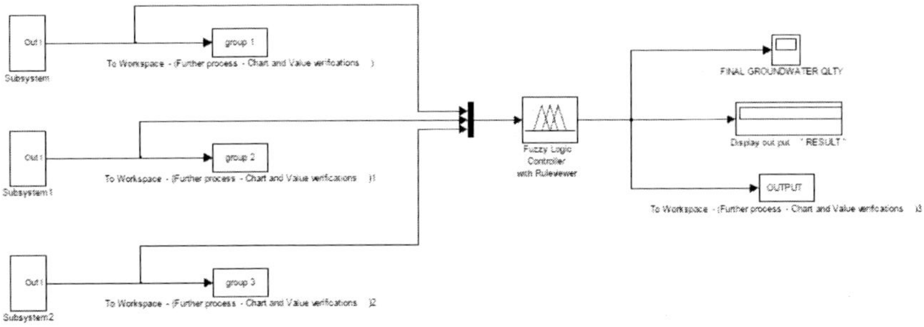

Figure 5. Block diagram for the fuzzy simulink process of FIP for water quality assessment.

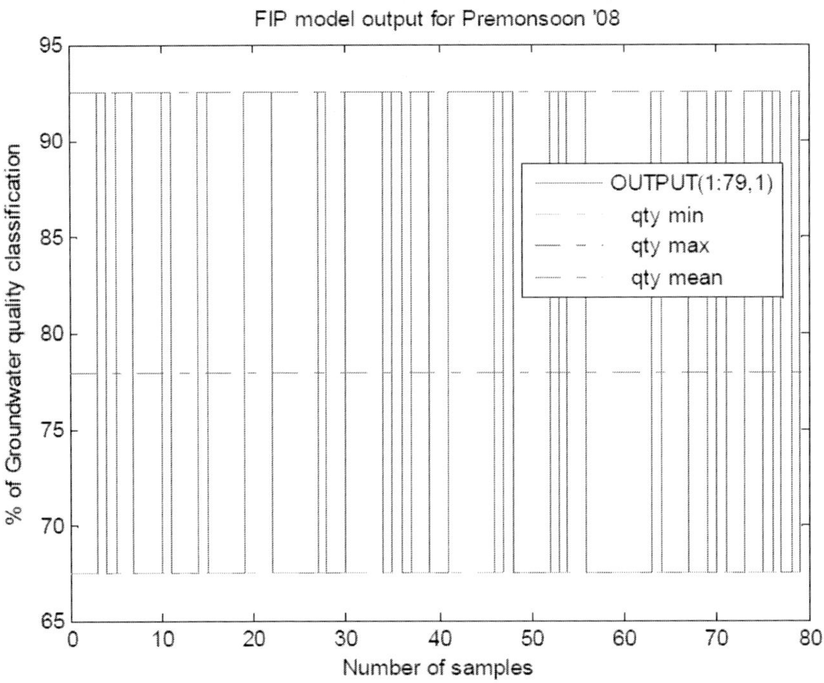

Figure 6. Subsurface water potable frequency showing Pre monsoon 2008.

3. RESULTS AND DISCUSSIONS

Physio-chemical Groundwater quality assessment by deterministic method for drinking groundwater usage on the basis of 8 water quality parameters were compared with the concentration in the water with point value prescribed limits [21,23]. In case Groundwater quality model approach, these 8 parameters were divided in the four categories on the basis of expert opinion having their importance with respect to drinking water quality criteria.

The major findings of the research study are:

1) The accuracy of chemical analysis was verified by ionic balance errors and was found Minimal.

2) The hydro chemical analyses revealed that water samples in the study area was characterized by hard to very hard, fresh to brackish and alkaline in nature. The highly turbid water may cause health risk as excessive turbidity can protect pathogenic microorganisms from the effects of disinfectants and also stimulate the growth of bacteria during storage.

3) Characteristic by pH values, most of the water samples were alkaline in nature which are well within permissible limit (6.5 – 8.5) and some of the samples have been found acceptable for usage and the ranges are between 6.5 and 9.2 meeting BIS standards of IS: 10500: 1991 and WHO (2006) guidelines.

4) Based on Electrical Conductivity (Ec) values measured all water samples Zone-I (Srirangam) are desirable (< 1 mS/cm) for potability.

5) The microbial results indicated that all the samples exceeded the acceptable limit of drinking water. Presence of Salmonella and Shigella was also detected.

6) Although, the water samples contain significant concentrations of calcium, magnesium, sulphate, alkalinity (HCO_3 + CO_3) and potassium ions, predominant presence of two ions, sodium and chloride were found in the samples.

7) It was found that 60% of drinking water samples were found desirable for drinking purpose and the remaining samples could also acceptable for drinking in the absence of alternate water sources.

8) The groundwater quality assessment for potability using fuzzy logic controller has been estimated for Pre monsoon period (March to July) for the years 2006 to 2008. Potability indices show that the groundwater quality of bore well No. 25 of Zone–II (Ariyamangalam)

was very low, as it is near to Municipal solid waste dumping yard. The lowest potability of 32.5% was found in bore well No. 73 of Zone–IV (K. Abisekapuram), located very near to sewer line and an open pond with stagnant water. The remaining bore wells in all the four zones records the potability between 67.5% and 92.5% and is illustrated in **Figure 7**.

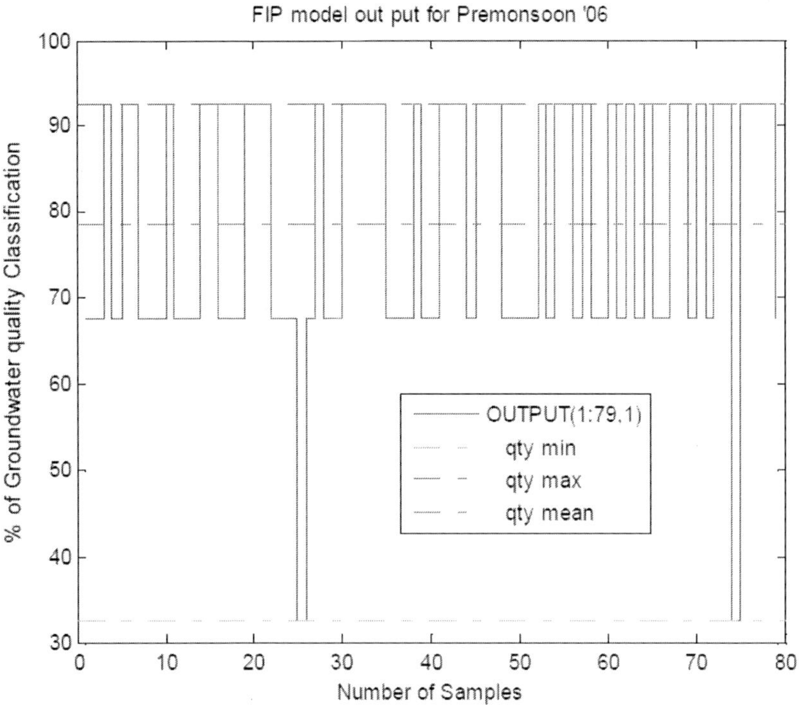

Figure 7. Subsurface water potable frequency during premonsoon periods 2006.

9) The groundwater quality assessment for potability using fuzzy logic controller has been estimated for Post monsoon period (August to November) for the years 2006 and 2008. The potability of 32.5% for the bore well No. 59, 62, 63 and 73 of Zone–IV (K. Abisekapuram). The potability indices show that the groundwater qualities of these Bore wells were located very near to an open ponds and stagnant waste water during the rainy season. The remaining bore wells have the potability between 67.5% and 92.5% and are illustrated in **Figure 8**.

10) Digital Surface Model (DSM) diagrams were constructed directly from major anions and cations, so as to consider common behaviors within

the study area. A typical DSM concentration based on Cl^- and HCO_3^- map is presented in **Figure 9**. There was high concentration in the Ariyamangalam zone makes the water non potable more than 50% of the sampled groundwater showed high chloride.

11) Pre-monsoon and post-monsoon contour potable water quality maps are constructed directly from major anions and cations from fuzzy Simulink model is shown in **Figure 10**. The contour maps of the equal concentrations based on standards is given in **Figure 10**.

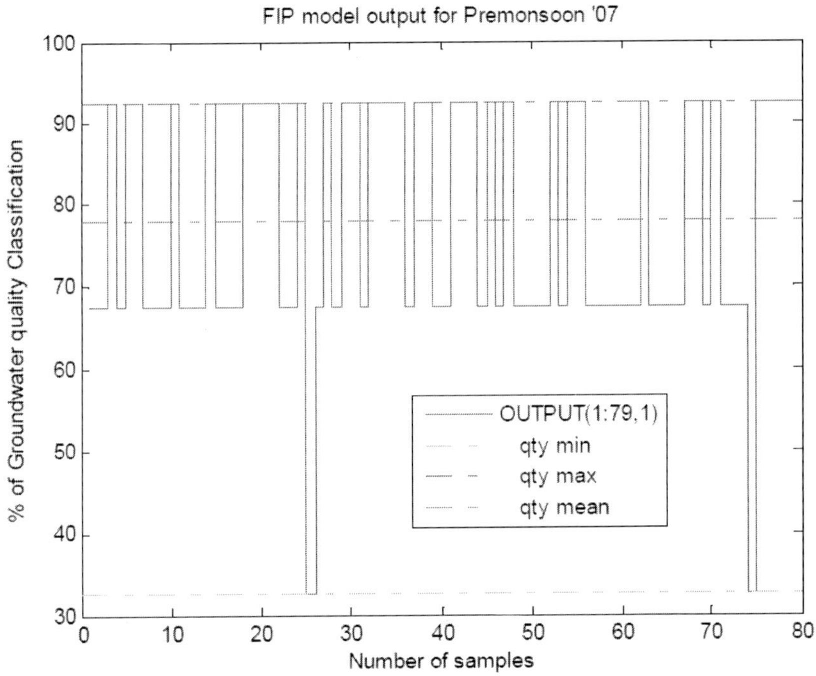

Figure 8. Subsurface water potable frequency during pre monsoon periods 2007.

The high risk was observed in the regions of Ariyamangalam zone consumption of ground water from the model during pre monsoon 2007. It was also noted that the market area and Ariyamangalam zone were the alarming zones of non potability regions during pre monsoon period 2006.

The high risks were noticed in some central parts of K. Abishekapuram towards Ariyamangalam due to the open pond and improper disposal of solid sluges.

4. CONCLUSIONS

The quality of the groundwater of the Tiruchirappalli city was monitored in 79 sampling wells for 3 years and major recorded data revealed that the concentrations of cations and anions were above the maximum, desirable for human consumption. The Electrical Conductivity was found to be the most significant parameter within input parameters used in the modeling. The developed model enabled well to test the data obtained from 79 samples of bore wells of Tiruchirappalli city.

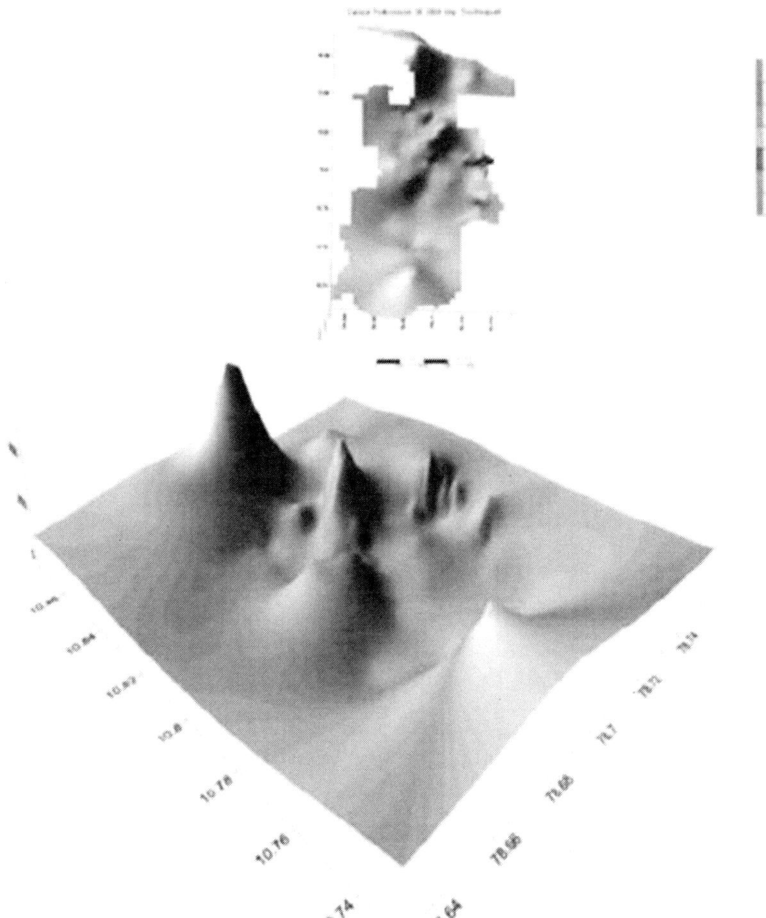

(a)

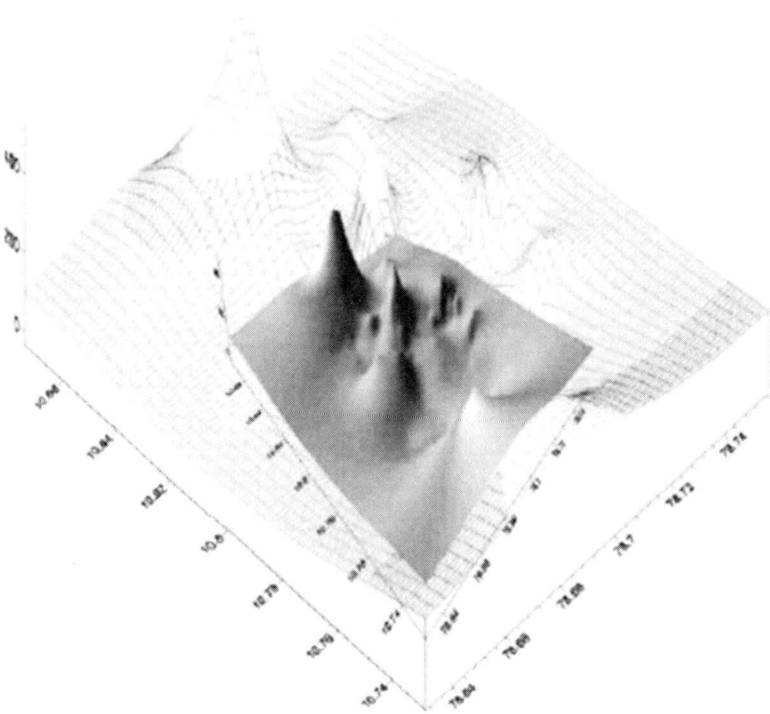

(b)

Figure 9. (a) DSM for chloride dispersion in Zone IV of Tiruchirappalli city; (b) DSM with Wire frame model for chloride dispersion in Zone IV of Tiruchirappalli city.

The groundwater in Tiruchirappalli meets all WHO drinking water standards with in the range of 67.5% to 92.5% for potable during pre monsoon condition of all the sampling durations.

As the sampling station of 24 and 73 were found in non potable condition due to vicinity of wastewater discharging areas and solid waste dumping sites.

During post monsoon all the sampling stations satisfies WHO [2] drinking water standards within in the range of 67.5% to 92.5%.

At the stations 59, 62 and 72 were reported with non potablity of 32.5% due to unhealthy environmental conditions of wastewater and local waste dumps near by the sampling points.

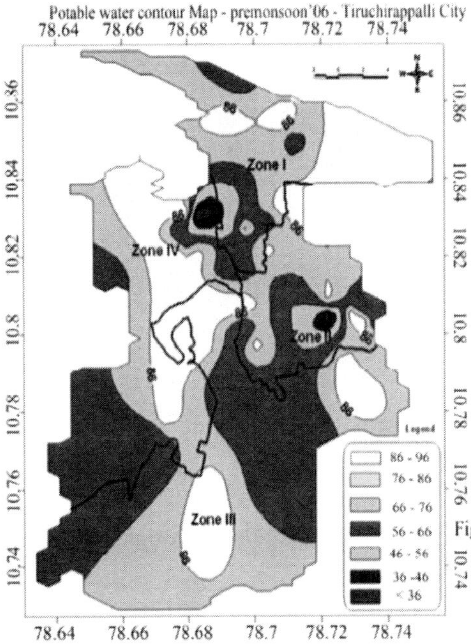

Potable Water Contour Map

Fuzzy Information Processing for Groundwater Quality Assessment

Figure .Potable Water Contour Map (in %)

Duing Pre monsoon 2006

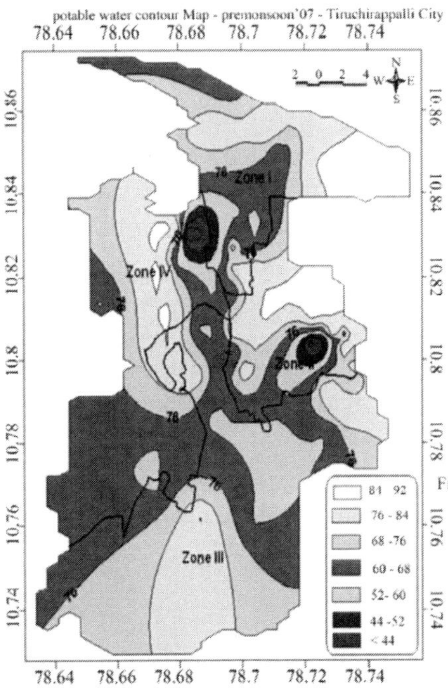

Potable Water Contour Map

Fuzzy Information Processing for Groundwater Quality

Figure. Potable Water Contour Map (in %)

During Pre monsoon 2007

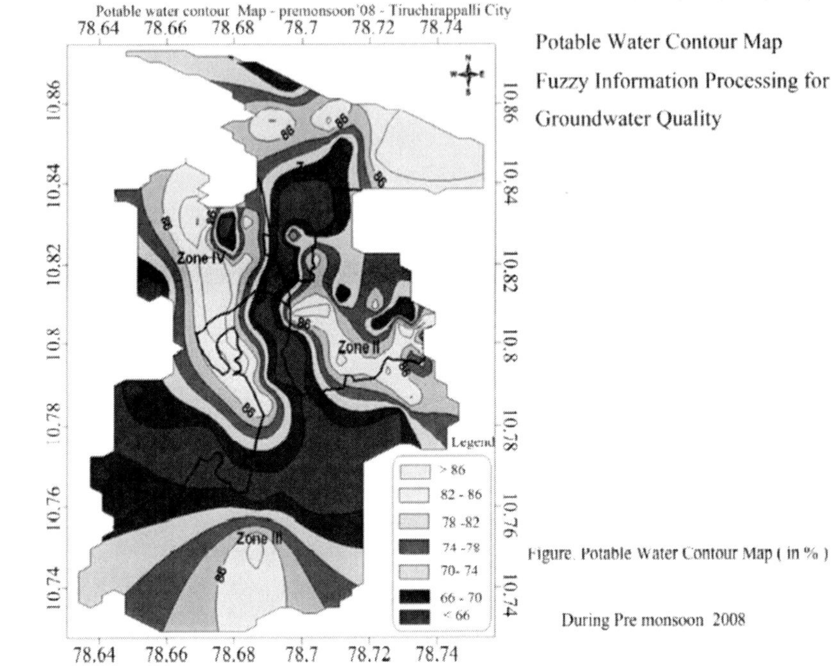

Figure 10. FIP potable water contour maps for pre monsoon 2006, 2007 and 2008.

Solid wastes including sledges were disposed, and without any pre treatment before dumping and no protection towards the subsurface water for potability.

In the previous study by the authors [13] the subsurface water quality in Ariyamangalam, Zone Tiruchirappalli City Corporation was seriously under threat by carbonates and sulphates near the sampling points of Ariyamangalam zone. Also contaminated by several pollutants as Ariyamangalam itself was currently polluted due to the waste dumping site and improper waste water vicinity.

Without immediate response, the subsurface water is currently degrading its consumption quantity and will not be potable in near future if the proper steps have not been taken care.

ACKNOWLEDGEMENTS

Natarajan Venkat Kumar, Samson Mathew, Ganapathiram Swaminathan The First author gratefully acknowledges the support by Fellowship under TEQIP, MHRD, and Govt. of India for his Doctoral study. The authors wish to express their sincere appreciation is also extended to the anonymous reviewers for their valuable comments to enhance the quality of this paper. Special thanks to Sarva Mangala Preveena for her constant support in preparation of this manuscript.

REFERENCES

1. Bureau of Indian Standard, "Indian Standard Specification for Drinking Water," BIS Publication No. IS (10501), New Delhi, 1991.
2. World Health Organization, "Guidelines for Drinking Water Quality Recommendation," Vol. 2, World Health Organization, Geneva, 2008.
3. M. F. Dahab, Y. W. Lee and I. Bogardi, "A Rule Based Fuzzy-Set Approach to Risk Analysis of Nitrate Contaminated Groundwater," Water Sciences Technology, Vol. 30, No. 7, 1994, pp. 45-52.
4. B. Dixon, "Prediction of Groundwater Vulnerability Using an Integrated GIS Based Neuro-Fuzzy Techniques," Journal of Spatial Hydrology, Vol. 4, No. 2, 2004, pp. 38-41.
5. R. Khaiwal and V. K. Garg, "Distribution of Fluoride in Groundwater and its Suitability Assessment for Drinking Purposes," International Journal of Environmental Health Research, Vol. 16, No. 2, 2006, pp. 163-166.
6. K. Schulz and B. Howe, "Uncertainty and Sensitivity Analysis of Water Transport Modeling in a Layered Soil Profile Using Fuzzy Set Theory," Journal of Hydroinformatics, Vol. 1, No. 2, 1999, pp. 127-138.
7. B. Dixon, H. D. Scott, J. C. Dixon and K. F. Steele, "Prediction of Aquifer Vulnerability to Pesticides Using Fuzzy-Rule Based Models of the Regional Scale," Physical Geography, Vol. 23, 2002, pp. 130-152.
8. B. Dixon, "Application of Neuro-Fuzzy Techniques in Predicting Groundwater Vulnerability: A GIS Based Sensitivity Analysis," Journal of Hydrology, Vol. 309, No. 1-4, 2005, pp. 17-38.
9. S. M. Preveena, M. H. Abdullah, A. Z. Aris and K. Bidin, "Groundwater Solution Techniques: Environmental Applications," Journal of Water Resource and Protection, Vol. 2, 2010, pp. 8-13.
10. Z. K. Sen, "Fuzzy Groundwater Classification Rule Derivation from Quality Maps," Water Quality Exposure Health, Vol. 1, No. 1, 2009, pp. 115-112.
11. American Public Health Association, "Standard Method for Examination of Water and Waste Water," 21st Edition, American Public Health Association, Washington, DC., 2005.
12. B. Shomar, S. A. Fkher and Alfred Yahya, "Assessment of Groundwater Quality in the Gaza Strip, Palestine Using GIS Mapping," Journal of Water Resource and Protection, Vol. 2, No. 2, 2010, pp. 93-104.
13. V. Kumar, N. S. Mathew and G. Swaminathan, "Fuzzy Information Processing for as Assessment of Groundwater Quality," International Journal of Soft Computing, Vol. 4, No. 1, 2009, pp. 1-9.
14. Z. Chen, G. H. Huan and A. Chakma, "Hybrid FuzzyStochastic Modeling Approach for Assessing Environmental Risks at Contaminated Groundwater Systems," Journal of Environmental Engineering, Vol. 129, 2003, pp. 79-88.

15. S. Liou and S. A. L. Wang, "Generalized Water Quality Index for Taiwan," Environmental Monitor Assessment, Vol. 9635-52, 2004.
16. S. Liou and S. L. Lo, "A Fuzzy Index Model for Tropic Status Evolution of Reservoir Waters," Water Research, Vol. 96, No. 1, 2004, pp. 35-52.
17. L. A. Zadeh, "Fuzzy Set," Information Control, Vol. 8, No. 3, 1965, pp. 338-353.
18. L. A. Zadeh, 'The Concept of a Linguistic Variable and its Application to Approximate Reasoning," Information Science, Vol. 8, 1975, pp. 199-249.
19. M. Jamshidi, "Tools for Intelligent Control: Fuzzy Controllers, Neural Networks and Genetic Algorithms," Philosophical Transactions of the Royal Society, Vol. 361, No. 1809, 2003, pp. 1781-1808.
20. C. Okoliand and S. D. Pawlowski, "The Delphi Method as a Research Tool an Example, Design Considerations and Applications," Information and Management, Vol. 42, No. 1, 2004, pp. 15-29.
21. V. K. Garg, S. Dahiya, A. Chaudhary and Deepshikha "Fluoride Distribution in Underground Waters of Jind District, Haryana, India," Ecology Environmental Conservation, Vol. 4, No. 172, 1998, pp. 19-23
22. E. M. Mamdani, "Advances in the Linguistic Synthesis of Fuzzy Controllers," International Journal of Man-Machine Studies, Vol. 8, No. 6, 1976, pp. 669-678.
23. N. Chang, H. W. Chen and S. K. King, "Identification of River Water Quality Using the Fuzzy Synthetic Evaluation Approach," Journal Environmental Management, Vol. 63, 2001, pp. 293-305.
24. C. O. Cude, "Water Quality Index: A Tool for Evaluation Water Quality Management Effectiveness," Journal of the American Water Resources Association, Vol. 37, No. 1, 2001, pp. 125-137.

CHAPTER 18

Spatiotemporal Analysis of Groundwater Recharge Trends and Variability in Northern Taiwan

Hsin-Fu Yeh[1*], Chen-Feng Yeh[1], Jhe-Wei Lee[2] and Cheng-Haw Lee[1]

[1] *Department of Resources Engineering, National Cheng Kung University, Tainan, Taiwan*

[2] *Department of Hydraulic and Ocean Engineering, National Cheng Kung University, Tainan, Taiwan*

ABSTRACT

In this study, the base flow estimation method was used to assess long-term changes of groundwater recharge in Northern Taiwan. The Mann-Kendall test was used to examine the characteristics of the trends. This was followed by trend slope calculation and change-point analysis. The annual groundwater recharge was found to exhibit a significant upward trend for the Fushan and Hengxi stations (Tamsui river basin). On the other hand, the Ximen Bridge station (Lanyang river basin) recorded a significant downward trend. Calculations showed that the rate of change for the Fengshan and Touqian river basins was small (less than 10%). However, that for the following stations was greater than 30%: Fushan, Hengxi, Ximen Bridge, and Niudou (also in the Lanyang river basin). The results of the change-point analysis further indicated a significant change-point for the annual recharge at Fushan, Hengxi, and Ximen Bridge stations in 1999,

1983, and 2001, respectively. The findings can be used for regional hydrological studies and as reference for water resource planning.

Keywords: groundwater recharge, base flow, Mann-Kendall test, Northern Taiwan

1. INTRODUCTION

The annual average precipitation in Taiwan is 2500 mm, 2.6 times higher than the global average values. In spite of the high precipitation per unit area, the annual precipitation per person is only 1/7 of the global average, due to both small land area and large population [1]. Analysis of environmental sustainability indicators revealed that Taiwan ranked 18th among 146 countries experiencing water shortages, with the average available water per person in Taiwan being 1740 m^3/year [2]. Climate changes and increasing temperature experienced in recent decades caused the precipitation in Taiwan to increase during the rainy season, while decreasing in the dry season, resulting in a drier dry season and a wetter wet season [3]. Consequently, stream flow in both dry and wet seasons varied significantly. Thus, it is urgent to assess and discuss groundwater recharge characteristics, toward effective water resource management in Taiwan.

When analyzing the hydrogeological model and groundwater system of a region, the study on recharging groundwater through precipitation is a very important but complex issue. Meteorological factors (e.g., intensity and delay of precipitation, temperature, humidity, and wind speed), thickness of the soil layer, prevailing groundwater level, topographical conditions on the surface, vegetation cover, and land uses, all have direct relationships with the system and must be considered [4–10]. It is very easy to measure precipitation and run-off amounts when analyzing water balance, but it is very difficult to quantify the recharging process. Its evaluation requires not only precipitation data, but also other factors, such as prevailing climatic conditions, soil type, soil moisture status, vegetation cover, and evapotranspiration conditions [11, 12]. The infiltration of precipitation into the soil leads to the recharging process and is an important factor determining circulation of the groundwater system and its recharge volume.

Groundwater recharge can be quantitatively estimated using two methods. The first is the water balance model [13–16], which is applicable for humid regions; the second is for arid regions and involves the use of tensiometers, tracers, infiltrometers, and other instruments on site to observe the movement of water in the unsaturated zone, before estimating the groundwater recharge for that area [17–19]. However, it is generally more difficult to implement the second method because of high costs and the need for long-term monitoring on site. The existing methods for estimating groundwater recharge at the regional level using the water balance model are further divided into two types: (i) precipitation, infiltration, run-off, evapotranspiration, and groundwater recharge are treated as components of an interrelated system, with the soil moisture status being that of an ever-changing soil water balance model [7, 20, 21]; and (ii) the hydrograph of a stream flow is used to estimate its base flow, with the latter being treated as the groundwater recharge (based on the assumption that the heterogeneous hydrogeological conditions within the catchment area are ignored) [22–26].

In this study, the base flow estimation method was used to determine groundwater recharge and the evapotranspiration from the unsaturated zone was not considered. The calculation method to determine the effective recharge is simple, and it neither requires any complicated hydrogeological models, nor factors such as weather conditions and soil types. In this study, the calculation method was used to assess long-term changes of groundwater recharge in Northern Taiwan, and the findings can serve as reference for the management of water resources.

2. METHODOLOGY

2.1. Base flow model

Based on the concept of water balance and the data from stream flow gauging stations in the main river basins of the catchment area, this study employed the stream flow PARTitioning (PART) program (a base flow analysis method) developed by the US Geological Survey [27]. This method utilized the stream flow data to separate the base flow, which is regarded as the groundwater recharge. However, for rainy and humid regions and steep mountainous areas, if the estimation of groundwater recharge is based solely on base flow separation, it will often lead to overestimation of the base flow during the wet season [7, 28]. Therefore, the calculation of a steady base flow separation is required. In this study, a method for steady base flow analysis was adopted for this calculation to subsequently estimate a reasonable groundwater recharge.

Using Grey theory, the steady base flow analysis obtains data trends with reference to the rearrangement and accumulation of data. The steady base flow analysis uses trends for a low-flow period, a steady base flow period, and an overestimated base flow period, as shown in **Figure 1**. These are obtained after the rearrangement and accumulation of the separated base flows, and then, the steady base flow period is linearly extrapolated to achieve the steady base flow.

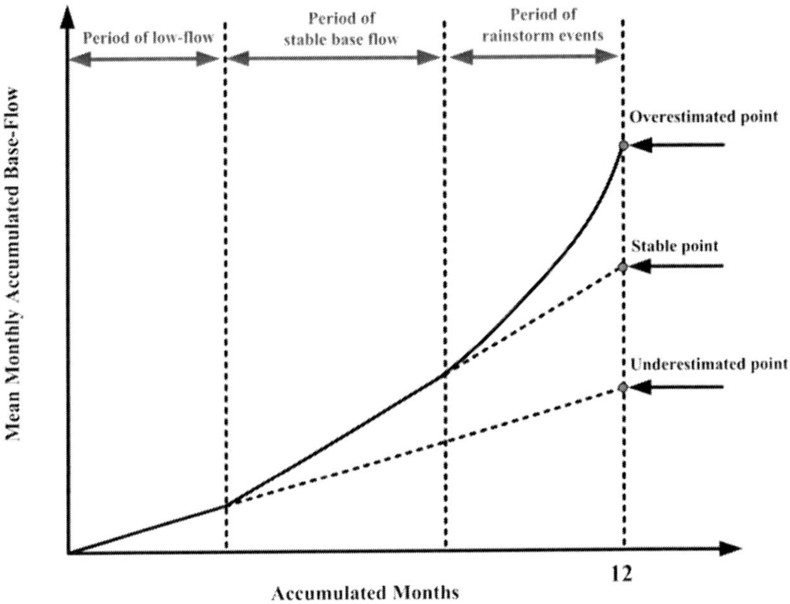

Figure 1. The diagram of the stable base flow analysis.

The steps in the analytical process are as follows:

1) Obtain the base flow of each month by base flow separation.
2) Sum the base flow per month over several years and then average the sum to achieve the long-term mean base flow on a month-by-month basis.
3) Sort the long-term mean base flow of each month in descending order and accumulate them to obtain the base flow accumulated per month and the trend of such base flows.
4) Determine the rising point of the base flow by the trend line of the accumulated base flow and obtain the steady base flow period.
5) Obtain the annual base flow, namely the annual groundwater recharge, by the extrapolation of the linear regression equation.

2.2. Mann-kendall test

Mann-Kendall (MK) [29, 30] test is a nonparametric method developed from Kendall's tau (τ). It can be used to test the relationship between two sets of data. The advantages of this method is that extreme values and missing data problems will not seriously affect the certification value. The MK test assesses the trend in a series via comparing the value of the series before and after to determine whether the series exhibits a specific degree of trend. The null hypothesis given that if there is not significant trend in the series, test statistic *S* is defined as follows:

$$Sign(X_j - X_i) = \begin{cases} +1 \,, X_j - X_i > 0 \\ 0 \,\,\,\, , X_j - X_i = 0 \\ -1 \,, X_j - X_i < 0 \end{cases} , S = \sum_{i=1}^{n-1} \sum_{j=i+1}^{n} Sign(X_j - X_i) \quad (1)$$

where $\{X1, X_2 X3, \ldots, Xn\}$ is stream flow data which is arranged in accordance with time $\{T_1, T_2, T_3, \ldots, Tn\}$. n is the number of data. When n is close to infinity, the probability of the S distribution curve will present as a normal distribution with a mean of 0. In addition, when n is more than 10, the variance of S can be substituted into the following approximate solution:

$$\sigma^2 = \frac{n(n-1)(2n+5)}{18} \quad (2)$$

In this study, long-term stream flow data are likely to be repeated in the data series; thus, Kendall modified the approximated solution Eq. (2) to Eq. (3).

$$\sigma^2 = \frac{1}{18}\left[n(n-1)(2n+5) - \sum_{u=1} u(u-1)(2u+5)\right] \quad (3)$$

where u is the duplicate value number of the data series.

Finally, the normalized statistical test S values becomes the Z value, as follows:

$$Z = \begin{cases} \frac{S-1}{\sigma} \,, S > 0 \\ 0 \quad\;\; , S = 0 \\ \frac{S+1}{\sigma} \,, S < 0 \end{cases}$$

When Z is a positive value, this indicates that the series is exhibiting an increasing trend; in contrast, when the value is negative, it indicates that the series has decreased. At this time, the obtained Z value should be tested by a significance test to assess whether the series is significant. Assuming a significance level of α, if $|Z| \geqq Z_\alpha$, the null hypothesis is rejected, which represents that the series has a significant trend, otherwise the series has no significant trend. In the study, the significance level is set as α=0.05. When $|Z| \geqq 1.96$, the series has a significant trend. When it is below this level, there is no significant trend.

2.3. Theil-sen slope

The Theil-Sen slope [31] is used to estimate the magnitude of the trend slope. The Theil-Sen slope estimation method is different from the slope values calculated using a linear regression, because it selects the median value, and therefore, the properties are less affected by extreme values. Thus, it is often used with the MK test. Slope β is defined as follows:

$$\beta = Median\left(\frac{X_j - X_i}{j-i}\right), foralli < j \tag{5}$$

$$X(t) = \beta t + C$$
$$X(t) = X_1 \sim X_n, t = 1 \sim n \tag{6}$$

2.4. Mann-whitney-pettit test

The Mann-Whitney-Pettit (MWP) test [32] can be used to search for significant change points in a data series. The definition of a change point is when a data series $\{X_1, X_2, \ldots, X_n\}$ has a change point X_t, Order $\{X_1, X_2, \ldots, X_t\}$ is $F_1(X)$ and $\{X_{t+1}, X_{t+2}, \ldots, X_n\}$ is $F_2(X)$, then $F_1(X) \neq F_2(X)$. The definition of $U_{t,n}$ is as shown in Eq. (7) shown. If there is not a change point in the data series, $|U_{t,n}|$ on the function of time, t will continue to rise, and there will be no turning point. On the contrary, if there is a change point, $|U_{t,n}||$ on the function of time t, there will be a decreasing turning point. In the same data series, the turning point may occur several times on behalf of this data series, and there may be more than one change point.

$$Sign(X_i - X_j) = \begin{cases} +1, & X_i - X_j > 0 \\ 0, & X_i - X_j = 0 \\ -1, & X_i - X_j < 0 \end{cases}, U_{t,n} = \sum_{i=1}^{t} \sum_{j=t+1}^{n} Sign(X_i - X_j) \tag{7}$$

$$K_n = Max\left|U_{t,n}\right|, 1 \leq t < n \tag{8}$$

To confirm that change points exist, Eq. (8) is used to calculate the extreme value of $|U_{t,n}|$ that is turning point as $Kn_{>}$. Equation (9) is used to calculate the probability of a change point. In this study, P = 0.95 is set the as confidence level, where P > 0.95 judges that the time is a significant changing point.

$$P = 1 - \exp\left(\frac{-6K_n{}^2}{n^2+n^3}\right) \tag{9}$$

However, in some data series, a change point may not exist by itself; thus, Eq. (10) is used to calculate each year's $P(t)$ value. The $P(t)$ value is identified when it is greater than the confidence level.

$$P(t) = 1 - \exp\left(\frac{-6\left|U_{t,n}\right|^2}{n^2+n^3}\right) \tag{10}$$

3. STUDY AREA

The northern region of Taiwan includes several administrative districts, namely Yilan County, Keelung City, New Taipei City, Taipei City, Taoyuan County, and Hsinchu City. There are five types of terrains in this region: plains (mainly distributed in the Taoyuan Alluvial Fan and the Hsinchu Plain in the west, as well as the Yilan River Delta in the east); hills (mainly the Chutung Hill in Keelung); tablelands (the Linkou Plateau); basins (the Taipei Basin); and mountainous areas (the Tatun Mountain range in the north and the Central Mountain range). With these varied topographic features, this region ranges significantly in height from the <100 m above mean sea level (AMSL) plain areas to the <500 m AMSL hilly areas and to the >1000 m AMSL mountainous areas. The main rivers in the region are the Tamsui River, Lanyang River, Fengshan River, and Touqian River, with drainage areas of 2726, 978, 250.1, and 565.9 km^2, respectively. The annual average overall stream flow in the region is less than 15.1 billion m^3, which is less than that of the central, southern, and eastern regions. The data for the past several years indicate that the stream flow of the northern region during the dry season is different from that of the wet season. The total stream flow of the dry season (from November to April) is approximately

5.62 billion m³; while during the wet season (from May to October), it is 9.48 billion m³ [33].

The long-term data on stream flow in Northern Taiwan were collected from the Water Resources Agency, Ministry of Economic Affairs. We selected stream flow stations that are not affected by artificial irrigation facilities and for which recorded data go as far back as 30 years. The eight gauging stations studied were Niudou and Ximen Bridge (Lanyang river basin); Fushan, Gaoyi, and Hengxi (Tamsui river basin); Xinpu (Fengshan river basin); and Neiwan and Shangping (Touqian river basin). Detailed information and geographical location of the various stations are shown in **Table 1** and **Figure 2**, respectively.

Table 1. Information on gauging stations in Northern Taiwan.

Basin	**Station**	**Area (km^2)**	**X-coordinate (TWD67[a])**	**Y-coordinate (TWD67[a])**	**Record years**
Lanyang River	Niu-Dou	446.7	306388.4	2726321	1979-2013
	Ximen Bridge	101.4	324454.8	2739377	1984-2013
Danshui River	Fu-Shan	160.4	298991.3	2742949.3	1953-2012
	Gaoyi	542.0	286029.3	2734394.2	1957-2002
	Hengchi	52.9	289452.4	2758619.3	1958-2012
Fengshan River	Hsin-Pu	208.1	255810.3	2746676	1970-2012
Touqian River	Nei-Wan	139.1	267503.3	2733084	1971-2012
	Shang-Ping	221.7	260738.5	2729330	1971-2012

[i] - Note: [a] is Taiwan triangulation points' coordinates.

4. RESULTS AND DISCUSSION

4.1. Results of base flow separation

After collecting the annual stream flow data of the stations, the PART program developed by the United States Geological Survey (USGS) was used for analysis and to calculate the base flow of each station. The results showed that the trends of the base flow and stream flow were consistent, with the former ranging from 60.9 to 284.9 cm/year. Since Taiwan is located in a subtropical humid zone with perennial precipitation, during the steady base flow period, the recharge derived from base flow separation might still be greater than the groundwater recharge and discharge situations reflected by stream flow.

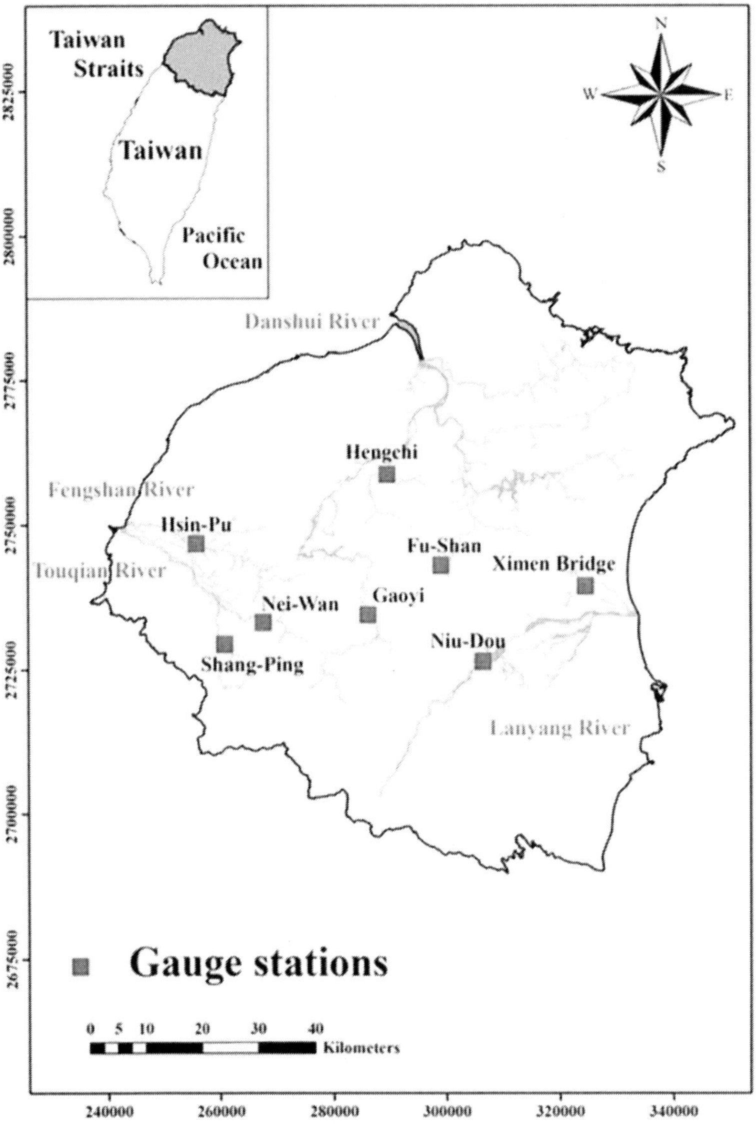

Figure 2. Spatial distribution of gauging stations in Northern Taiwan.

From a short-term perspective, base flow is affected by the amount of precipitation for that year. However, the amount of water discharged from the groundwater system should be a constant value over the long-term. Hence, stable base flow analysis was used to examine and estimate the groundwater recharge. Lee et al. [7] pointed out that the calculated base flow is likely to be overestimated when the base flow estimation method is used to assess stream flow data. This will in turn affect the calculations of the groundwater recharge. After estimating the base flow for all eight stations, evaluation was made using stable base flow analysis. This provided the depth of annual groundwater recharge for each station. The results are shown in **Table 2**. The depth of annual groundwater recharge for the northern region was 15.6–160.65 cm/year. The maximum and minimum values were recorded at the Fushan and Niudou stations, respectively.

Table 2. Results of the depth of annual groundwater recharge in Northern Taiwan.

Basin	Station	Record years	Recharge depth (cm/year)
Lanyang River	Niu-Dou	1979-2010	15.60
	Ximen Bridge	1983-2012	142.89
Danshui River	Fu-Shan	1953-2012	160.65
	Gaoyi	1957-2002	144.07
	Hengchi	1958-2012	50.55
Fengshan River	Hsin-Pu	1970-2012	20.56
Touqian River	Nei-Wan	1971-2012	52.23
	Shang-Ping	1971-2012	55.83

4.2. Analysis of groundwater recharge

After calculating the depth of groundwater recharge for the respective stations using stable base flow analysis, the results were multiplied by the area of the water catchment. This gave the annual groundwater recharge for each station and the long-term average amount (**Table 3**). Annual groundwater recharge for the northern region worked out to be between 2.67×10^7 and $7.81 \times 10^8 m^3$/year. Both largest and smallest volumes were found in the Tamsui river basin, at the Gaoyi and Hengxi stations, respectively.

Table 3. Results of annual groundwater recharge in Northern Taiwan.

Basin	Station	Record years	area (km^2)	Recharge Depth (cm/year)	annual recharge (m^3/year)
Lanyang River	Niu-Dou	1979-2010	446.7	15.60	6.97×10^7
	Ximen Bridge	1983-2012	101.4	142.89	1.45×10^8
Danshui River	Fu-Shan	1953-2012	160.4	160.65	2.58×10^8
	Gaoyi	1957-2002	542.0	144.07	7.81×10^8
	Hengchi	1958-2012	52.9	50.55	2.67×10^7
Fengshan River	Hsin-Pu	1970-2012	208.1	20.56	4.28×10^7
Touqian River	Nei-Wan	1971-2012	139.1	52.23	7.26×10^7
	Shang-Ping	1971-2012	221.7	55.83	1.24×10^8

4.3. Analysis of trends in groundwater recharge

The Mann-Kendall test was used to analyze the characteristics of the long-term annual recharge trends for Northern Taiwan and to examine the distribution of significant trends. The test results are shown in **Table 4**. The significant level α = 5% was adopted as the standard, meaning that $|Z_{\alpha/2}|$ = 1.96 was used to verify that a trend was of significance. Three river flow stations were found to exhibit significant trends, namely Fushan, Hengxi, and Ximen Bridge stations. It was a significant upward trend for the first two stations, but a significant downward trend for the third station (**Table 4**). Overall, among the eight stations within the study area, only two exhibited downward trends: Gaoyi station at upstream Tamsui River and Ximen Bridge at downstream Lanyang River. The spatial distribution of trends for the various stations is shown in **Figure 3**. In addition, the Theil-Sen estimation method was used to calculate the trend slope. An upward and downward trend is indicated by a value larger and smaller than zero, respectively. The trend line can also be calculated using the trend slope values and annual recharge volume for the individual years. **Equation (10)** was then used to calculate the rate of change and the results are shown in **Figure 4**.

Table 4. Results of significant trend, slope, and relative change in Northern Taiwan.

Basin	Station	Record years	Mann-Kendall test result	Slope estimator	Relative change
Lanyang River	Niu-Dou	1979-2010	1.768	0.155	30.9%
	Ximen Bridge	1983-2012	-3.604*	-5.254	-106.6%
Danshui River	Fu-Shan	1953-2012	2.991*	1.198	44.0%
	Gaoyi	1957-2002	-0.095	-0.026	-0.8%
	Hengchi	1958-2012	2.195*	0.363	33.7%
Fengshan River	Hsin-Pu	1970-2012	0.415	0.049	10.0%
Touqian River	Nei-Wan	1971-2012	0.398	0.087	7.0%
	Shang-Ping	1971-2012	0.356	0.105	7.9%

[i] - Note: *indicates the significant trends. The positive values represent increasing trends, and the negative ones represent decreasing trends.

It was observed that the rate of change for the annual recharge was small at the Fengshan and Touqian river basins: that for the Xinpu, Neiwan, and Shangping stations was all less than 10%. For the Tamsui river basin, with the exception of Gaoyi, the rate of change for the other stations was greater than 30%. In particular, the increases for Fushan and Hengxi stations were 44.0 and 33.7%, respectively. The rate of change was also greater than 30% for the Niudou and Ximen Bridge stations. However, it was an increase of 30.9% for the former, but a decrease of 106.6% for the latter. The slopes of the significant trends in stream flow are shown in **Figure 4**.

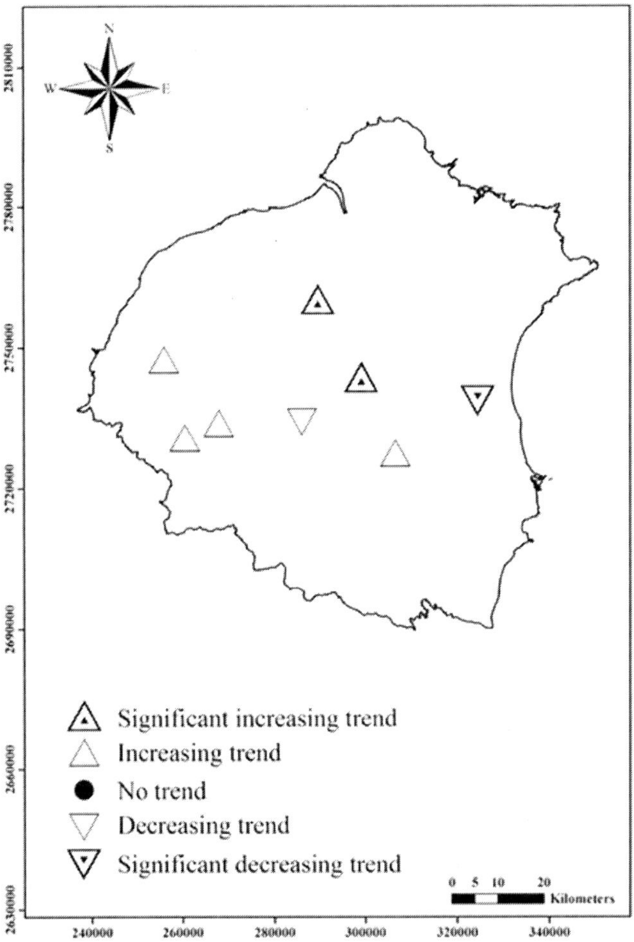

Figure 3. Map showing spatial variation in trends in annual recharge in Northern Taiwan.

The change-point analysis results revealed that change points occurred at three gauging stations: Fushan, Hengxi, and Ximen Bridge (**Table 5** and **Figure 5**). The change point for the Fushan station occurred in 1999. The average annual groundwater recharge before and after the change point was 23.1×10^2 and 35.2×10^7 m^3/year, respectively. The amount exhibited an upward trend post-1999, and the rate of increase was 46.9%. For the Hengxi station, the change point occurred in 1983, with the before and after volume being 2.2 × 107 and 3.3×10^7 m^3year, respectively. There was an overall upward trend as well, and the rate of increase was 41.7%. The change point for the Ximen Bridge station occurred in 2001. The before and after volumes there were 19.6×10^7 and 6.7×10^7 m^3/year, respectively. The overall trend after the change point declined by as much as 89.4%. Among the three aforementioned gauging stations, the magnitude of change at the Ximen Bridge station was the largest.

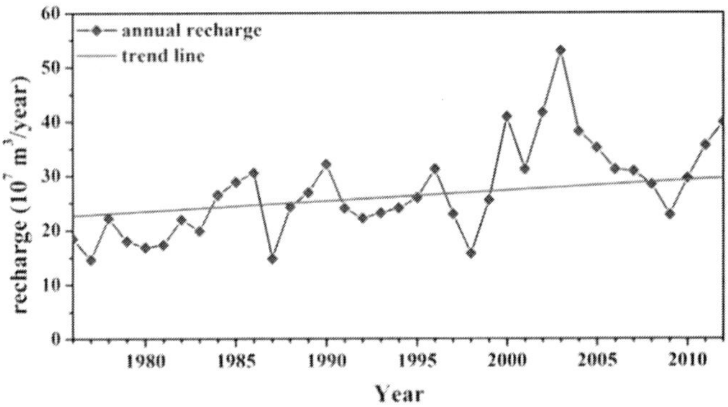

(a) Fu-Shan station

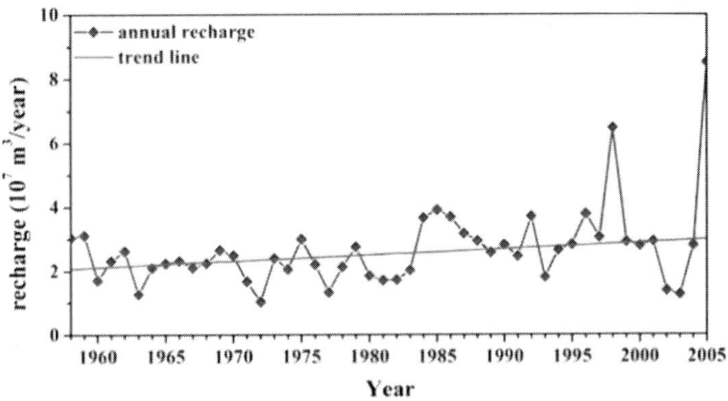

(b) Hengchi station

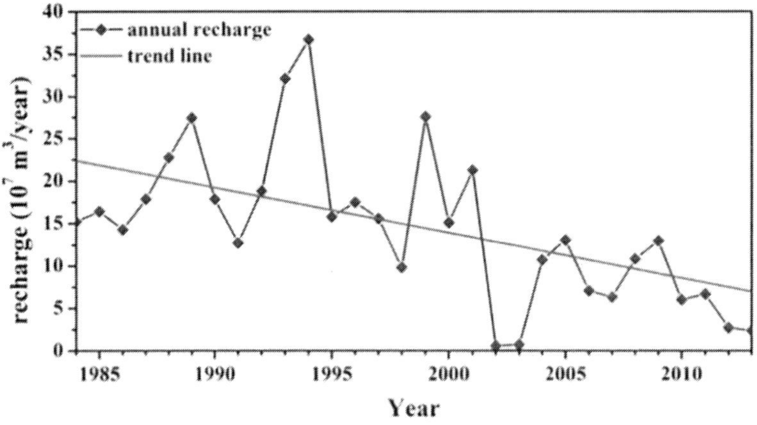

(c) Ximen Bridge station

Figure 4. The significant trend line of gauging stations in Northern Taiwan. (a). Fu-Shan station, (b) Hengchi station, (c) Ximen Bridge station

Table 5. Results of change points using cumulative deviations and Mann-Whitney-Pettitt test.

Basin	**Station**	**Change point (year)**	**P (Mann-Whitney-Pettitt)**	**Mean recharge (107 m^3/year)**		**Relative change at the change point**
			before change point	**after change point**		
Lanyang River	Niu-Dou	-	0.8423	-	-	-
	Ximen Bridge	2001	0.9999	19.6	6.7	-89.4%
Danshui River	Fu-Shan	1999	0.9995	23.1	35.2	46.9%
	Gaoyi	-	0.8553	-	-	-
	Hengchi	1983	0.9990	2.2	3.3	41.7%
Fengshan River	Hsin-Pu	-	1.0660	-	-	-
Touqian River	Nei-Wan	-	0.8012	-	-	-
	Shang-Ping	-	0.7013	-	-	-

[i] - Note: The number in bold indicates a statistically significant difference.

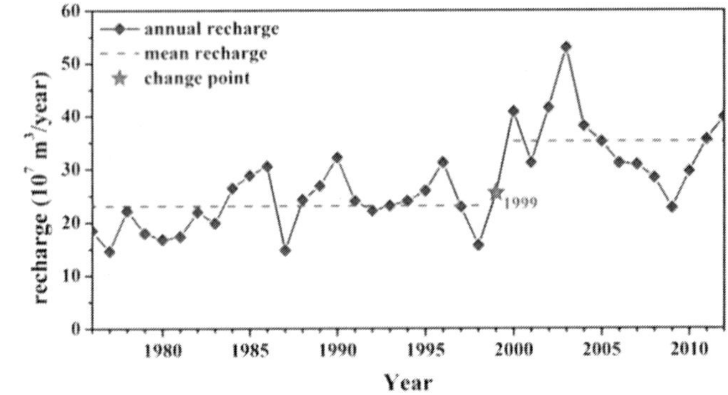

(a) Fu-Shan station

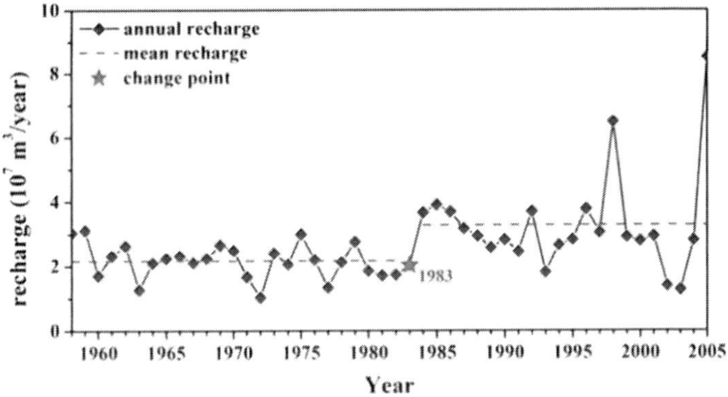

(b) Hengchi station

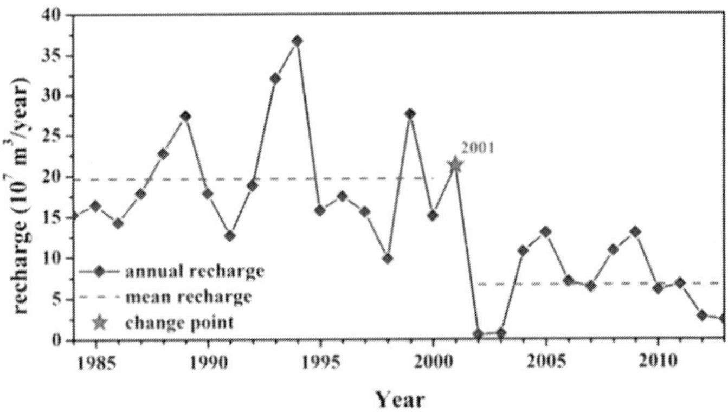

(c) Ximen Bridge station

Figure 5. The results of change points of significant trend stations. (a)Fu-Shan station, (b) Hengchi station, (c)Ximen Bridge station

5. CONCLUSIONS

In this study, the base flow estimation method and stable base flow analysis were used to evaluate the long-term stream flow data of eight stations in Northern Taiwan. The range was between $2.67 \times 10^7$and 7.81×10^8 m^3/year based on the cumulative annual recharge. The characteristics of the trends were examined using the Mann-Kendall test. Further, trend slope calculation and change-point analysis were carried out. The results provided an understanding of the trends for the eight stations and indicated that only two had downward trends: Gaoyi station at the upstream of the Tamsui River and Ximen Bridge at the downstream of the Lanyang River. The rate of change for the Fengshan and Touqian river basins was relatively small. Separately, the rate of change for the annual groundwater recharge was greater than 30% for all stations in the Lanyang river basin, while Gaoyi station was the only exception in the Tamsui river basin. The results of the change-point analysis showed that the change point for the Fushan, Hengxi, and Ximen Bridge stations occurred in 1999, 1983, and 2001, respectively. The average annual groundwater recharge for the first two stations exhibited an upward trend before and after the change point (46.9 and 41.7%, respectively), while that for the last station decreased by as much as 89.4%. The findings can be used for regional hydrological studies and as reference for water resources planning.

ACKNOWLEDGEMENTS

The authors are grateful for the support of the Research Project of the Ministry of Science and Technology (MOST 103-2221-E-006-205-MY2).

REFERENCES

1. Chung YD, Kuo CC, Chen CS. The temporal variation of regional rainfall characteristics in Taiwan. Journal of Chinese Agricultural Engineering. 2009;55:1–18. (in Chinese)
2. Esty D, Srebotnjak T, Goodall M, Andonov B, Campbell K, Gregg K, Kim C, Li Q, Martinez M, Townsend J, Zhang B, Levy M, de Sherbinin A, Anderson B, Saltelli A, Saisana M, Nardo M, Dahl A. Environmental sustainability index: benchmarking national environmental stewardship. Yale Center for Environmental Law & Policy: New Haven, USA; 2005.
3. Hsu HH, Chen CT, Lu MM, Chen YM, Chou C, Wu YC. Climate change in Taiwan: scientific report 2011; National Science and Technology Center for Disaster Reduction: New Taipei City, Taiwan.
4. Scanlon BR, Healy RW, Cook PG. Choosing appropriate techniques for quantifying groundwater recharge. Hydrogeology Journal. 2002;10:18–39.
5. Keese KE, Scanlon BR, Reedy RC. Assessing controls on diffuse groundwater recharge using unsaturated flow modeling. Water Resources Research. 2005;41:W06010.

6. Döll P, Fiedler K. Global-scale modeling of groundwater recharge. Hydrology and Earth System Sciences. 2008;12:863–885.
7. Lee CH, Yeh HF,Chen JF. Estimation of groundwater recharge using soil moisture budget method and base-flow model. Environmental Geology. 2008;54:1787–1797.
8. Kim JH, Jackson RB. A global analysis of groundwater recharge for vegetation, climate, and soils. Vadose Zone Journal. 2012;11:188–222.
9. Ibrahim M, Favreau G, Scanlon BR, Seidel JL, Le Coz M, Demarty J, Cappelaere B. Long-term increase in diffuse groundwater recharge following expansion of rainfed cultivation in the Sahel, West Africa. Hydrogeology Journal. 2014;22:1293–1305.
10. Sanford WE, Nelms DL, Pope JP, Selnick DL. Estimating mean long-term hydrologic budget components for watersheds and counties: an application to the commonwealth of Virginia, USA. Hydrology Current Research. 2015;6:191.
11. Wang TJ, Istanbulluoglu E, Lenters J, Scott D. On the role of groundwater and soil texture in the regional water balance: an investigation of the Nebraska Sand Hills, USA. Water Resources Research. 2009;45:W10413.
12. Taylor RG, Scanlon B, Doll P, et al. Ground water and climate change. Nature Climate Change. 2013;3:322–329.
13. Dripps WR, Bradbury KR. A simple daily soil-water balance model for estimating the spatial and temporal distribution of groundwater recharge in temperate humid areas. Hydrogeology Journal. 2007;15:433–444.
14. Westenbroek SM, Kelson VA, Dripps WR, Hunt RJ, Bradbury KR. SWB-a modified Thornthwaithe-Mather soil-water-balance code for estimating groundwater recharge. U.S. Geological Survey. Techniques and Methods. 2010;6–A31.60p.
15. Mair A, Hagedorn B, Tillery S, El-Kadi AI, Westenbroek S, Ha K, Koh GW. Temporal and spatial variability of groundwater recharge on Jeju Island, Korea. Journal of Hydrology. 2013;501:213–226.
16. Beigi E, Tsai FTC. GIS-based water budget framework for high-resolution groundwater recharge estimation of large-scale humid regions. Journal of Hydrologic Engineering. 2014;19:05014004.
17. Cuthbert MO, Mackay R, Nimmo JR. Linking soil moisture balance and source-responsive models to estimate diffuse and preferential components of groundwater recharge. Hydrology and Earth System Sciences. 2013;17:1003–1019.
18. Bekele E, Patterson B, Toze S, Furness A, Higginson S, Shackleton M. Aquifer residence times for recycled water estimated using chemical tracers and the propagation of temperature signals at a managed aquifer recharge site in Australia. Hydrogeology Journal. 2014;22:1383–1401.
19. Vásquez V, Thomsen A, Iversen BV, Jensen R, Ringgaard R, Schelde K. Integrating lysimeter drainage and eddy covariance flux measurements in a groundwater recharge model. Hydrological Sciences Journal. 2015;60:1520–1537.
20. Bakundukize C, Camp MV, Walraevens K. Estimation of groundwater recharge in Bugesera region (Burundi) using soil moisture budget approach. Geologica Belgica. 2011;14:85–102.
21. Oke, MO, Martins O, Idowu OA. Determination of rainfall-recharge relationship in River Ona basin using soil moisture balance and water fluctuation methods. International Journal of Water Resources and Environmental Engineering. 2014;6:1–11.
22. Nathan RJ, McMahon TA. Evaluation of automated techniques for base flow and recession analysis. Water Resources Research. 1990;26:1465–1473.
23. Rutledge AT, Daniel III CC. Testing an automated method to estimate ground-water recharge from streamflow records. Ground Water. 1994;32:180–189.
24. Mau DP, Winter TC. Estimating ground-water recharge from streamflow hydrographs for a small mountain watershed in a temperate humid climate, New Hampshire, USA. Ground Water. 1997;35:291–304.
25. Lee CH, Chen WP, Lee RH. Estimation of groundwater recharge using water balance coupled with base-flow-record estimation and stable-base-flow analysis. Environmental Geology. 2006;51:73–82.

26. Yeh HF, Lin HI, Lee ST, Chang MH, Hsu KC, Lee CH. GIS and SBF for estimating groundwater recharge of a mountainous basin in Wu River watershed, Taiwan. Journal of Earth System Science. 2014;123:503–516.
27. Rutledge AT. Computer programs for describing the recession of ground-water discharge and for estimating mean ground-water recharge and discharge from streamflow records-update. U.S. Geological Survey Water-Resources Investigations Report 98–4148;1998. 43p.
28. Rutledge AT. Considerations for use of the RORA program to estimate ground-water recharge from streamflow records. U.S. Geological Survey. 2000;Open-File Report 00-156, 44p.
29. Mann HB. Non-parametric test against trend. Econometrica. 1945;13:245–259.
30. Kemdall MG. Rank correlation methods. Charles Griffin, London; 1975.
31. Sen PK. Estimates of the regression coefficient based on Kendall's tau. Journal of the American Statistical Association. 1968;63:1379–1389.
32. Pettit AN. A non-parametric approach to the change point problem. Applied Statistics. 1979;28:126–135.
33. Water Resources Agency (WRA). Hydrological year book of Taiwan. Ministry of Economic Affairs: Taipei, Taiwan; 2013.

INDEX